FOG FOR 5G AND IoT

WILEY SERIES ON INFORMATION AND COMMUNICATION TECHNOLOGY

Series Editors: T. Russell Hsing, Vincent K. N. Lau, and Mung Chiang

A complete list of the titles in this series appears at the end of this volume.

FOG FOR 5G AND IoT

Edited by

Mung Chiang
Arthur LeGrand Doty Professor of Electrical Engineering, Princeton University, Princeton, NJ, USA

Bharath Balasubramanian
Senior Inventive Scientist, ATT Labs Research, Bedminster, NJ, USA

Flavio Bonomi
Founder and CEO, Nebbiolo Technologies, Milpitas, CA, USA

This edition first published 2017

Registered Office
John Wiley & Sons, Inc., 111 River Street, Hoboken, NJ 07030, USA

Editorial Office
111 River Street, Hoboken, NJ 07030, USA

For details of our global editorial offices, customer services, and more information about Wiley products visit us at www.wiley.com.

Wiley also publishes its books in a variety of electronic formats and by print-on-demand. Some content that appears in standard print versions of this book may not be available in other formats.

Library of Congress Cataloging-in-Publication Data

Names: Chiang, Mung, editor. | Balasubramanian, Bharath, editor. | Bonomi, Flavio, editor.
Title: Fog for 5G and IoT / edited by Mung Chiang, Bharath Balasubramanian, Flavio Bonomi.
Description: Hoboken, NJ, USA : John Wiley & Sons Inc., 2017. | Includes bibliographical references and index.
Identifiers: LCCN 2016042091| ISBN 9781119187134 (cloth) | ISBN 9781119187172 (epub) | ISBN 9781119187158 (epdf)
Subjects: LCSH: Electronic data processing–Distributed processing. | Distributed shared memory. | Storage area networks (Computer networks) | Mobile computing. | Internet of things. | Cloud computing.
Classification: LCC QA76.9.D5 F636 2017 | DDC 004.67/82–dc23
LC record available at https://lccn.loc.gov/2016042091

Cover image: Cultura/Seb Oliver/Gettyimages
Cover design by Wiley

Set in 10/12pt Times by SPi Global, Pondicherry, India

10 9 8 7 6 5 4 3 2 1

CONTENTS

CONTRIBUTORS

MOSTAFA AMMAR, School of Computer Science, College of Computing, Georgia Institute of Technology, Atlanta, GA, USA

HELDER ANTUNES, Corporate Strategic Innovations Group, Cisco Systems, Inc., San Jose, CA, USA

A. SALMAN AVESTIMEHR, Department of Electrical Engineering, University of Southern California, Los Angeles, CA, USA

BHARATH BALASUBRAMANIAN, ATT Labs Research, Bedminster, NJ, USA

SUMAN BANERJEE, Department of Computer Sciences, University of Wisconsin-Madison, Madison, WI, USA

FLAVIO BONOMI, Nebbiolo Technologies, Inc., Milpitas, CA, USA

S.-H. GARY CHAN, Department of Computer Science and Engineering, The Hong Kong University of Science and Technology, Clear Water Bay, Hong Kong

XU CHEN, School of ECEE, Arizona State University, Tempe, AZ, USA

MUNG CHIANG, EDGE Labs; Department of Electrical Engineering, Princeton University, Princeton, NJ, USA

SANGTAE HA, Department of Computer Science, University of Colorado at Boulder, Boulder, CO, USA

KARIM HABAK, School of Computer Science, College of Computing, Georgia Institute of Technology, Atlanta, GA, USA

ROBERT J. HALL, AT&T Labs Research, Bedminster, NJ, USA

KHALED A. HARRAS, Computer Science Department, School of Computer Science, Carnegie Mellon University, Doha, Qatar

CARLEE JOE-WONG, Electrical and Computer Engineering, Carnegie Mellon University, Silicon Valley, CA, USA

STEVEN Y. KO, University at Buffalo, The State University of New York, Buffalo, NY, USA

PENG LIU, Pennsylvania State University, State College, PA; Department of Computer Sciences, University of Wisconsin-Madison, Madison, WI, USA

ZHENMING LIU, Department of Computer Science, College of William and Mary, Williamsburg, VA, USA

ZHI LIU, Global Information and Telecommunication Institute, Waseda University, Tokyo, Japan

SATYAJAYANT MISRA, Department of Computer Science, New Mexico State University, Las Cruces, NM, USA

ANDREAS F. MOLISCH, Department of Electrical Engineering, University of Southern California, Los Angeles, CA, USA

ASHISH PATRO, Department of Computer Sciences, University of Wisconsin-Madison, Madison, WI, USA

STEFAN POLEDNA, TTTech Computertechnik AG, Wien, Austria

CONG SHI, School of Computer Science, College of Computing, Georgia Institute of Technology, Atlanta, GA, USA; Square, Inc., San Francisco, CA, USA

WILFRIED STEINER, TTTech Computertechnik AG, Wien, Austria

DEEPAK S. TURAGA, IBM T. J. Watson Research Center, Yorktown, New York, NY, USA

MIHAELA VAN DER SCHAAR, Electrical Engineering Department, University of California at Los Angeles, Los Angeles, CA, USA

DALE WILLIS, Department of Computer Sciences, University of Wisconsin-Madison, Madison, WI, USA

FELIX MING FAI WONG, Yelp Inc., San Francisco, CA, USA

ELLEN W. ZEGURA, School of Computer Science, College of Computing, Georgia Institute of Technology, Atlanta, GA, USA

BO ZHANG, Department of Computer Science and Engineering, The Hong Kong University of Science and Technology, Clear Water Bay, Hong Kong

JUNSHAN ZHANG, School of ECEE, Arizona State University, Tempe, AZ, USA

TAO ZHANG, Corporate Strategic Innovation Group, Cisco Systems, Inc., San Jose, CA, USA

YI ZHENG, Corporate Strategic Innovation Group, Cisco Systems, Inc., San Jose, CA, USA

RAYMOND ZHENG, Corporate Strategic Innovation Group, Cisco Systems, Inc., San Jose, CA, USA

Introduction

BHARATH BALASUBRAMANIAN,[1] MUNG CHIANG,[2] and FLAVIO BONOMI[3]

[1] *ATT Labs Research, Bedminster, NJ, USA*
[2] *EDGE Labs, Princeton University, Princeton, NJ, USA*
[3] *Nebbiolo Technologies, Inc., Milpitas, CA, USA*

The past 15 years have seen the rise of the cloud, along with rapid increase in Internet backbone traffic and more sophisticated cellular core networks. There are three different types of clouds: (i) data centers, (ii) backbone IP networks, and (iii) cellular core networks, responsible for computation, storage, communication, and network management. Now the functions of these three types of clouds are descending to be among or near the end users, as the "fog." Empowered by the latest chips, radios, and sensors, the edge devices today are capable of performing complex functions including computation, storage, sensing, and network management. In this book, we explore the evolving notion of the *fog architecture* that incorporates networking, computing, and storage.

Architecture is about the division of labor in modularization: who does what, at what timescale, and how to glue them back together. The division of labor between layers, between control plane and data plane, and between cloud and fog [1] in turn supports various application domains. We take the following as a working definition of the fog architecture: it is an architecture for the cloud-to-things (C2T) continuum that uses one or a collaborative multitude of end-user clients or near-user edge devices to carry out a substantial amount of storage, communication, and control, configuration, measurement, and management. Engineering artifacts that may use the fog architecture include 5G, home/personal networking, embedded AI, and the Internet of things (IoT) [2].

In Figure I.1, we highlight that fog can refer to an architecture for computing, storage, control, or communication network, and that as a network architecture it may support a variety of applications. We contrast between the fog architecture and the current practice of the cloud along the following three dimensions:

1. Carry out a substantial amount of storage at or near the end user (rather than stored primarily in large-scale data centers).
2. Carry out a substantial amount of communication at or near the end user (rather than all routed through the backbone network).

Fog for 5G and IoT, First Edition. Edited by Mung Chiang, Bharath Balasubramanian, and Flavio Bonomi.

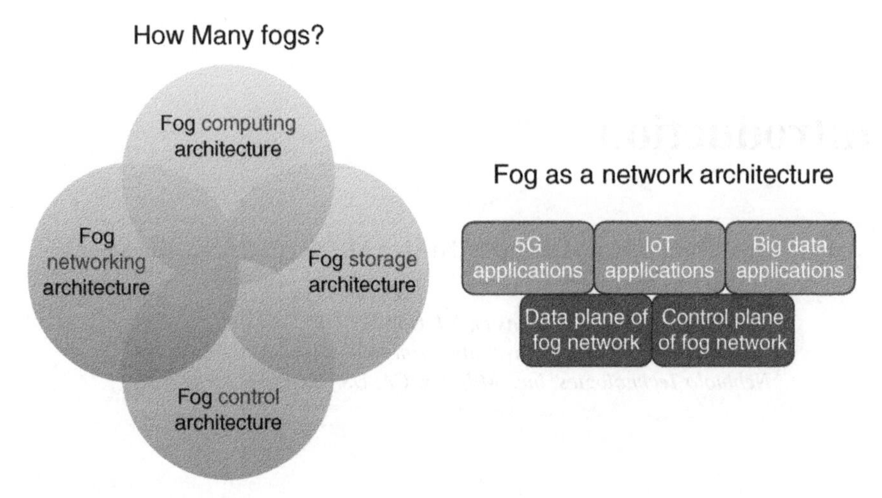

Figure I.1 Fog architectures and applications. Supported by such architectures.

3. Carry out a substantial amount of computing and management, including network measurement, control, and configuration, at or near the end user (rather than controlled primarily by gateways such as those in the LTE core).

Why would we be interested in the fog view now? There are four main reasons summarized as CEAL. Many examples in recent publications, across mobile and landline, and from physical layer beamforming to application layer edge analytics have started leveraging these advantages [3–8]:

1. *Cognition: Awareness of Client-Centric Objectives.* Following the end-to-end principle, some of the applications can be best enabled by knowing the requirements on the clients. This is especially true when privacy and reliability cannot be trusted in the cloud or when security is enhanced by shortening the extent over which communication is carried out.
2. *Efficiency: Pooling of Local Resources.* There are typically hundreds of gigabytes sitting idle on tablets, laptops, and set-top boxes in a household every evening, across a table in a conference room, or among the passengers of a public transit system. Similarly, idle processing power, sensing ability, and wireless connectivity on the edge may be pooled within a fog network.
3. *Agility: Rapid Innovation and Affordable Scaling.* It is usually much faster and cheaper to experiment with client and edge devices. Rather than waiting for vendors of large boxes inside the network to adopt an innovation, in the fog world a small team may take advantages of smartphone API and SDK, the proliferation of mobile apps, and offer a networking service through its own API.

4. *Latency: Real-Time Processing and Cyber–Physical System Control.* Edge data analytics, as well as the actions it enables through control loops, often have stringent time requirement and can only be carried out on the edge or the "things", here and now. This is particularly essential for Tactile Internet: the vision of millisecond reaction time on networks that enable virtual–reality-type interfaces between humans and devices.

We further elaborate on the previous potential advantages of fog. Client and edge devices have increasing strength and capabilities. For instance, the original iPhone had a single core 412 MHz ARM processor with 128MB RAM and 8GB storage space. The iPhone 5S on the other hand carries a dual-core 1.3 GHz Apple A7 processor with 1GB RAM, 64 GB storage space, and enhanced GPU capabilities. Intel's mobile chip Atom and Nvidia's Tegra too promise near similar specifications. The increase in strength and capabilities implies complex functionality such as CPU/GPU intensive gaming, powerful location/context tracking sensors, and enhanced storage. Further, as suggested in [9], these interconnected edge devices will play a crucial role in orchestrating the IoT. Edge devices including mobile phones and wearable devices use a rich variety of sensors including gyroscopes, accelerometers, and odometers to monitor the environment around them. This enables the crucial notion of exploiting context both personal in terms of location and physical/psychological characteristics and context in the communal sense of how devices are interacting with other devices around them.

As the need for cloud-based services increases, the amount of data traffic generated in the core networks is increasing at an alarming rate. Cisco predicts that cloud traffic will increase almost four to five times over the next 5 years [10]. Further, they predict that cloud IP traffic will account for nearly two-thirds of all data center traffic by 2017. Can the fog alleviate some of this by satisfying application needs locally? For example, can part of cloud storage be moved closer to the user with edge/client devices acting as micro-data centers? Can videos be cached efficiently at the edge devices to reduce accesses to the cloud? Or more broadly, can edge devices perform an active role in orchestrating both data plane-based cloud services and control plane-based core network services?

Accesses to the cloud often span geographically distant entities with round-trip times of nearly 150–200 ms. Access latency is a crucial factor in the end-user experience with studies showing that a 20% decrease in RTTs results in a 15% decrease in page load time [11]. A significant way to decrease the RTT for content access is to place as much of the content physically close to the end user as possible. While decreasing latency is beneficial to all services, it may be a necessity for many services in the future. For example, services involving augmented reality applications may not tolerate latencies of more than 10–20 ms [12]. Hence, any computation/processing for these kind of services need to be performed locally. Fog services may play a significant part in addressing this challenge.

The fog R&D will leverage past experience in sensor networks, peer-to-peer systems, and mobile ad hoc networks while incorporating the latest advances in devices, systems, and data science to reshape the "balance of power" in the ecosystem between powerful data centers and the edge devices. Toward that end, this book serves as the first introduction to the evolving fog architecture, compiling work traversing many different areas that fit into this paradigm.

In this book, we will encounter many use cases and applications that in many ways are not necessarily new and revolutionary and have been conceived in the context of distributed computing, networking, and storage systems. Computing resources have been always distributed in homes, in factories, along roads and highways, in cities, and in their shopping centers. The field of pervasive or ubiquitous computing has been active for a long time. Networking has always deployed switches, routers, and middleboxes at the edge. Caching media and data at the edge has been fundamental to the evolution of Web services and video delivery.

As is typical of any emergent area of R & D, many of the themes in the fog architecture are not completely new and instead are evolved versions of accumulated transformations in the past decade or two:

- Compared with peer-to-peer (P2P) networks in the mid-2000s, fog is not just about content sharing (or data plane as a whole) but also network measurement, control and configuration, and service definition.
- Compared with mobile ad hoc network (MANET) research a decade ago, we have much more powerful and diverse off-the-shelf edge devices and applications now, together with the structure/hierarchy that comes with cellular/broadband networks.
- Compared with generic edge networking in the past, fog networking provides a new layer of meaning to the end-to-end principle: not only do edge devices optimize among themselves, but also they collectively measure and control the rest of the network.

Along with two other network architecture themes, ICN and SDN, each with a longer history, the fog is revisiting the foundation of how to think about and engineer networks, that is, how to optimize network functions: who does what and how to glue them back together:

- *Information-Centric Networks.* Redefine functions (to operate on digital objects rather than just bytes)
- *Software-Defined Networks.* Virtualize functions (through a centralized control plane)
- *Fog Networks.* Relocate functions (closer to the end users along the C2T continuum)

While fog networks do not have to have any virtualization or to be information centric, one could also imagine an information-centric, software-defined fog network (since these three branches are not orthogonal).

With its adoption of the most modern concepts developed in the IT domain and at the same time with its need to satisfy the requirements of the operational technology (OT) domains, such as time-sensitive and deterministic behaviors in networking, computing and storage, sensor and actuator support and aggregation, and sometimes even safety support, the fog is a perfect conduit for the highly promising convergence of IT and OT in many key IoT verticals. In this perspective, the fog not only builds on and incorporates many of the traditional relevant technologies from sensor and ad hoc network, ubiquitous computing, distributed storage, etc. but also manifests in a timely manner new and specific characteristics coming from the IT and OT convergence behind IoT.

As the cloud catalyzed, consolidated, and evolved a range of existing technologies and approaches, the fog is catalyzing, consolidating, and evolving a range of edge technologies and approaches in a creative and rich mix, at this special transition time into IoT. Complementing the swarm of endpoints and the cloud, the fog will enable the seamless deployment of distributed applications, responding to the needs of critical use cases in a broad array of verticals. For example, some of the early work on fog architecture and functionality was driven by specific applications in connected vehicle and transportation, smart grid, the support of distributed analytics, and the improvement of Web services and video delivery [9, 13, 14].

I.1 SUMMARY OF CHAPTERS

Following the above paragraphs, the chapters in this edited volume are divided into three broad sections. In the first four chapters, we describe work that presents techniques to enable communication and management of the devices in a fog network involving their interaction with the cloud, management of their bandwidth requirements, and prescriptions on how the edge devices can often work together to fulfill their requirements. The next natural step is to understand how to perform the two fundamental components of many applications on the edge: storage and computation. We focus on this aspect in the following three chapters. And finally, we focus on the applications that will be enabled on top of the fog infrastructure and the challenges in realizing them.

Communication and Management In the first chapter the authors present a unique edge computing framework, called *ParaDrop*, that allows developers to leverage one of the most stable and persistent computing resources in the end customer premises: the gateway (e.g., the Wi-Fi access point or home set-top box). Based on a platform that allows the deployment of containers on these edge devices, the authors show how interesting applications such as security cameras and environment sensors can be deployed on these devices. While the first chapter focuses on an operating system agnostic container-based approach, the second chapter posits that the underlying operating system on these devices too should evolve to support fog computing and networking. In a broad analysis, the authors focus on four important aspects: *why* do these systems need to provide better properties to support the fog, *where* do they need to improve, *what* are the exact properties that need to be provided, and finally *how* can they provide these better properties?

To enable rich communication in the fog, bandwidth needs have to be addressed. Following the philosophy of fog networking, why can't the power of edge devices be used to leverage this? In the second chapter, the authors present a home-user-based bandwidth management solution to cope with the growing demand for bandwidth, with a novel technique that puts more intelligence in both the home gateways and the end-user devices. They show that using a two-level system, one based on the gateways "buying bandwidth" from the ISPs within a fixed budget driven by incentives and the other based on end-user prioritization of applications, much better utilization of network bandwidth can be achieved.

The following chapter addresses this question from the point of view of peer-to-peer communication among devices. They present a game theory-based mechanism that end-user devices like tablets and cell phones can use to cooperate with one another and act as relays for each other's network traffic, thereby boosting network capability. An important aspect of fog management and communication is that of addressing the potentially thousands and maybe even millions of fog–IoT devices.

In the final chapter, the author contends that traditional IP-based addressing will not always work for field IoT devices working in a fog environment, interacting with cloud servers or among themselves. This is primarily due to factors such as device mobility, spatial density of devices, and gaps in coverage. As an alternative, they propose a technique of geographic addressing where communication protocols allow devices to specify the destination devices based on their geographic location rather than IP address.

Computation and Storage Following the first section of chapters on communication and management of fog devices, we move on to two important platform functions: storage and caching for video delivery in fog networks and techniques for fog computation. The first chapter in this section presents caching schemes for video on demand (VoD), especially to optimize the last wireless hop in video delivery. While most CDN-based systems focus on caching at the edge of the *network*, the authors here focus on caching in *edge devices* such as Femto helper nodes (similar to Femto base stations) and the end-user devices themselves.

The second chapter, on the other hand, shifts the focus from VoD to live streaming, a use case with very different requirements but similar potential uses of the fog paradigm. The authors discuss a technique through which the end-user devices collaborate to deliver live streams to each other, operating as a wireless fog. They focus on a crucial problem in such systems—that of errors due to lossy wireless links—and present a store–recover–forward strategy for wireless multihop fog networks that combines traditional store and forward techniques with network coding.

In the final chapter of this section, we move from storage to general-purpose computation in fog. Similar to other chapters in this book, the authors posit that mobile devices have now become far more powerful and can hence perform several computations locally, with carefully planned fog architectures. They focus on two such designs: femto cloud, in which they discuss a general purpose architecture of a computational platform for mobile devices, and Serendipity, in which they consider a more severe version of the same problem in which devices are highly mobile and often tasks need to be off-loaded to one another.

Applications Having set the foundation with the previous section on the platform requirements and innovations, we finally move on to applications built on the fog architecture. In the first chapter in this section, the authors provide a close look at the challenges facing the connected car, an IoT use case that is increasingly prominent these days. In particular, they focus on the electrical architecture that will enable this application and describe how fog computing with its virtualization techniques, platform unification of several concerns such as security and management will help alleviate these challenges.

In the following chapter, the authors provide a detailed analysis of distributed stream processing systems and online learning frameworks with a view to building what they term a *smarter planet*. In their vision of smart planet, they envisage a world in which users are constantly gathering data from their surroundings, processing this data, performing meaningful analysis, and taking decisions based on this analysis. The main challenge however is that given the potentially huge number of low-power sensors and the mobility of the users, all this data analysis needs to be heavily distributed through its life cycle. The combination of potent-distributed learning frameworks and fog computing that will provide the platform capabilities for such frameworks can bring forth the vision of the smarter planet.

Finally, we end the book with a chapter on how fog computing can help address the crucial needs of security in IoT devices. The authors start with the question: what is so different about IoT security as opposed to standard enterprise security, and what needs to change? They then go on answering these questions and identify IoT concerns ranging from the incredibly large number of such devices to the need for keeping them regularly updated with regard to security information. Crucially, they focus on how the fog paradigm can help address many of these concerns by providing frameworks and platforms to alleviate the load on the IoT devices and perform functions such as endpoint authentication and security updates.

The electronic supplemental content to support use of this book is available online at https://booksupport.wiley.com

I.2 ACKNOWLEDGMENTS

This book would not have been possible without help from numerous people, and we wish to sincerely thank all of them.

In particular, Dr. Jiasi Chen, Dr. Michael Wang, Dr. Christopher Brinton, Dr. Srinivas Narayana, Dr. Zhe Huang, and Dr. Zhenming Liu provided valuable feedback on the individual chapters of the book. The publisher, John Wiley and Sons, made a thorough effort to get the book curated and published. We are grateful to the support from funding agencies of National Science Foundation under the fog research grants. Last but not the least, the book will ultimately stand on its contents, and we are grateful to all the chapter authors for their technical contributions and never-ending enthusiasm in writing this book.

REFERENCES

1. Mung Chiang, Steven H. Low, A. Robert Calderbank, and John C. Doyle. Layering as optimization decomposition: A mathematical theory of network architectures. In *Proceedings of the IEEE*, volume 95, pages 255–312, January 2007.
2. Mung Chiang and Tuo Zhang, Fog and IoT: an overview of research opportunities. *IEEE Journal of Internet of Things*, 3(6), December 2016.
3. Abhijnan Chakraborty, Vishnu Navda, Venkata N. Padmanabhan, and Ramachandran Ramjee. Coordinating cellular background transfers using load sense. In *Proceedings of the 19th Annual International Conference on Mobile Computing & Networking*, MobiCom '13, pages 63–74, New York, NY, USA, 2013. ACM.
4. Ehsan Aryafar, Alireza Keshavarz-Haddad, Michael Wang, and Mung Chiang. Rat selection games in hetnets. In *INFOCOM*, pages 998–1006 April 14–19, 2013. IEEE Turin, Italy.
5. Luca Canzian and Mihaela van der Schaar. Real-time stream mining: Online knowledge extraction using classifier networks. *IEEE Network*, 29(5):10–16, 2015.
6. Jae Yoon Chung, Carlee Joe-Wong, Sangtae Ha, James Won-Ki Hong, and Mung Chiang. Cyrus: Towards client-defined cloud storage. In *Proceedings of the 10th European Conference on Computer Systems*, EuroSys '15, pages 17:1–17:16, New York, NY, USA, 2015. ACM.
7. Felix Ming Fai Wong, Carlee Joe-Wong, Sangtae Ha, Zhenming Liu, and Mung Chiang. Mind your own bandwidth: An edge solution to peak-hour broadband congestion. *CoRR*, abs/1312.7844, 2013.
8. Yongjiu Du, Ehsan Aryafar, Joseph Camp, and Mung Chiang. iBeam: Intelligent client-side multi-user beamforming in wireless networks. In *2014 IEEE Conference on Computer Communications, INFOCOM 2014*, pages 817–825, Toronto, Canada, April 27–May 2, 2014. IEEE.
9. Flavio Bonomi, Rodolfo Milito, Jiang Zhu, and Sateesh Addepalli. Fog computing and its role in the internet of things. In *Proceedings of the First Edition of the MCC Workshop on Mobile Cloud Computing*, MCC '12, pages 13–16, New York, NY, USA, 2012. ACM.
10. Cisco Global Cloud Index: Forecast and Methodology. http://www.intercomms.net/issue-21/pdfs/articles/cisco.pdf (accessed September 12, 2016).
11. Latency: The New Web Performance Bottleneck. https://www.igvita.com/2012/07/19/latency-the-new-web-performance-bottleneck/ (accessed September 12, 2016).
12. W. Pasman, Arjen Van Der Schaaf, R.L. Lagendijk, and Frederik W. Jansen. Low latency rendering and positioning for mobile augmented reality. In *Proceedings Vision Modeling and Visualization '99*, pages 309–315, 1999.

13. Flavio Bonomi. Cloud and fog computing: Trade-offs and applications. In *EON-2011 Workshop, at the International Symposium on Computer Architecture (ISCA 2011)*, San Jose, USA, June 4–8, 2011.

14. Xiaoqing Zhu, Douglas S. Chan, Hao Hu, Mythili S. Prabhu, Elango Ganesan, and Flavio Bonomi. Improving video performance with edge servers in the fog computing architecture. *Intel Technology Journal* 19(1):202–224, 2015.

PART I
Communication and Management of Fog

1 ParaDrop: An Edge Computing Platform in Home Gateways

SUMAN BANERJEE,[1] PENG LIU,[1,2] ASHISH PATRO,[1] and DALE WILLIS[1]

[1]*Department of Computer Sciences, University of Wisconsin-Madison, Madison, WI, USA*
[2]*Pennsylvania State University, State College, PA, USA*

1.1 INTRODUCTION

The last decade has seen a rapid diversification of computing platforms, devices, and services. For example, desktops used to be the primary computing platform until the turn of the century. Since then, laptops and more recently handheld devices such as laptops and tablets have been widely adopted. Wearable devices and the Internet of things (IoT) are the latest trends in this space. This has also led to widespread adoption of the "cloud" as a ubiquitous platform for supporting applications and services across these different devices.

Simultaneously, cloud computing platforms, such as Amazon EC2 and Google App Engine, have become a popular approach to provide ubiquitous access to services across different user devices. Third-party developers have come to rely on cloud computing platforms to provide high quality services to their end users, since they are reliable, always on, and robust. Netflix and Dropbox are examples of popular cloud-based services. Cloud services require developers to host services, applications, and data on off-site data centers. But, due to application-specific reasons, a growing number of high quality services restrict computational tasks to be colocated with the end user. For example, latency-sensitive applications require the backend service to be located to a user's current location. Over the years, a number of research threads have proposed that a better end-user experience is possible if the computation is performed close to the end user. This is typically referred to as "edge computing" and comes in various flavors including: cyber foraging [1], cloudlets [2], and more recently fog computing [3].

Fog for 5G and IoT, First Edition. Edited by Mung Chiang, Bharath Balasubramanian, and Flavio Bonomi.

This chapter presents a unique edge computing framework, called ParaDrop, which allows developers to leverage one of the last bastions of persistent computing resources in the end customer premises: the gateway (e.g., the Wi-Fi access point (AP) or home set-top box). Using this platform, which has been fully implemented on commodity gateways, developers can design virtually isolated compute containers to provide a persistent computational presence in the proximity of the end user. The compute containers retain user state and also move with the users as the latter changes their points of attachment. We demonstrate the capabilities of this platform by demonstrating useful third-party applications, which utilize the ParaDrop framework. The ParaDrop framework also allows for multitenancy through virtualization, dynamic installation through the developer API, and tight resource control through a managed policy design.

1.1.1 Enabling Multitenant Wireless Gateways and Applications through ParaDrop

A decade or two ago, the desktop computer was the only reliable computing platform within the home where third-party applications could reliably and persistently run. However diverse mobile devices, such as smartphones and tablets, have deprecated the desktop computer since, and today persistent third-party applications are often run in remote cloud-based servers. While cloud-based third-party services have many advantages, the rise of edge computing concepts stems from the observation that many services can benefit from a persistent computing platform, right in the end-user premises.

With end-user devices going mobile, there is one remaining device that provides all the capabilities developers require for their services, as well as the proximity expected from an edge computational framework. The gateway—which could be a home Wi-Fi AP or a cable set-top box provided by a network operator—is a platform that is continuously on and due to its pervasiveness is a primary entry point into the end-user premises for such third-party services.

We want to push computation onto the home gateways (e.g., Wi-Fi APs and cable set-top boxes) for the following reasons:

- The home gateways can handle it—modern home gateways are much more powerful than what they need to be for their networking workload. What is more if you are not running a Web server out of the house, your gateway sits dormant majority of the time (when no one is home using it).
- Utilizing computational resources in the home gateway gives us a footprint within the home to devices that are starved for computational resources, namely, IoT devices. Using ParaDrop, developers can piggyback their IoT devices onto the AP without the need for cloud services OR a dedicated desktop!
- Every household connected to the Internet by definition must contain an Internet gateway somewhere in the house. With these devices sitting around, we can use them to their full potential.

- Pervasive Hardware: Our world is quickly moving toward households only having mobile devices (tablets and laptops) in the home that are not always on or always connected. Developers can no longer rely on pushing software into the home without also developing their own hardware too.

A Developer-Centric Framework. In this chapter, we examine the requirements of services in order to build an edge computing platform, which enables developers to provide services to the end user in place of a cloud computing platform. A focus on edge computation would require developers to think differently about their application development process; however we believe there are many benefits to a distributed platform such as ParaDrop. The developer has remained our focus in the design and implementation of our platform. Thus, we have implemented ParaDrop to include a fully featured API for development, with a focus on a centrally managed framework. Through virtualization, ParaDrop enables each developer access to resources in a way as to completely isolate all services on the gateway. A tightly controlled resource policy has been developed, which allows fair performance between all services.

1.1.2 ParaDrop Capabilities

ParaDrop takes advantage of the fact that resources of the gateway are underutilized most of the time. Thus each service, referred to as a chute (as in parachute), borrows CPU time, unused memory, and extra disk space from the gateway. This allows vendors an unexplored opportunity to provide added value to their services through the close proximity footprint of the gateway.

Figure 1.1 shows ParaDrop system running on real hardware, the “Wi-Fi home gateway,” along with two services to motivate our platform: “security camera” and “environment sensors.” ParaDrop has been implemented on PC engines ALIX 2D2 single board computer running OpenWrt “Barrier Breaker” on an AMD Geode 500 MHz processor with 256 MB of RAM. This low-end hardware platform was chosen to showcase ParaDrop’s capabilities with existing gateway hardware.

We have emulated two third-party developers who have migrated their services to the ParaDrop platform to showcase the potential of ParaDrop. Each of these services contains a fully implemented set of applications to capture, process, store, and visualize the data from their wireless sensors within a virtually isolated environment. The first service is a wireless environmental sensor designed as part of the Emonix research platform [4], which we refer to as “EnvSense.” The second service is a wireless security camera based on a commercially available D-Link DCS 931L webcam, which we call “SecCam.” Leveraging the ParaDrop platform, the two developer services allow us to motivate the following characteristics of ParaDrop:

- *Privacy.* Many sensors and even webcams today rely on the cloud as the only storage mechanism for generated data. Leveraging the ParaDrop platform, the end user no longer must rely on cloud storage for the data generated by their private devices and instead can borrow disk space available in the gateway for such data.

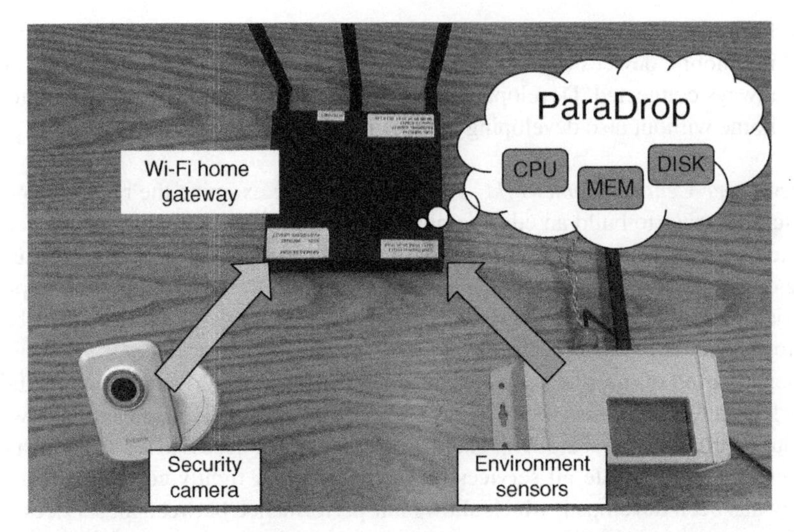

Figure 1.1 The fully implemented ParaDrop platform on the Wi-Fi home gateway, which shares its resources with two wireless devices including a security camera and environment sensor.

- *Low Latency.* Many simple processing tasks required by sensors are performed in the cloud today. By moving these simple processing tasks onto gateway hardware, one hop away from the sensor itself, a reliable low-latency service can be implemented by the developer.
- *Proprietary Friendly.* From a developer's perspective, the cloud is the best option to deploy their proprietary software because it is under their complete control. Using ParaDrop, a developer can package up the same software binaries and deploy them within the gateway to execute in a virtualized environment, which is still under their complete control.
- *Local Networking Context.* In the typical service implemented by a developer, the data is consumed only by the end user yet stored in the cloud. This requires data generated by a security camera in the home to travel out to a server somewhere in the Internet and upon the end user's request travel back from this server into the end-user device for viewing. Utilizing the ParaDrop platform, a developer can ensure that only data requested by the end user is transmitted through Internet paths to the end-user device.
- *Internet Disconnectivity.* Finally, as services become more heterogeneous, they will move away from simple "nice to have" features into mission critical, life saving services. While generally accepted as unlikely, a disconnection from the Internet makes a cloud-based sensor completely useless and is unacceptable for services such as health monitoring. In this case, a developer could leverage the always-on nature of the gateway to process data from these sensors, even when the Internet seems to be down.

1.2 IMPLEMENTING SERVICES FOR THE PARADROP PLATFORM

The primary component of ParaDrop is the virtual machine called a chute (short for parachute) because the framework uses it to install services across different APs. Each developer can deploy many chutes (Figure 1.2) to their AP, thanks to a low-overhead virtualization technology: Linux containers (LXC). These chutes allow for fully isolated use of computational resources on the AP. As you design and implement services on your AP, you can, and should, separate these services into unique chutes. Figure 1.3 shows an example chute configuration specified in the *Chute.struct* file.

There are several primary concerns of the ParaDrop platform including installation procedure, API, and networking configuration.

Dynamic Installation. In order to allow end users to easily add services to their gateway, each service should have the ability to be dynamically installed. This process is possible through the virtualization environment of each chute. When an end user wishes to add a service to their home, they simply register an account with the developer. Using the ParaDrop API, the developer links the user's account with their gateway. If the service utilizes a wireless device, the gateway can fully integrate with the device without any interference from the end user.

ParaDrop API. The focus of ParaDrop is to enable third-party developers to provide high quality services to their users. In order to enable this, a seamless API was developed, based on a RESTful paradigm, which allows the developer to have complete control over the configuration of their chutes.

Developers can use the API to query and monitor the status of the Paradrop platform:

- Persistent State: Users (type, permissions, etc.), chutes (description, resource requirements, etc.), and gateways (configuration, accessories, location, etc.)
- Real-Time State: Running status of chutes and gateway

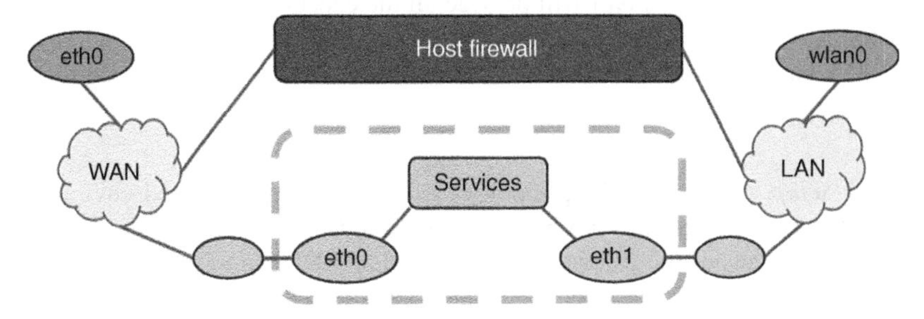

Figure 1.2 The dashed box shows the block diagram representation of a "chute" installed on a ParaDrop-enabled access point. Each chute hosts a stand-alone service and has its own network subnet.

```
"disk": {
  "size": 123456
},
"net": {
  "wan": {
  "type": "wan",
  "intfName": "eth0",
  "ipaddr": "10.100.10.1",
  "netmask": "255.255.255.0"
},
"wifi": {
  "type": "wifi",
  "intfName": "eth1",
  "ipaddr": "10.100.11.1",
  "netmask": "255.255.255.0",
  "ssid": "Virtual0",
  "encryption": "psk2",
  "key": "wifi1234"
}
```

Figure 1.3 An example Chute.struct file, which is used to specify the key configuration parameters of a chute that hosts a stand-alone service. Parameters such as CPU, memory, disk requirements, and network configurations are specified as JSON key–value pairs. ParaDrop provides chute configuration templates to developers, which can customized based on application requirements.

Developers can also use the API to control the system:

- Publish a chute to the store or remove a chute from the store.
- Register/unregister a gateway.
- Install, start, and revoke a chute on one or many gateways.

As services evolve, the API will provide all the capabilities required without the need for modification to the configuration software. This is possible through the use of a JSON-based data back end, which allows abstract configuration and control over each chute.

Network Setup. The networking topology of a dynamic, virtualized environment controlled by several entities is very complex. In order to maintain control over the networking aspects of the gateway, we leveraged an SDN paradigm. All configuration related to networking between the chutes and the gateway is handled through a cloud service, which is interfaced by the developers and network operators. The use of SDN is what allows developers to transparently redirect the user's request to their Web services from within the gateway.

Resource Policy. The multitenancy aspects of ParaDrop require tight policy control over the gateway and its limited resources. Currently the major resources controlled by ParaDrop include CPU, memory, and networking. Using the API, the developer specifies the type of resources they require depending on the services they implement. Through the management interface, the network operator can dynamically adjust the resources provided to each chute. These resources are adjusted first by a request sent to the chute, and, if not acted upon, then by force through the virtualization framework tools.

1.3 DEVELOP SERVICES FOR PARADROP

IoT is becoming a huge part of the networking world. Yet many IoT devices rely on back end services that must traverse the Internet to utilize their full potential. Using ParaDrop, we can pull that intelligence back into the AP.

1.3.1 A Security Camera Service Using ParaDrop

In this section, we present a walk-through about using a Wi-Fi-based video camera with a ParaDrop AP to implement a security camera service called SecCam.

The SecCam service is based on a commercially available wireless IP camera, where we took the role of developer to fully implement the service.

For this service, we require networking interfaces to communicate with the webcam and the Internet, as well as ample storage for images. To augment storage resources on ParaDrop gateways, we add a flash card to the gateway device, which provides GBs of storage.

The applications for SecCam allow for motion detection from the webcam, user-defined alerts, and visualization of the detected images. The motion detection component is a Python-based program with user-defined characteristics such as threshold of motion, time of day, and rate of detection. Visualization of the motion is implemented as a PHP-based Web page, which is hosted within the SecCam chute.

This example in the section creates a chute for the "SecCam" service with the following end result:

- *Create the SecCam SSID.* This SSID provides an isolated Wi-Fi network and subnet to the security cameras. This is designed so that devices purchased by end users do not have to be programmed when they arrive at the house (they can be flashed with a default SSID and password by the company). This subnet will not have internet access and any network traffic be consumed by the chute.
- *Image Capture Service.* The service will run a simple Python program to capture images from an IP camera, calculate differences to detect motion, and store those images to disk. The images stored to disk will then be visualized using a Web server, which runs inside the chute.

```
"disk": {
  "size": 123456
},
"net": {
  "wan": {
  "type": "wan",
  "intfName": "eth0",
  "ipaddr": "10.100.10.1",
  "netmask": "255.255.255.0"
},
"wifi": {
  "type": "wifi",
  "intfName": "eth1",
  "ipaddr": "10.100.11.1",
  "netmask": "255.255.255.0",
  "ssid": "SecCam",
  "encryption": "psk2",
  "key": "noOneCanHackThis"
}
```

Figure 1.4 The primary Chute.struct component for the SecCam chute.

```
{
  "name":"www",
  "path":"/srv/www",
  "location":"@paradrop.server(seccam/srv.tar.gz)",
  "sha1":"526bb8cb52458aad4043c56980cd238551b46b7e",
  "todo":"EXTRACT"
}, {
  "name":"root",
  "path":"/root",
  "sha1":"1633ea1d6351929cc2c8717d1611dcb41681b585",
  "location":"@paradrop.server(seccam/seccam.py)"
}
```

Figure 1.5 The Chute.files component lists the files required for the SecCam chute.

1.3.1.1 Defining the SecCam Chute *Chute.struct.* As discussed earlier, we first need to define the primary *Chute.struct* component first for our awesome SecCam chute (Figure 1.4).

Chute.files. For a chute, the *Chute.files* component lists any files that must exist on the chute's disk in order for it to operate properly. This can include things like bash scripts, Python programs, PHP code, etc.

The rules in Figure 1.5 show files required for our SecCam application. The "www" attribute specifies Web server PHP code to download *seccam/srv.tar.gz* from an *examples* directory on the ParaDrop server to the chute's root file system (FS). Similarly the "root" attribute downloads seccam.py to /root. The "sha1" values let the code running on ParaDrop to verify it properly downloading the code into the chute before it launches.

Chute.resource. As much as possible, ParaDrop tries to be a lean virtualized platform (hence our use of LXC over more traditional virtualization methods). For this reason, we explicitly make the developer define and be aware of the resources they will require for their chute.

These resources are broken down into three categories:

1. *CPU.* The CPU shares devoted to this chute, in most cases the default value, will be fine; if you know the chute will not perform CPU intensive tasks or you want to lower the priority of the tasks it will perform, you can lower the CPU value, by default it is 1024 (meaning equal sharing between all chutes).
2. *Memory.* The AP we have implemented for ParaDrop contains 2 GB of DDR3 memory, so compared with a typical AP memory will not be hard to come by. The default value for memory should typically be fine, but keep in mind: the memory value is a hard limit; if you define it to be too low, your chute's kernel may not even fully boot due to out-of-memory (OOM) issues.
3. *Networking.* The final resource to be defined for chutes is any network throughput requirements of the chute. These are specified in kbps for both upload and download for each interface in the chute. If you are designing a chute with low priority but its use is primarily a virtual router, rather than lowering the CPU resources (which will not greatly affect throughput rates), you should lower the overall throughput provided to the interface instead.

Figure 1.6 shows the *Chute.resource* component for the SecCam chute. We choose the default CPU and memory configuration and specify a high-bandwidth limit to allow high-volume video traffic from the Wi-Fi camera.

Chute.runtime. The *Chute.runtime* component specifies what operations will be performed within the chute itself. We refer to these as the runtime rules (Figure 1.7). The webhosting runtime attribute creates an instance of uhttpd with the arguments specified. The DHCP server runtime macro sets up a default DHCP server inside the chute so that future security cameras can connect to it properly.

Chute.traffic. In many situations, the chute you are implementing will need to interface with devices that for any number of reasons may not be associated to your

```
"cpu": "@resource.cpu.DEFAULT",
"memory": '@resource.memory.DEFAULT',
"wan": {"down": 25000, "up": 10000},
"wifi": {"down": 25000, "up": 10000}
```

Figure 1.6 The Chute.resource component specifies the resource consumption limits for the SecCam chute.

```
{
  "name": "webhosting",
  "program": "uhttpd",
  "args": "-p 80 -i .php=/usr/bin/php-cgi -h /srv/www"
}, {
  "name": "DHCP Server",
  "program": "@net.runtime.dhcpserver"
}
```

Figure 1.7 The Chute.runtime component for the SecCam chute.

```
{
  "name": "Web",
  "description": "Allows the chute to provide a webserver
                  on WAN",
  "rule": "@net.traffic.redirect(@net.host.lan:*:5000,
           wifi:10.100.13.1:80)"
}, {
  "name": "HostSSH",
  "description": "Allows the host stack access to SSH",
  "rule": "@net.traffic.redirect(@net.host.lan:*:5001,
           wifi:10.100.13.1:22)"
}
```

Figure 1.8 The Chute.traffic component allows users to access data within the SecCam chute.

chute's network directly (via a Wi-Fi interface). In these cases for security purposes, the ParaDrop platform allows the developer to implement traffic rules. These rules are implemented in the host networking stack's firewall rules and allow for things like a computer on the host LAN network to access a particular port within a deployed chute (called port forwarding in firewall land).

For the SecCam application, the images are stored within the chute but need to be accessible to users on the LAN network. The Web rule allows the user connected on the LAN network to access Web pages hosted by a uhttpd Web server running inside a chute. The host SSH rule allows the user to SSH into the chute from his laptop (mainly for debugging) connected to the LAN network by using the default ParaDrop SSID (Figure 1.8).

1.3.2 An Environmental Sensor Service Using ParaDrop

Since the wireless environmental sensor was fully implemented as a part of the Emonix research platform, we only need to migrate the service, rather than rewrite it to fit ParaDrop platform. The original service runs in a cloud server to collect data from the sensors, process and store the data, and visualize the data. After identifying the resources required to run the service, we can develop a chute for it so that the service can run in ParaDrop gateways, which are close to the sensors. As the steps to develop a chute for it are the same as the SecCam application, we do not discuss them in detail here.

REFERENCES

1. R. Balan, J. Flinn, M. Satyanarayanan, S. Sinnamohideen, and H.-I. Yang. The case for cyber foraging. In *Proceedings of the 10th Workshop on ACM SIGOPS European Workshop*, EW 10, pages 87–92, New York, NY, USA, 2002. ACM.
2. M. Satyanarayanan, P. Bahl, R. Caceres, and N. Davies. The case for vm-based cloudlets in mobile computing. *IEEE Pervasive Computing*, 8(4):14–23, 2009.
3. F. Bonomi, R. Milito, J. Zhu, and S. Addepalli. Fog computing and its role in the internet of things. In *Proceedings of the First Edition of the MCC Workshop on Mobile Cloud Computing*, MCC '12, pages 13–16, New York, NY, USA, 2012. ACM.
4. N. Klingensmith, D. Willis, and S. Banerjee. A distributed energy monitoring and analytics platform and its use cases. In *Proceedings of the Fifth ACM Workshop on Embedded Systems For Energy-Efficient Buildings*, BuildSys'13, pages 5:1–5:8, New York, NY, USA, 2013. ACM.

2 Mind Your Own Bandwidth

CARLEE JOE-WONG,[1] SANGTAE HA,[2] ZHENMING LIU,[3]
FELIX MING FAI WONG,[4] and MUNG CHIANG[5]

[1]*Electrical and Computer Engineering, Carnegie Mellon University, Silicon Valley, CA, USA*
[2]*Department of Computer Science, University of Colorado at Boulder, Boulder, CO, USA*
[3]*Department of Computer Science, College of William and Mary, Williamsburg, VA, USA*
[4]*Yelp Inc., San Francisco, CA, USA*
[5]*Department of Electrical Engineering, Princeton University, Princeton, NJ, USA*

2.1 INTRODUCTION

The growing popularity of the Internet, and particularly streaming and cloud services like Dropbox and Netflix, has caused a dramatic increase in data usage since 2010 [1]. Internet service providers (ISPs) are thus confronting a difficult question: how should they cope with this growing demand for data as it threatens to overwhelm their available network capacity?

An obvious answer to this question would be to simply expand network capacity to meet users' demands. However, demand for data is growing at such a fast rate that the necessary capacity increases and would require prohibitive amounts of investment [2]. Thus, instead of expanding the bandwidth supply, many ISPs have turned to managing user demand. By limiting bandwidth demand at any given time to lie below the network capacity threshold, ISPs can prevent their networks from becoming over-congested. However, demand restriction is not easy. Simple restriction policies, such as throttling heavy users, can effectively reduce network congestion, but they can unfairly target heavy users. Moreover, simple throttling does not account for the fact that different types of applications require different amounts of data. For instance, throttling streaming videos can significantly degrade users' quality of experience (QoE) while throttling a peer-to-peer download merely delays the time until a user receives a file.

To account for user QoE, ISPs can try to infer users' QoE by classifying their data traffic into different applications and provisioning bandwidth accordingly (e.g.,

Fog for 5G and IoT, First Edition. Edited by Mung Chiang, Bharath Balasubramanian, and Flavio Bonomi.

streaming media gets more Mbps than file downloads). Yet such targeted bandwidth restriction can violate users' privacy, as users may not want their providers to know what types of applications they use.

To preserve users' privacy, ISPs could instead introduce distributed, user-based demand shaping. In these types of systems, users would self-throttle their network traffic, using only the amount that they need, at times when they need it. Since users would be in charge of the throttling, they could take their individual QoE into account, so that users who needed more bandwidth would receive it. However, without suitable incentives, users will not moderate their demand. And giving users these incentives is difficult for an ISP; since the ISP does not know exactly which applications the user needs, it does not know users' exact valuations for network bandwidth and cannot tailor the offered incentives accordingly.

In this chapter, we propose a distributed, user-based incentive mechanism and demonstrate its benefits for both ISPs and users. Users spend virtual "QoE credits" to purchase higher bandwidth from the ISP; since individual users make their own spending decisions, these can take into account individual user QoE. At the same time, the ISP has a degree of control over the total number of credits that users can spend (i.e., the credit budgets for individual users), in order to ensure that users are treated fairly. Within each user's access point, the user can decide how to allocate his or her purchased bandwidth among the different applications and devices that are currently connected to the access point.

2.1.1 Leveraging the Fog

Our user-based system is an example of the *fog networking* paradigm, which shifts communication and control of networked services to the network edge. A fog-like architecture can bring many benefits, for example, allowing users to better control the services they receive while maintaining privacy about their service preferences. Other advantages include scalability; since much of the network functionality is shifted to user devices, expanding the service to other users simply requires their devices to connect to the network. Any entities remaining in the network core will likely scale more easily than before, as they now provide less complex functionalities.

Fog networking can be applied to a broad range of systems, ranging from distributed storage and computing to network bandwidth allocation. In this chapter, we focus on the particular case of broadband cable networks, which have experienced significant congestion in recent years [3]. Our system pushes congestion management to home gateways, that is, combinations of a broadband modem and an in-home Wi-Fi access point. Though often treated as black boxes by their users, we show that these gateways can be configured to implement the fine-granular bandwidth control required by our system.

2.1.2 A Home Solution to a Home Problem

We propose to allocate bandwidth in a two-level hierarchy, as shown in Figure 2.1. Bandwidth is first allocated among home gateways (Level 1) and then allocated locally among the users and devices connected to each gateway (Level 2).

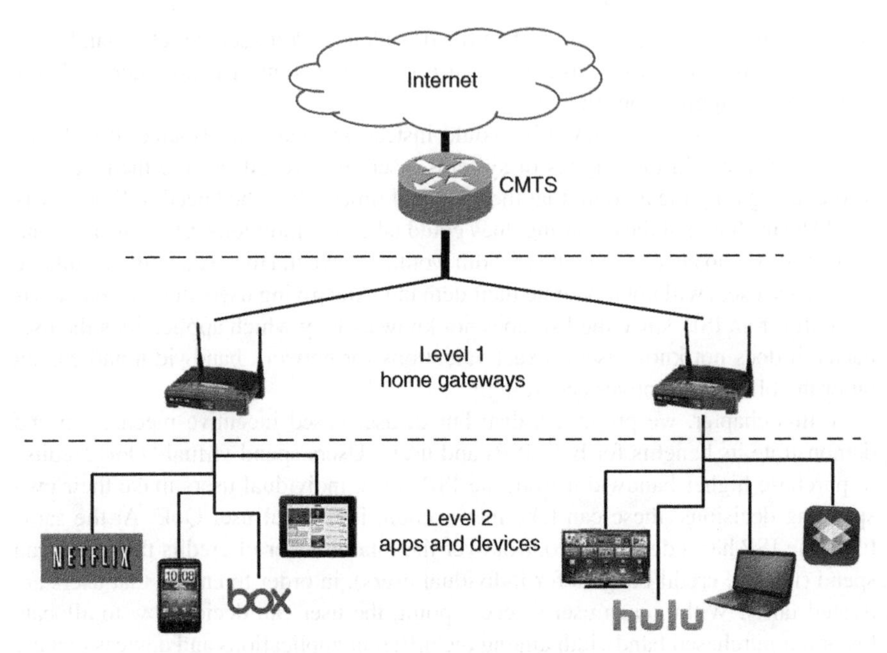

Figure 2.1 Hierarchical edge-based bandwidth allocation.

Level 1: Our Level 1 solution distributes *QoE credits* [4] to the gateways. Each gateway uses its credits to "purchase" guaranteed bandwidth rates at congested times, giving it an incentive to moderate network usage due to its limited credit budget. We limit the total bandwidth demand to the network capacity by fixing the total number of credits in the system and recirculating credits to gateways as they are spent.

The ISP divides the day into a series of discrete time periods, for example, each lasting an hour, and designates some as "congested." At such times, traffic is divided into two classes: a first-tier class that gateways must purchase with credits and a second-tier class that requires no credits but is always of lower priority.[1] This scheme ensures that the network is fully utilized if there is sufficient demand yet still encourages gateways to spend credits at different times.

At the start of the period, each gateway first decides how many of its credits to spend, that is, how much guaranteed bandwidth to purchase.[2] A central server in the ISP's network records the total credits spent by each gateway and redistributes the appropriate number of credits to each gateway in the next time period. Each gateway updates its budget by deducting the credits spent in the previous time period and

[1]By spending credits, users essentially sort their first-class traffic into customized classes defined by the different guaranteed bandwidths purchased.

[2]In practice, an automated agent acting on behalf of the gateway's users makes this decision, with possible manual overrides.

adding the number of credits received. In the next time period, each gateway then knows its updated budget and can again choose how many credits to spend.

In addition to the privacy and scalability advantages of a fog-like architecture, our specific credit redistribution scheme enables us to realize several additional benefits:

- *Fairness Across Gateways.* The credits circulated back to each gateway at a given time depend on other gateways' behavior. Thus, gateways that spend a lot of credits in one time period will have fewer to spend later, ensuring that every gateway will be able to use a fair portion of the bandwidth over time.
- *User-Driven QoE Optimization.* Each gateway spends credits so as to maximize its own overall satisfaction or QoE and is free to allocate this purchased bandwidth among its apps and devices. Our credit redistribution mechanism ensures that gateways' credit spending choices optimize the collective social welfare, that is, *all* gateways' satisfaction, over time.
- *Incremental Deployability.* Since credit spending decisions are made locally by each gateway, our system can be easily deployed via modified home gateways and does not require substantive changes to ISPs' network architecture. Moreover, the number of total credits is fixed, so we can incrementally deploy the solution by starting with a small number of credits and introducing more as more gateways begin to participate.

Level 2: Our Level 2 solution allows gateway users to prioritize different apps and automatically allocates bandwidth accordingly.[3] One user, for instance, might prioritize streaming music, while another might prioritize file transfers. The gateway then divides its purchased bandwidth among these apps according to their priorities. We focus on *elephant traffic*, which tends to be non-bursty and amenable to bandwidth throttling. Since the allocation runs locally at each gateway, the user has full control over these decisions at a session level.

Prioritizing different applications requires both automatically classifying sessions entering the gateway into different apps and enforcing rate limits for each app. While standard mechanisms are available for doing so in a router, they generally require static priority configurations and server-side support when prioritizing downlink traffic. We thus develop our own classification and rate limiting solutions, which run locally on each gateway, and prototype them in a modified home gateway router.

We briefly discuss related works in Section 2.2 before introducing our credit redistribution algorithm in Section 2.3. We show that gateways' total satisfaction is maximized in equilibrium and that each gateway receives a fair amount of bandwidth over time. In Section 2.4, we present practical algorithms for each gateway to decide how many credits to spend and to distribute bandwidth among its apps and devices. We discuss our prototype implementation on a home gateway router in Section 2.5 and present simulation and implementation results of an example scenario in Sections 2.6 and 2.7. We conclude in Section 2.8.

[3]Should explicit prioritization prove too complex for average users, we can introduce default priorities for different types of apps and devices.

2.2 RELATED WORK

As the need for customized data services has grown, there has been much recent work on developing smart home gateways with plain Linux/Windows or open-source router software like OpenWrt. Smart home gateways have been used for network measurement [5,6], providing intuitive interfaces for home network management [7, 8] and better QoE provisioning [9,10]. One such gateway uses weighted fair queueing to allocate uplink traffic according to manual QoE feedback [11]. Another uses traffic classification to sort flows into a fixed set of QoE classes, each of which is given a different priority at the gateway [12]. However, we are not aware of any work in *coordinating* bandwidth usage across households.

Using pricing to manage network congestion is a long-studied research area [2], and recent work has considered practical user interfaces for responding to data prices [13, 14]. Our work differs in targeting broadband users on flat-fee service plans, prompting us to use a credit scheme instead of extra fees for prioritized access. Other credit-based schemes have been proposed for flow admission control [15] and for regulating access to higher-quality service [16], but these have remained theoretical proposals, due to users' reluctance to manually make complex token bidding decisions. We present a complete solution, from algorithms to implementation, for a specific problem of peak-hour broadband access. Moreover, our solution leverages user-specified QoE indicators to optimize traffic according to users' needs; while some works have introduced ways for users to give QoE feedback [17, 18], none of them have used this information to adjust bandwidth allocations.

From an implementation perspective, we develop a new incoming rate limiting tool, as off-the-shelf tools (e.g., Linux `tc`) are insufficient for our application. The congestion manager (CM) project [19] shares similar goals of reducing congestion at the network edge, but we propose a QoE credit scheme to incentivize users to reduce usage, while CM provides an API for applications to adapt to varying network conditions and requires sender-side support. Receiver-side rate control is mostly done through explicitly controlling the receive window [20] or the receive socket buffer [21], for example, to implement low-priority transfers [22] and prioritize traffic [23,24]. Our solution does not modify client devices or track the number of active connections and their round trip times (RTTs). It also avoids interfering with Linux's buffer autotuning mechanism [25]. Our approach of implicit window control is similar to that of Trickle [26], but they serve different goals. Trickle is designed for non-root users to *voluntarily* rate limit their applications, while we aim to impose mandatory rate limits that are transparent to users.

2.3 CREDIT DISTRIBUTION AND OPTIMAL SPENDING

In this section, we describe the bandwidth allocation at the higher level of Figure 2.1. We first describe our system of credits for purchasing bandwidth (Section 2.3.1) and show that it satisfies several fairness properties. We then show that even if each gateway selfishly maximizes its own satisfaction, the total satisfaction across all gateways can be maximized (Section 2.3.2). All proofs are in Appendices 2.A.1–2.A.6.

2.3.1 Credit Distribution

We divide congested times of the day into discrete time periods, for example, of a half-hour duration, and allow gateways to "purchase" bandwidth in each period. The spent credits are redistributed at the end of each period. Users' credit budgets at the end of each day carry over into the next day, so our model is not affected by these time gaps in credit spending.

We suppose that a fixed number $B = \beta C$ of credits is shared by n different gateways, where C is the network capacity in Gbps and β an over-provisioning factor chosen by the ISP. We consider $T + 1$ time periods indexed by $t = 0, 1, \ldots, T$, for example, $T + 1$ periods per week. We use b_{it} to denote the budget, that is, number of credits owned, of gateway i at time t, and we suppose that the total credits are initially distributed equally across gateways, that is, $b_{i0} = B/n$ for all i. For brevity, in the remainder of the paper, we use "budget" to mean "credit budget" or the number of credits available to the gateway at a given time. We use x_{it} to denote the number of credits used by gateway i in time period t. We then update each gateway i's budget as

$$b_{i,t+1} = b_{it} - x_{it} + \frac{1}{n-1} \sum_{j \neq i} x_{jt}, \tag{2.1}$$

where we sum over all gateways j except gateway i. Each gateway i is constrained by $0 \leq x_{it} \leq b_{it}$: it cannot spend negative credits, and the number of credits spent cannot exceed its budget. This credit redistribution scheme conserves the total number of credits for all times t:

Lemma 2.1 *At any time t, the number of credits distributed among gateways is fixed, that is, $\sum_{i=1}^{n} b_{it} = B = \beta C$.*

Since users cannot spend more than their budgets, their total bandwidth purchases are therefore limited to at most βC.

Heavy gateways are prevented from hogging the network, as a large x_{jt} (i.e., large usage by gateway j at time t) simply means that the other gateways $i \neq j$ will receive larger budgets in the time interval $t + 1$. This natural fluctuation in credit budgets enforces a form of *fairness across gateways*. In fact, if this redistribution leads back to a previous budget allocation, all gateways spend the same number of credits:

Lemma 2.2 *Suppose that for some times s and t, $b_{is} = b_{it}$ for all gateways i, for example, $s = 0$ and $b_{it} = B/n$. Then each gateway spends the same number of credits between times s and t: for all gateways i and j, $\sum_{\tau=s}^{t-1} x_{i\tau} = \sum_{\tau=s}^{t-1} x_{j\tau}$.*

Using this result, we can more generally bind the difference in the number of credits gateways can spend:

Proposition 2.1 *At any time t, for any two gateways i and j,* $\left|\sum_{s=0}^{t} x_{is} - \sum_{s=0}^{t} x_{js}\right| \leq B(n-1)/n$. *Thus, the time-averaged difference in spending*

$$\lim_{t\to\infty} \frac{1}{t}\left|\sum_{s=0}^{t} x_{is} - \sum_{s=0}^{t} x_{js}\right| \leq \lim_{t\to\infty} \frac{B(n-1)}{nt} = 0. \tag{2.2}$$

Over time, fairness is enforced in the sense that all gateways can spend approximately the same number of credits.

Though these fairness results limit heavy gateways from hogging the network, gateways with less usage may conversely "hoard" credits, hurting other gateways' budgets. To limit hoarding, we cap each gateway's budget at a maximal value of $\overline{B}$, with $B/n < \overline{B} \leq B$.[4] For instance, the ISP might choose $\overline{B} = B/(n-m+1)$, where m is the minimum number of gateways on the network at any given time. The $n-m$ inactive gateways at that time can then hoard at most $B(n-m)/(n-m+1)$ credits, letting active gateways use the remaining $B/(n-m+1)$ credits.

To enforce this budget cap, the excess budget $\left(b_{it} - x_{it} + \sum_{j\neq i} x_{jt}/(n-1)\right) - \overline{B}$ of any gateway i exceeding the cap is evenly distributed among all gateways below the cap. Should these credits push any gateway over the cap, the resulting excess is evenly redistributed to the remaining gateways until all budgets are below the cap. Since $\overline{B} > B/n$ and we reallocate to fewer gateways after each iteration, this process converges after at most $n-1$ iterations. We expect that users will rarely reach the budget cap, as even without the cap, no single gateway can hoard all available credits:

Proposition 2.2 *Let* $\alpha = (n-2)/(n-1)$ *and suppose that a given gateway i uses at least* ϵ *bandwidth every p periods, where* $B/n > \epsilon \geq 0$ *and p may denote, for example, 1 day. Then at any time t, gateway i's budget is*

$$b_{it} \leq \frac{B}{n}\alpha^{t+1} + B\left(1-\alpha^{t+1}\right) - \epsilon\left(\frac{\alpha^{p} - \alpha^{p\left(1+\left\lfloor \frac{t+1}{p}\right\rfloor\right)}}{1-\alpha^{p}}\right)$$

$$\to B - \frac{\epsilon\alpha^{p}}{1-\alpha^{p}} \tag{2.3}$$

as $t \to \infty$. *Thus, if* $\epsilon > 0$, $b_{it} < B$. *Moreover, at any fixed time t, at most one gateway can have a budget of zero credits.*

For instance, if a gateway spends ϵ credits at each time, then as $t \to \infty$, $b_{it} \leq B - \epsilon(n-2)$; if ϵ is relatively large, a gateway hoards fewer credits, since these are redistributed among others once spent. Conversely, a gateway that spends very little can asymptotically hoard almost B credits. More broadly, if a number m of gateways are inactive in a network for a certain number of time periods s, then we can bound the number of credits these m gateways accumulate:

[4] If $\overline{B} = B/n$, then we would have $b_{it} = B/n$ for all gateways i at all times t. We therefore take $\overline{B} > B/n$ to ensure that there is a feasible set of budgets $\{b_{it}\}$ with each $b_{it} \leq \overline{B}$.

Proposition 2.3 *Suppose that m gateways are inactive from times 0 to* $s-1$ *(i.e., for s periods). Then the number of credits that these gateways can accumulate by time s is given by*

$$\sum_{i=1}^{m} b_{is} - b_{i0} \leq \left(1 - \left(\frac{n-2}{n-1}\right)^{s}\right)\left(B - \sum_{i=1}^{m} b_{i0}\right) \tag{2.4}$$

where we index the inactive gateways by $i = 1, 2, \ldots, m$.

2.3.2 Optimal Credit Spending

Given the previous credit distribution scheme, each gateway must decide how many credits to spend in each period. To formalize this mathematically, let U_{it} denote gateway i's utility as a function of the guaranteed bandwidth x_{it} in time interval t. Though gateways may increase their utilities with second-tier traffic, we do not consider this traffic in our formulation. Second-tier bandwidth is difficult to predict: gateways could only obtain historical information on its availability by regularly sending such traffic, which they are unlikely to do.

We consider a finite time horizon T, for example, 1 week, since the utility functions cannot be reliably known far into the future. Each gateway i then optimizes its total utility from the current time s to $s+T$:

$$\max_{x_{it}} \sum_{t=s}^{s+T} U_{it}\left(x_{it}\right), \quad \text{s.t. } 0 \leq x_{it} \leq b_{it}, \ \forall t. \tag{2.5}$$

Here the budgets b_{it} are calculated using the credit redistribution scheme (2.1), with appropriate adjustments to enforce the budget limit $\overline{B}$. For ease of analysis, we do not model these budget caps here. In practice, the ISP can cap gateways' budgets for each time period during the credit redistribution.

We first note that the budget expressions (2.1) can be used to rewrite the inequality $x_{it} \leq b_{it}$ as the linear function

$$\sum_{\tau=s}^{t} x_{i\tau} - \sum_{j\neq i}\sum_{\tau=s}^{t-1} \frac{x_{j\tau}}{n-1} \leq b_{i0}. \tag{2.6}$$

Thus, if the U_{it} are concave functions, then given the amount spent by other gateways, $x_{j\tau}$, (2.5) is a convex optimization problem with linear constraints.[5]

Since each gateway chooses its own x_{it} to solve (2.5), these joint optimization problems may be viewed in a game theoretic sense: each gateway is making a decision that affects the utilities of other gateways. From this perspective, the game has a Nash equilibrium at the social optimum:

[5]The assumption of concavity, that is, $U''_{it}(x_{it}) < 0$, may be justified with the economic principle of diminishing marginal utility as bandwidth increases.

Proposition 2.4 *Consider the global optimization problem*

$$\max_{x_{it}} \sum_{i=1}^{n} \sum_{t=s}^{s+T} U_{it}\left(x_{it}\right), \quad \text{s.t. } 0 \leq x_{it} \leq b_{it}, \ \forall i, t \tag{2.7}$$

with the credit redistribution (2.1) and strictly concave U_{it}. Then an optimal solution $\{x_{it}^*\}$ to (2.7) is a Nash equilibrium.

While Proposition 2.4's result is encouraging from a system standpoint, in practice this Nash equilibrium may never be achieved. Since the gateways do not know each others' utility functions, they do not know how many credits will be spent and redistributed at future times, making the future credit budgets unknown parameters in each gateway's optimization problem. These must be estimated based on historical observations, which we discuss in the next section.

2.4 AN ONLINE BANDWIDTH ALLOCATION ALGORITHM

We now consider a gateway's actions at both levels of bandwidth allocation. We first give an algorithm to decide credit spending (Level 1) and then show how the purchased bandwidth can be divided at the gateway (Level 2). Using Algorithm 2.1, each gateway iteratively estimates the future credits redistributed, chooses how many credits to spend, prioritizes apps, and updates its credit estimates. We assume the gateway's automated agent knows its users' utility functions.

2.4.1 Estimating Other Gateways' Spending

To be consistent with (2.5)'s finite time horizon, we suppose that gateways employ a *sliding window* optimization. At any given time s, gateway i chooses rates for the next T periods $s, \ldots, s+T-1$ so as to maximize its utility for those periods. At time $s+1$, the gateway updates its estimates of future credits redistributed and optimizes over the next T periods.

We use *scenario optimization* to estimate the number of credits each gateway will receive in the future.[6] Scenario optimization considers a finite set S_i of possible scenarios for each gateway i, associating each scenario $\sigma \in S_i$ with a probability π_σ that it will take place. Computing the credit redistribution and optimal spending x_{it} for each σ then yields a probability distribution of the possible credits spent. In our case, a "scenario" is a set of utility functions $\{U_{jt}\}$ for the other gateways. We parameterize these scenarios by noting that gateways' utilities depend on the application used, for example, streaming versus downloading files. We consider K different applications and define $u_k(x)$ as the (predetermined) utility from an application of type k (e.g., $k=1$ corresponds to streaming, $k=2$ to file downloads, etc.). We thus take

$$U_{jt} = \gamma_{jt} \sum_{k=1}^{K} p_{jt}^{k} u_k \tag{2.8}$$

[6]This technique is often used in finance to solve optimization problems with stochastic constraints that are hard to predict, for example, market dynamics [27].

Algorithm 2.1 Gateway Spending Decisions

$s \leftarrow 1$ {s tracks the current time.}
while $s > 0$ **do**
  if $s > 1$ **then**
    Update estimate of future amounts redistributed using Algorithm 2.2.
  end if
  Calculate $\sum_{j \neq i} x_{jt}/(n-1)$ for $t = s, \ldots, s + T - 1$.
  Solve (2.5) with budget constraints (2.9) given $\sum_{j \neq i} x_{jt}/(n-1)$.
  Choose the application priorities μ_k by solving (2.10).
  $s \leftarrow s + 1$
end while

for each gateway j, where γ_{jt} is a scaling factor specified by individual gateways. The variable p_{jt}^k denotes the (estimated) probability that gateway j optimizes its usage with the utility function u_k, for example, if app k is used the most at time t.

With this utility definition, we can define a scenario σ by the coefficients $\gamma_{jt}(\sigma)$ and $p_{jt}^k(\sigma)$ of gateways' utility functions. Since gateway i cannot distinguish between other gateways, it need only estimate their behavior in aggregate. These gateways can be thus viewed as one "gateway" j by adding their utility functions and budget constraints. Gateway j then maximizes

$$U_{jt} = \sum_{t=1}^{T} \gamma_{jt}(\sigma) \sum_{k=1}^{K} p_{jt}^k(\sigma) u_k$$

subject to the budget constraints $0 \leq x_{jt} \leq b_{jt}$, where the coefficients $\gamma_{jt}(\sigma) p_{jt}^k(\sigma)$ represent the added coefficients for all gateways $\neq i$. Since gateway i cannot know the accuracy of its or gateway j's estimates of future usage, for the purpose of estimation, we assume that both gateways' future usage estimates are correct. Thus, following Proposition 2.4, all gateways choose their usage so as to maximize the collective utility $\sum_t \left(U_{jt} + U_{it}\right)$ subject to the budget constraints. This optimization may be solved to calculate the credits $\sum_{j \neq i} x_{jt}(\sigma)/(n-1)$ redistributed to user i at each time t in scenario σ.

To improve our credit estimates, at each time t we update the scenario probabilities π_σ by comparing the observed number of credits redistributed at time $t-1$, denoted by $\sum_{j \neq i} \overline{x}_{j,t-1}/(n-1)$, with the estimated amount redistributed $\sum_{j \neq i} x_{j,t-1}(\sigma)/(n-1)$ for each $\sigma \in S_i$. We suppose that gateways' behavior is sufficiently periodic (e.g., over $T = 1$ week) for the π_σ at times t and $t + T$ to be the same.

We use $P\left(\sum_{j \neq i} \overline{x}_{j,t-1} = \sum_{j \neq i} x_{j,t-1}(\sigma)\right)$ to denote the probability that, given $\sum_{j \neq i} \overline{x}_{j,t-1}/(n-1)$ at time t, gateways $\neq i$ use scenario σ's utility function at time t. We can calculate these probabilities by measuring the L_2 discrepancy between the estimated and observed credits redistributed:

Algorithm 2.2 Estimating Credit Redistribution

$s \leftarrow 1$ {s tracks the current time.}
while $s > 0$ **do**
  for all gateways $i = 1, \ldots, n$ **do** {this loop may be run in parallel}
    Choose scenarios S_i.
    for each scenario $\sigma \in S$ **do**
      Calculate the predicted amount redistributed $\sum_{j \neq i} x_{jt}(\sigma)/(n-1)$ for $t = s, \ldots, s+T-1$, assuming other gateways know x_{it} for all t.
      if $s > 1$ **then**
        Update probability π_σ using Bayes' Rule.
      end if
    end for
  end for
end while

$$P\left(\sum_{j\neq i} \bar{x}_{j,t-1} = \sum_{j\neq i} x_{j,t-1}(\sigma)\right) = \frac{1}{|S_i|-1}\left(1 - \frac{\left(\sum_{j\neq i} \bar{x}_{j,t-1} - \sum_{j\neq i} x_{j,t-1}(\sigma)\right)^2}{\sum_{l=1}^{|S_i|}\left(\sum_{j\neq i} \bar{x}_{j,t-1} - \sum_{j\neq i} x_{j,t-1}(l)\right)^2}\right).$$

We then update the scenario probabilities π_σ using Bayes' rule and use the new π_σ in Algorithm 2.2.

2.4.2 Online Spending Decisions and App Prioritization

Algorithm 2.1 shows how the credits spent in different scenarios are incorporated into choosing a gateway's rates x_{it} and application priorities. Each gateway constrains its spending depending on the estimated redistributed credits: for instance, a conservative gateway might choose the x_{it} so that the budget constraints $0 \leq x_{it} \leq b_{it}$ hold for all scenarios. In the discussion that follows, we suppose that gateways constrain the x_{it} so that (2.6) holds in expectation:

$$\sum_{\tau=s}^{t} x_{i\tau} - \sum_{\sigma\in S_i} \pi_\sigma \left(\sum_{j\neq i}\sum_{\tau=s}^{t-1} \frac{x_{j\tau}(\sigma)}{n-1}\right) \leq b_{i0}. \tag{2.9}$$

We also constrain $b_{it} \leq \overline{B}$, that is, the expected budget at a given time cannot exceed the budget cap; users would rather spend more credits to remain under the budget cap than be forced to redistribute excess credits to other gateways.

Each gateway can further improve its own experience with its Level 2 allocation, dividing the purchased bandwidth among its apps. It does so by assigning priorities to different devices and applications, so that higher-priority apps receive more bandwidth. Since users cannot be expected to manually specify priorities in each time period, we introduce an automated algorithm that leverages the gateway's known utility functions (2.8) to optimally set application priorities.

We consider the K application categories in (2.8) and use μ_k to represent each category k's priority. Since the applications that are active at a given time may change during a period, for example, if a user starts or stops watching a video, we define an app's priority in relative terms: for any apps k_1 and k_2, $\mu_{k_1}/\mu_{k_2} = y_{k_1}/y_{k_2}$, where y_k is the bandwidth allocated to application k and $\sum_k y_k = x_{it}$, ensuring that all the purchased bandwidth is used. We normalize the priorities to sum to 1: $\sum_k \mu_k = 1$.

Since it is difficult to predict which apps will be active at a given instant of time, we choose the app priorities μ_k according to a "worst-case scenario," in which all apps are simultaneously active. In this case, each app k receives $y_k = \mu_k x_{it}$ bandwidth, and we choose the μ_k to maximize total utility:

$$\max_{\mu_k} \sum_{j=1}^{K} u_k \left(\mu_k x_{it} \right), \quad \text{s.t.} \sum_{k=1}^{m} \mu_k = 1. \tag{2.10}$$

Since each function u_k is assumed to be concave and the constraint is linear in the μ_k, (2.10) is a convex optimization problem and may be solved rapidly with standard methods.

2.5 DESIGN AND IMPLEMENTATION

Figure 2.2 summarizes the architecture of our system. It consists of four modules: (i) When traffic goes through the gateway for forwarding, it is passed to a device and application classifier to identify the traffic type and priority. (ii) All traffic is

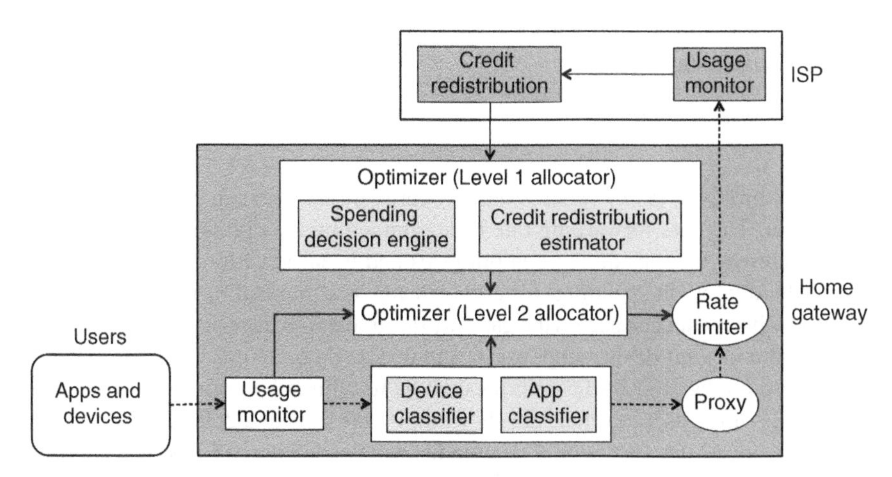

Figure 2.2 System architecture. Dashed lines represent traffic flow, and solid lines represent rate and credit information.

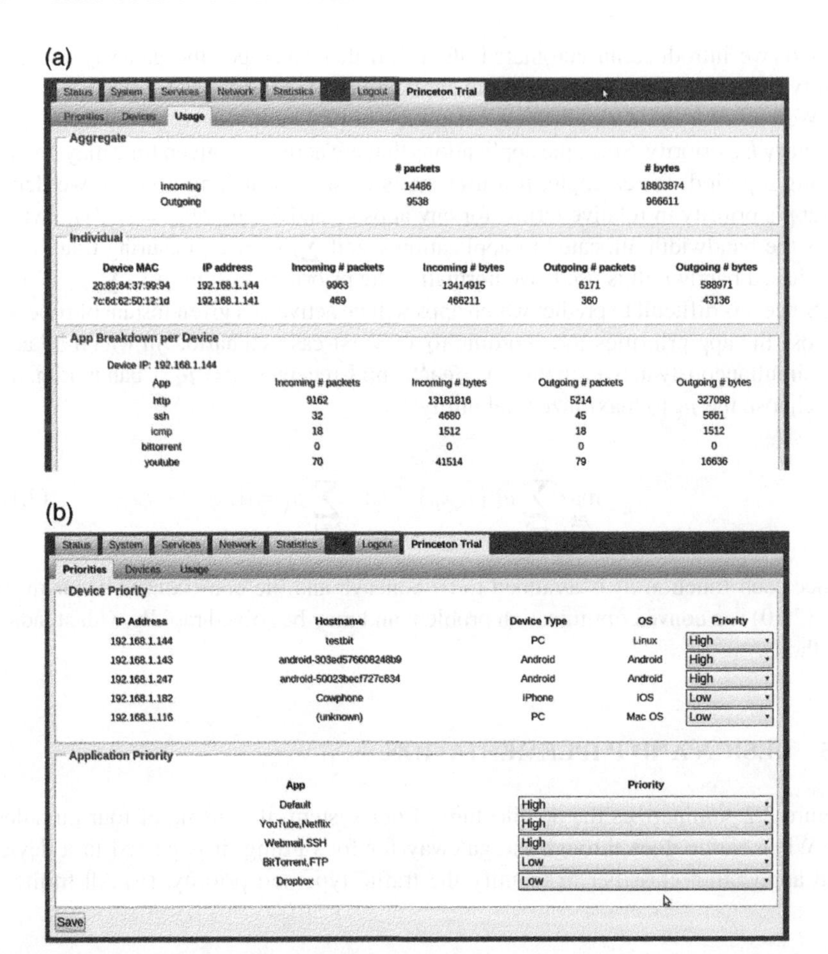

Figure 2.3 Screenshots of the web interface. (a) Usage tracking and (b) traffic priorities and device/OS classification.

redirected through a proxy process that forwards traffic between client devices and the Internet. The data forwarding rate is determined by the optimizer (L2 allocator) in each gateway by considering app priorities and is enforced by a rate limiter. (iii) The bandwidth (credit spending) for each gateway is computed by the optimizer (L1 allocator). (iv) A user can access the gateway through a web interface to view its usage (at aggregate or joint device–app levels) and update its preferences, that is, when to spend more credits and traffic priorities, so as to adjust the optimizer's decisions. Screenshots of the user interface are shown in Figure 2.3a and 2.3b.

We implement our system in a commodity wireless router, a Cisco E2100L with an Atheros 9130 MIPS-based 400 MHz processor, 64 MB memory, and 8 MB flash storage. We replaced the factory default firmware with OpenWrt, a Linux distribution commonly used for embedded devices. The implementation poses two significant challenges:

Traffic and Device Classification. Standard approaches for classifying traffic from different devices include port-based protocol detection and operating system (OS) fingerprinting. However, different apps can run on the same protocol, for example, videos streamed in HTTP, and many device types run on the same OS, for example, most smartphones run on a variant of Linux. Moreover, home gateways have only limited computational resources, but both these approaches require significant computational overhead.

Rate Limiting and Prioritization. We can limit a session's bandwidth rate by directly setting its TCP-advertised window size [23,24]. However, doing so requires knowing each connection's RTT and the number of active connections, both of which can be difficult to estimate in practice.

2.5.1 Traffic and Device Classification

To build a low-overhead classifier, we integrate a kernel-level `netfilter` module that inspects the first several packets of a connection for application matching. If a match is found, the classifier marks the connection with a mark to be queried at userspace by our proxy processes through `netlink`.

Our classifier module performs traffic classification *above* layer 7, that is, it can differentiate YouTube and Netflix, through a combination of content matching, byte tracking, and protocol fingerprinting. We classify devices and OSes by using the same module to monitor HTTP traffic and inspect user-agent header strings for device information. This approach is practically effective due to the prevalence of devices using HTTP traffic.

2.5.2 Rate Limiting Engine

Our goal is to (i) enforce an aggregate rate limit over multiple connections and (ii) enforce prioritization, that is, which gets higher bandwidth, among the connections given the aggregate limit. In this paper we only consider throttling *incoming* traffic, because the other direction can be easily and accurately done using standard token bucket-based traffic shaping tools.

Transparent Proxy. During the establishment of a connection between a client device and a server, it is intercepted at the gateway and redirected to the proxy process running in the gateway. Then the proxy establishes a new connection to the server on behalf of the client and forwards traffic between the two (proxy-server and client-proxy) connections. We use the Linux `splice()` function to achieve zero copying, that is, all data are handled in kernel space.

Implicit Receive Window Control. TCP's flow control mechanism allows the receiver of a connection to advertise a receive window to the sender so that incoming traffic does not overwhelm the receiver's buffer. While originally set to match the available receiver buffer space, the receive window can be artificially set to limit bandwidth using the relation $\texttt{cwnd} = \texttt{rate} \times \texttt{RTT}$: given a maximum `rate` and

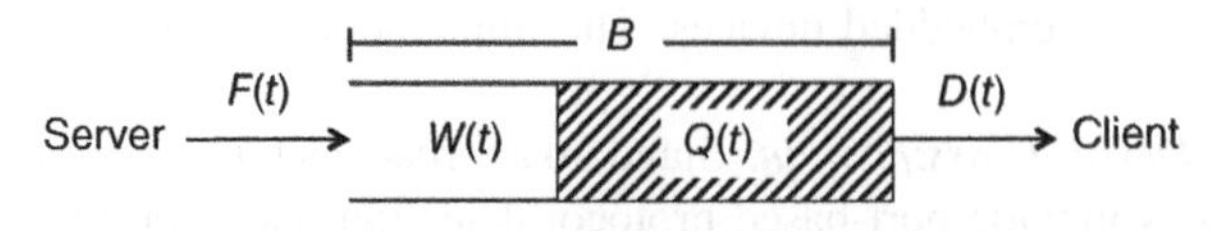

Figure 2.4 Receive buffer model.

measured RTT, the receive window can be set to no greater than rate × RTT. We opt for an *adaptive* approach such that the proxy does not need to know the RTT or compute the exact window size.

To illustrate our approach we consider a one connection case. As data from the server arrive at the proxy, they are queued at the proxy's receive buffer until the proxy issues a `recv()` on the proxy-server socket to process and clear them (at the same time the proxy issues a `send()` on the client–proxy socket to forward the data to the client). Note that if we modulate the frequency and the size of `recv()`'s, we modulate the size of the receive buffer and effectively the sending rate.

More specifically, we consider the model in Figure 2.4: the queue is the proxy's receive buffer, B is the receive buffer size,[7] and at time t, $F(t)$ is the fill rate (sending rate, which the proxy cannot directly control), $D(t)$ is the drain rate (how frequent the proxy issues `recv()`'s), $Q(t)$ is the queue length, and $W(t) = B - Q(t)$ is the advertised window size.

Suppose updates happen at intervals of Δt. The window update equation is then

$$W(t + \Delta t) = W(t) + \left[D(t) - F(t)\right]^{+} \Delta t \tag{2.11}$$

and taking a fluid approximation by setting $\Delta t \to 0$, we have

$$\dot{W}(t) = \left[D(t) - F(t)\right]^{+}. \tag{2.12}$$

Our rate limiting goal is equivalent to getting $F(t) = R$ for large enough t through controlling $D(t)$. By setting $D(t) = R$ at all t, it is not difficult to verify from (2.12) that at equilibrium[8] we have $F^{*}(t) = R$ and $W^{*}(t) = R \times \mathrm{RTT}$.

2.5.3 Traffic Prioritization Engine

When there are multiple connections, the proxy spawns multiple threads such that each thread serves one connection, and we aim to limit the aggregate rate R over all connections. To allocate bandwidth fairly among the connections, we coordinate socket reads of these threads through a time division multiplexing scheme, using a thread mutex, we create a virtual time resource such that each socket read is associated with an exclusively held time slot of length proportional to the number of bytes read. Although more complicated socket read scheduling mechanisms can be considered,

[7] The receive buffer size can change with time due to Linux's buffer autotuning mechanism, but these changes do not affect our algorithm.

[8] Note that if we throttle a connection through TCP flow control, a static equilibrium can indeed be achieved because the rate is now limited by the receive window rather than self-induced congestion, that is, the usual sawtooth $W(t)$ time evolution no longer occurs.

Algorithm 2.3 Pseudocode of Incoming Rate Control

```
Input: R, b, α_i,
  server_fd: socket of server connection,
  client_fd: socket of client connection,
  mutex: thread mutex shared by all connections
  while connection open do
    bytes_read = recv(server_fd, b)
    bytes_per_write = α_i × bytes_read
    while not all bytes_read written to client do
      send(client_fd, bytes_per_write)
      lock(mutex)
          sleep(bytes_per_write/R)
      unlock(mutex)
    end while
  end while
```

for simplicity we leave the scheduling to the OS, and from experiments, we observe that the time slots are shared fairly.

For traffic prioritization we assign a relative priority parameter $\alpha_i \in (0, 1]$ for every connection i such that for n *busy* connections, that is, each has a sufficiently large backlog, we want the sum of their rates R_i to be $\sum_{i=1}^{n} R_i = R$, and $R_i/R_j = \alpha_i/\alpha_j$ for $i, j = 1, \ldots, n$.

We achieve the desired prioritization through truncated reads. When the proxy issues a socket read, it needs to specify a maximum block size b to read (we set it as the page size of the processor architecture), and for a busy connection this limit b is always reached. If connection i is of lower priority with $\alpha_i < 1$, we truncate this block limit by setting it to be $\alpha_i b$. Since each access to a time slot is associated with a server, socket read (equivalently, a client socket write) of $\alpha_i b$ bytes and time slots are fairly distributed across connections, the achieved client rate D_i (equivalent to R_i) scales with α_i.

By virtue of statistical multiplexing, our rate allocation mechanism does not require the number of busy connections n, which is difficult to track in practice; hence it can readily accommodate new connections. To accommodate bursty connections, the proxy first queries the receive buffer for the number of pending bytes. If it is above b, then it does a truncated read as described previously; otherwise it does not. The pseudocode of a proxy thread is shown in Algorithm 2.3.

2.6 EXPERIMENTAL RESULTS

2.6.1 Rate Limiting

We compare our approach with the standard Linux traffic policing approach using the `tc` command with two different choices of the burst parameter. Two experiments are performed using `iperf`. In the first one, we fix network RTT to be 100 ms and vary the rate limit from 1 to 15 Mbps to observe the actual rate achieved. Figure 2.5a shows that our approach results in more accurate rate limiting (<4% error in each setting). While it appears that increasing the burst parameter helps in improving rate limiting accuracy, we note the values chosen are rather large (a typical value is 10k, while we use 50k and 200k) and may harm network stability. The sensitivity of the results of `tc` with respect to the parameters suggests the need for careful parameter tuning, which is undesirable given the diversity of network environments.

The first experiment hints that traffic policing, or using packet drops to signal the sender to reduce its rate, is too drastic as a rate control mechanism. Our second experiment confirms this observation. We fix the rate limit at 8 Mbps and burst parameter at 50k and vary network RTT from 20 to 100 ms. Figures 2.5b and 2.5c shows that `tc` results in significantly more packet retransmissions and higher jitter. This result shows that our approach is indeed more graceful in rate limiting.

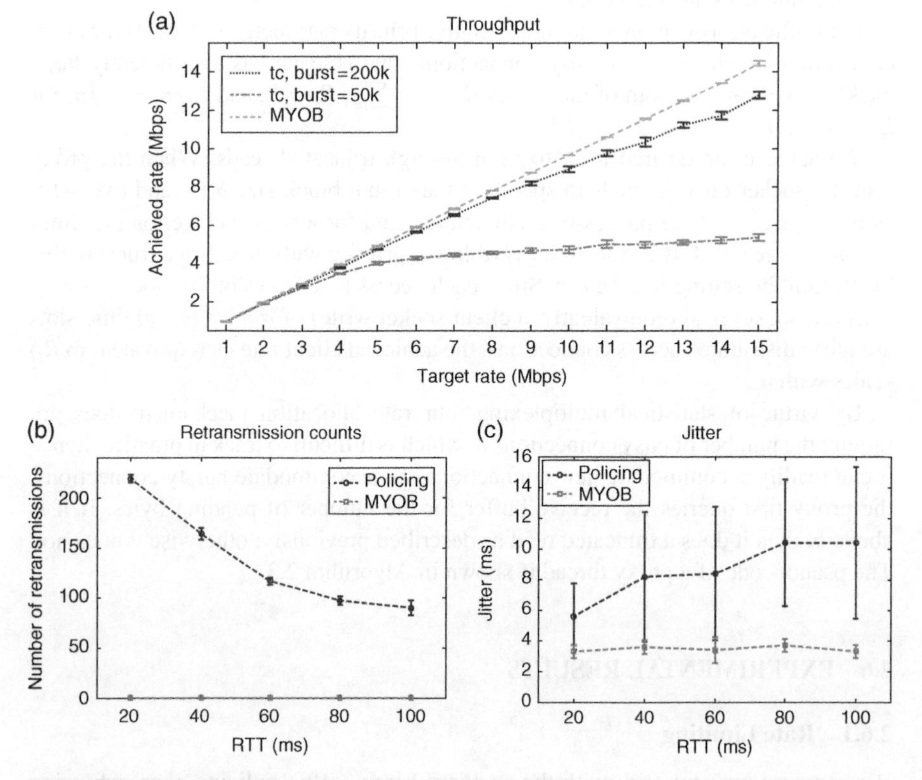

Figure 2.5 Our rate limiting algorithm is (a) more accurate than `tc` and (b, c) more graceful than rate limiting. We average all results over 10 runs, 60 seconds each, and show 95% confidence intervals.

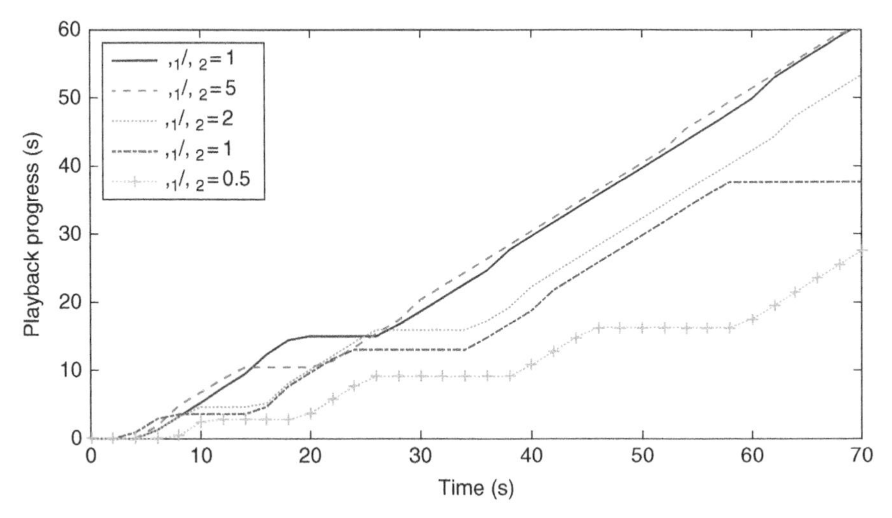

Figure 2.6 YouTube playback performance improves as α_1/α_2 increases and YouTube receives higher prioritization over `wget`.

2.6.2 Traffic Prioritization

Consider a scenario with two users, one watching a 720p YouTube video stream and the other downloading a large file with `wget`, competing for a limited bandwidth of 2 Mbps. We vary the priorities of the two types of traffic and observe the effect on video playback.

Let α_1 and α_2 be the priorities of YouTube and `wget`, respectively. With α_1 fixed, we vary α_2 and measure the amount of video played over time.[9] Note that there are two base cases: the case $\alpha_1/\alpha_2 = \infty$ corresponds to YouTube traffic without `wget` interference and is the best possible result we can expect and the case $\alpha_1/\alpha_2 = 1$ is equivalent to no prioritization. Figure 2.6 shows the results. When $\alpha_1/\alpha_2 > 1$, that is, YouTube has higher priority and playback performance (inversely related to the duration of pauses or flat regions in a curve) and is strictly better than the no prioritization case. Also, performance improves with increasing α_1/α_2 ratio. Not only is our system able to do fine-grained traffic classification with two types of traffic running under HTTP, but our traffic prioritization algorithm also produces noticeable improvement in user experience.

2.7 GATEWAY SHARING RESULTS

To demonstrate our sharing framework's efficacy, we simulate the behavior of 16 gateways sharing a cable link. We compare our credit-based allocation to *equal sharing*, in which the ISP divides its capacity into slots with a minimally

[9] We create a video-embedded webpage with a Javascript snippet that periodically queries the YouTube API for playback progress.

acceptable level of bandwidth, for example, 1 Mbps, and assigns them to gateways in a round-robin fashion until the network capacity is reached. This approach, which is similar to current practices in that gateways are all treated equally, risks inefficiency: gateways may gain little additional utility from the full bandwidth of their assigned slots, but cannot redistribute any excess bandwidth to gateways that would benefit more. Our credit-based approach addresses this disadvantage, and we show in our simulations that it significantly improves gateway utilities while enforcing a fair rate allocation. We then evaluate our online algorithm for users' credit spending decisions (Algorithm 2.1). We find that all gateways optimize their credit spending to achieve fair, near-optimal utilities despite their uncertain future budgets.

Simulation Setup. We suppose that credit-based sharing is enforced in the congested hours between 6 pm and midnight, with half-hour time slots. Users at each gateway are assumed to make their credit spending decisions based on their probability of using four types of applications: streaming, social networking, file downloads, and web browsing. We use the utility functions

$$u_1(x) = \frac{2(25x)^{1-\alpha_1}}{1-\alpha_1}$$

$$u_2(x) = \frac{(25x)^{1-\alpha_2}}{1-\alpha_2}$$

$$u_3(x) = \left(\frac{1}{\alpha_3 - 1} + \frac{(25x+1)^{1-\alpha_3}}{1-\alpha_3}\right)$$

$$u_4(x) = 15\left(\frac{1}{\alpha_4 - 1} + \frac{(25x+1)^{1-\alpha_4}}{1-\alpha_4}\right)$$

in (2.8) to respectively model the utility received from each application, where $(\alpha_1, \alpha_2, \alpha_3, \alpha_4) = (0.7, 0.5, 0.2, 3)$. The probabilities p_{it}^k of using each application are adapted from a recent measurement study of per app usage over time for iOS, Android, Windows, and Mac smartphones and computers [28]. Table 2.1 shows the devices at each gateway. We choose coefficients $\gamma_i(t)$ to be larger in the evening, as is consistent with observed data usage [28], and add random fluctuations to model period-to-period variations in each gateway's behavior.

TABLE 2.1 Number of Devices at Each Gateway

Gateways	iPhones	Androids	Windows Laptops	Mac Laptops
1,4,9,13	1	1	1	1
2,6,10,14	2	0	2	1
3,7,11,15	1	1	1	2
4,8,12,16	2	0	1	1

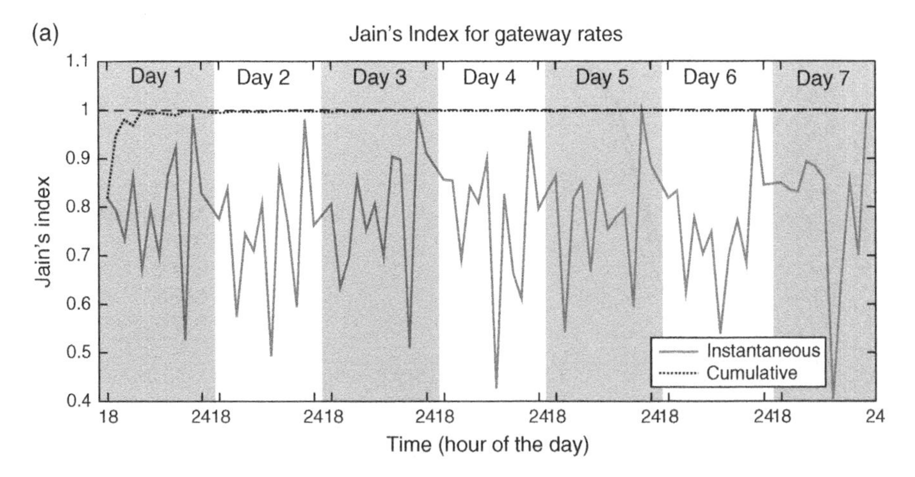

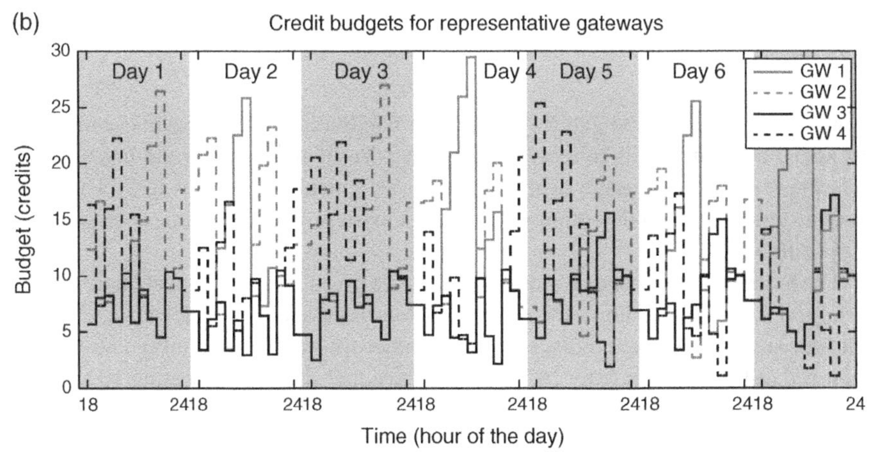

Figure 2.7 With our credit sharing scheme, all gateways (a) achieve comparable cumulative rates by (b) actively saving and spending credits at different times.

We assume a budget of $B = 160$ total credits, with each credit representing 1 Mbps.[10] The budget b_{it} for each gateway i is capped at 32 credits at any given time. In addition to the purchased bandwidth, we suppose that gateways send a random amount of traffic over the second tier, which is capped at the network capacity. We consider 1 week of credit redistributions and bandwidth allocations.

Globally Optimal Solution. We first compute the globally optimal rates, that is, those that maximize (2.7). To show that the overall rate allocation is fair, we compute Jain's index over the gateways' rates, including second-tier traffic, at each time in Figure 2.7a. Jain's index is relatively low at some times, indicating a large variation

[10]Though 160 Mbps is a relatively small bottleneck bandwidth, we limit the number of users and link capacity in order to better illustrate the effect of QoE credit allocation on individual users.

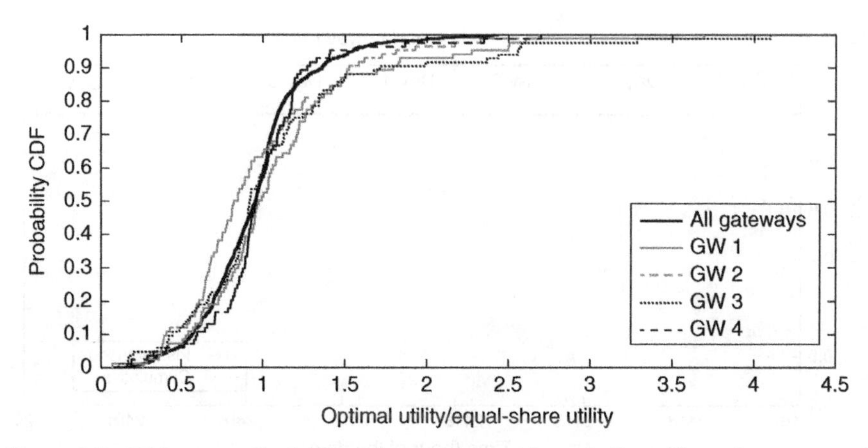

Figure 2.8 With our credit sharing scheme, users achieve similar utility gains over equal sharing over 1 week.

in gateways' rates: some gateways use little bandwidth to save credits, while others spend a lot of credits to receive large rates. Yet if we compute the index for all gateways' *cumulative* usage over time, its value quickly converges to 1. The gateways receive comparable cumulative rates, consistent with the fairness property of Proposition 2.1.

The large variability in gateway allocations at a given time can be seen more clearly in Figure 2.7b, which shows the budgets of four representative gateways over time. All four sometimes save credits to spend at other times. This flexibility causes total achieved utility to increase by 29.7% relative to equal sharing (allocating 10 Mbps to each gateway at all times). Figure 2.8 shows the cumulative density function (CDF) of the ratio of gateway utilities under credit allocation and equal sharing at different times. We plot the CDF over all gateways and times as well as the CDF over all times for each gateway shown in Figure 2.7b. All of the CDFs are comparable, indicating that credit-based allocation benefits all gateways' utilities. While the gateways reduce their utility nearly half of the time, the utility more than doubles in some periods.

Online Solution. We next compare the globally optimal utilities with those obtained when the gateways follow Algorithm 2.1. To perform the credit estimation, we use four scenarios, in which all other gateways are assumed to use only streaming, only social networking, and so on. Each gateway assumes (falsely) that the other gateways' $\gamma_i(t)$ coefficients are the same, and the probabilities π_σ of each scenario are initialized to be uniform. After learning the scenario distribution for only the simulation's first 4 days, the algorithm recovers most (84.7%) of the optimal utility for the remaining 3 days.

As with the optimal solution, at any given time, gateways' rates can be very different: Jain's indices in Figure 2.9a for all gateways' usage at a given time can be quite low. However, all gateways achieve similar cumulative rates: Jain's index of the

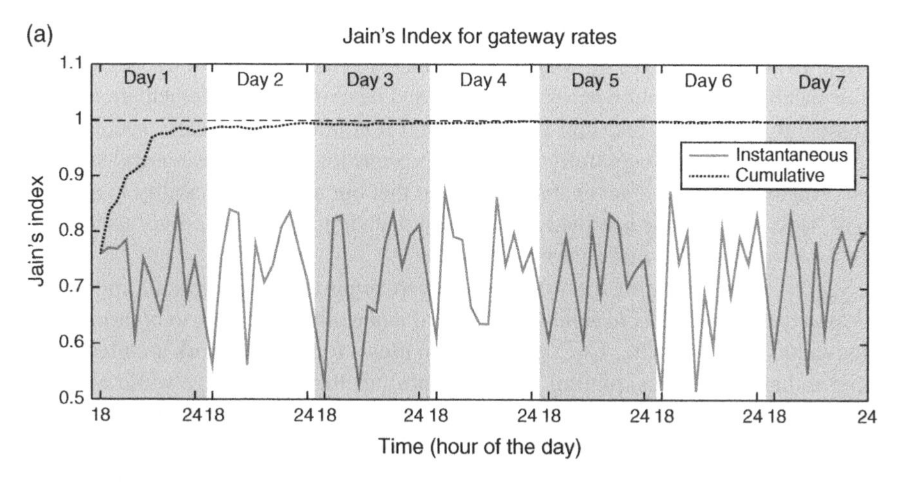

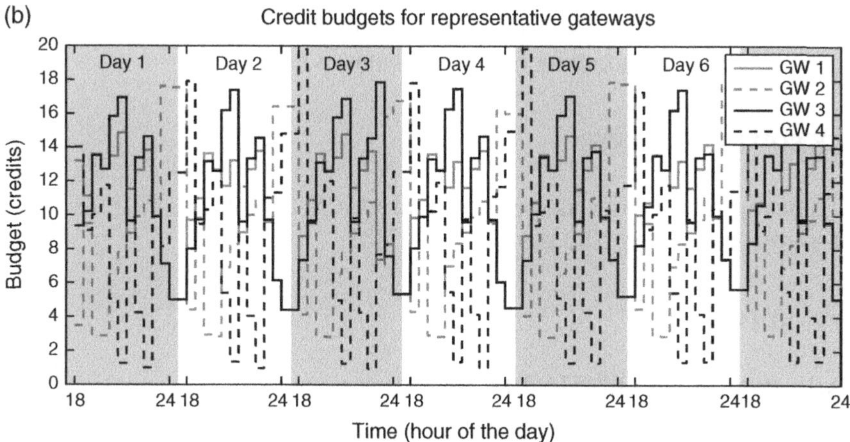

Figure 2.9 Despite their uncertain future budgets, with our online algorithm gateways (a) achieve comparable cumulative rates by (b) saving and spending credits at different times over 1 week.

cumulative rates quickly converges to 1. Indeed, the budgets (and thus spending) of four representative gateways (Figure 2.9b) vary over time, as with the optimal solution (Figure 2.7b). Incentivizing gateways to delay some of their usage significantly improves users' overall satisfaction and utility.

2.8 CONCLUDING REMARKS

In this paper, we propose to solve peak-hour broadband network congestion problems by pushing congestion management to the network edge. Our solution uses two levels of bandwidth allocation: in Level 1, home gateways purchase bandwidth on a shared link using QoE credits, and in Level 2, they divide the purchased bandwidth among their apps and devices. We show analytically that our credit distribution scheme yields a fair bandwidth allocation across gateways and describe our implementation of the bandwidth purchasing and app prioritization on commodity wireless routers. Our implementation can successfully enforce app priorities and increase users' satisfaction. Finally, we show in an example scenario that our algorithm's ability to adapt to users' QoE yields a fair bandwidth allocation that significantly improves user utility relative to a baseline equal-sharing scheme.

Successfully managing network congestion requires an integrated, end-to-end solution that allows ISPs to limit user demand while allowing users to optimize their bandwidth usage for better QoE. Our solution uses a fog-like network architecture to achieve these objectives, putting users in control of their data usage while allowing some control from the ISP in the form of bandwidth credits. While we implement this solution for cable networks, our methodology is applicable to other access technologies, for example, cellular, that involve shared medium access. Such technologies, wireless and wired, will increasingly need new congestion management mechanisms as user demand for bandwidth continues to grow.

ACKNOWLEDGMENTS

An earlier version of this work appeared in IEEE/ACM IWQoS 2015 [29]. This work was partly supported by NSF grant CNS-1525435.

APPENDIX 2.A

2.A.1 PROOF OF LEMMA 2.1

We proceed by induction: at time $t = 0$, clearly the sum of gateways' budgets $\sum_i b_{i0} = B$ from the budget initialization. Supposing that $\sum_{i=1}^{n} b_{it} = B$ at time t, we then calculate

$$\sum_{i=1}^{n} b_{i,t+1} = \sum_{i=1}^{n} \left(b_{it} - x_{it} + \sum_{j \neq i} \frac{x_{jt}}{n-1} \right)$$

$$= B - \sum_{i=1}^{n} x_{it} + \sum_{i=1}^{n} \frac{(n-1)x_{it}}{n-1} = B.$$

2.A.2 PROOF OF LEMMA 2.2

We first note that (2.1) is equivalent to the statement that

$$b_{it} = b_{is} + \sum_{\tau=s}^{t-1} \left(-x_{i\tau} + \sum_{j \neq i} x_{j\tau}/(n-1) \right).$$

Then if $b_{is} = b_{it}$ for all gateways i, we obtain the system of equations

$$\sum_{\tau=s}^{t-1} x_{i\tau} = \sum_{\tau=s}^{t-1} \sum_{j \neq i} \frac{x_{j\tau}}{n-1}. \tag{2.A.1}$$

It suffices to show that (2.A.1) implies the proposition.

We proceed by induction on n. If $n = 2$, then clearly (2.A.1) is exactly our desired result, since $n - 1 = 1$. We now suppose that the proposition holds for $n = m$ and show that it holds for $n = m + 1$. From (2.A.1), we have

$$\sum_{\tau=s}^{t-1} x_{1\tau} = \sum_{\tau=s}^{t-1} \sum_{j=2}^{n} \frac{x_{j\tau}}{n-1}.$$

Substituting this equality into (2.A.1) for $i > 1$, we have for all such i,

$$\sum_{\tau=s}^{t-1} x_{i\tau} = \sum_{\tau=s}^{t-1} \sum_{j=2}^{n} \frac{x_{j\tau}}{(n-1)^2} + \sum_{\tau=s}^{t-1} \sum_{j \neq i, j>1} \frac{x_{j\tau}}{n-1}.$$

Thus, we have upon rearranging that

$$\left(1 - \frac{1}{(n-1)^2}\right) \sum_{\tau=s}^{t-1} x_{i\tau} = \left(\frac{1}{(n-1)^2} + \frac{1}{n-1}\right) \sum_{\tau=s}^{t-1} \sum_{j \neq i, j>1} x_{j\tau}.$$

Simplifying, we obtain

$$\sum_{\tau=s}^{t-1} x_{i\tau} = \sum_{\tau=s}^{t-1} \sum_{j \neq i, j>1} \frac{x_{j\tau}}{n-2}$$

for all $i > 1$. By induction, this implies that $\sum_{\tau=s}^{t-1} x_{j\tau} = \sum_{\tau=s}^{t-1} x_{k\tau}$ for all $j, k > 1$, and the proposition follows upon solving for $\sum_{\tau=s}^{t-1} x_{1\tau}$.

2.A.3 PROOF OF PROPOSITION 2.1

We first show that given a distribution of budgets $\{b_{it}\}$ at a fixed time t, there exists a set of gateway spending decisions $\{x_{it}\}$ such that $b_{i,t+1} = B/n$ for all gateways i. Suppose that each gateway i spends $x_{it} = b_{it}(n-1)/n$ credits at time t. Then Lemma 2.1's budget conservation allows us to conclude that gateway i's budget at time $t + 1$ is

$$b_{i,t+1} = b_{it} - \frac{b_{it}(n-1)}{n} + \sum_{j \neq i} \frac{b_{jt}(n-1)}{n(n-1)} = \sum_{i=1}^{n} \frac{b_{it}}{n} = \frac{B}{n}.$$

We now observe that since each $b_{i0} = B/n$, we can apply Lemma 2.2 to conclude that

$$\sum_{s=0}^{t+1} x_{is} = \sum_{s=0}^{t} x_{is} + \frac{b_{it}(n-1)}{n} = \sum_{s=0}^{t} x_{js} + \frac{b_{jt}(n-1)}{n} = \sum_{s=0}^{t+1} x_{js}$$

for all gateways i and j. We then rearrange this equation to find the first part of the proposition:

$$\left| \sum_{s=0}^{t} x_{is} - \sum_{s=0}^{t} x_{js} \right| = \left| b_{jt} - b_{it} \right| \frac{n-1}{n} \leq \frac{B(n-1)}{n}.$$

The time average follows immediately upon dividing by t and taking limits as $st \to \infty$.

2.A.4 PROOF OF PROPOSITION 2.2

To prove the first part of the proposition, we note that if each $x_{it} = 0$, then (2.1) yields

$$b_{i,t+1} = b_{it} - x_{it} + \sum_{j \neq i} \frac{x_{jt}}{n-1} \leq \frac{B}{n-1} + b_{it} \frac{n-2}{n-1} - x_{it},$$

where the inequality comes from each gateway's budget constraint $\sum_{j \neq i} x_{jt} \leq \sum_{j \neq i} b_{jt} = B - b_{it}$. Thus, at time $t+1$, we have

$$\begin{aligned} b_{i,t+1} &= \sum_{\tau=0}^{t} \left(\frac{B}{n-1} - x_{i\tau} \right) \left(\frac{n-2}{n-1} \right)^{\tau} + \frac{B}{n} \left(\frac{n-2}{n-1} \right)^{t+1} \\ &\leq \frac{B}{n} \alpha^{t+1} + B \left(1 - \alpha^{t+1} \right) - \epsilon \sum_{\tau=1}^{\left\lfloor \frac{t+1}{p} \right\rfloor} \alpha^{p\tau} \\ &= \frac{B}{n} \alpha^{t+1} + B \left(1 - \alpha^{t+1} \right) - \epsilon \left(\frac{\alpha^{p} - \alpha^{p\left(1 + \left\lfloor \frac{t+1}{p} \right\rfloor\right)}}{1 - \alpha^{p}} \right) \end{aligned}$$

as desired, using the fact that $\sum_{\tau=s}^{s+n} x_{i\tau} \geq \epsilon$ at any time s. We obtain (2.3) by taking $t \to \infty$, substituting for $\alpha = (n-2)/(n-1)$, and simplifying.

To prove the second part of the proposition, suppose that gateways i and k both have zero budgets at time $t+1$, that is, $b_{i,t+1} = b_{k,t+1} = 0$, but that $b_{it} > 0$. Since each $b_{i0} = B/n > 0$, such a time t must exist. But then from (2.1), $b_{i,t+1} = b_{it} - x_{it} + \sum_{j \neq i} x_{jt}/(n-1) = 0$, and since each $x_{jt} \geq 0$, we have $x_{it} = b_{it} > 0$. But then $b_{k,t+1} = b_{kt} - x_{kt} + \sum_{j \neq k} x_{jt}/(n-1)$, and since $x_{kt} \leq b_{kt}$, we have $b_{k,t+1} > 0$, which is a contradiction. Thus, at most one gateway can have zero budget in any given time period.

2.A.5 PROOF OF PROPOSITION 2.3

We first note that at each time $t < s$,

$$\begin{aligned}\sum_{i=1}^{m} b_{i,t+1} &\leq \sum_{i=1}^{m} b_{it} + \sum_{j>i} \frac{x_{it}}{n-1} \\ &\leq \sum_{i=1}^{m} b_{it} + \frac{B - \sum_{i=1}^{m} b_{it}}{n-1} \\ &= \frac{B}{n-1} + \left(\frac{n-2}{n-1}\right) \sum_{i=1}^{m} b_{it}.\end{aligned}$$

An inductive argument then shows that

$$\sum_{i=1}^{m} b_{i,s} \leq \frac{B}{n-1} \left(\sum_{\tau=0}^{s-1} \left(\frac{n-2}{n-1}\right)^{s} \right) + \left(\frac{n-2}{n-1}\right)^{s} \sum_{i=1}^{m} b_{i0}.$$

Expanding the sums and subtracting $\sum_{i=1}^{m} b_{i0}$ then yield the proposition.

2.A.6 PROOF OF PROPOSITION 2.4

Suppose that $\{x_{it}^*\}$ solve (2.7), and let λ_{it} denote the corresponding Lagrange multiplier for the constraint $0 \leq x_{it} \leq b_{it}$, with v_{it} the multiplier for the constraint $x_{it} \geq 0$. Since the U_{it} are strictly concave, it suffices to show that these multipliers satisfy the Karush–Kuhn–Tucker (KKT) conditions for (2.5), augmented by all gateways' constraints:

$$\max_{x_{it}} \sum_{t=s}^{s+T} U_{it}\left(x_{it}\right), \quad \text{s.t. } 0 \leq x_{it} \leq b_{it}, \ \forall i, t.$$

Since the budget constraints $0 \leq x_{it} \leq b_{it}$ are identical to those of (2.7), it suffices to show that

$$\frac{dU_{it}}{dx_{it}} - \sum_{j=1}^{n} \sum_{\tau=t}^{s+T} \lambda_{i\tau} + \sum_{j \neq i} \sum_{\tau=t}^{s+T-1} \frac{\lambda_{j\tau}}{n-1} + v_{it} = 0, \qquad (2.A.2)$$

where we use (2.6) to sum over the appropriate multipliers $\lambda_{i\tau}$. However, this equation is just one of the KKT conditions for (2.7): the only change between (2.7) and (2.5) is the addition of utility terms $U_{jt}(x_{jt})$, which are additively decoupled from gateway i's spending decisions x_{it}. Thus, (2.A.2) must be satisfied by the x_{it}^* and multipliers λ_{it}, v_{it}. Each gateway i is thus optimizing its own utility, given other gateways' credit spending decisions x_{jt}^*.

REFERENCES

1. Cisco Systems. Cisco visual networking index: Forecast and methodology, 2014–2019, August 2015. http://tinyurl.com/VNI2014 (accessed September 20, 2016).
2. Soumya Sen, Carlee Joe-Wong, Sangtae Ha, and Mung Chiang. A survey of smart data pricing: Past proposals, current plans, and future trends. *ACM Computing Surveys*, 46(2):15:1–15:37, 2013.
3. Comcast. Learn How Network Congestion Management Affects Your Internet Use, 2016. https://customer.xfinity.com/help-and-support/internet/network-management-information (accessed December 20, 2016).
4. Frank P. Kelly, Aman K. Maulloo, and David H. K. Tan. Rate control for communication networks: Shadow prices, proportional fairness, and stability. *Journal of Operational Research Society*, 49:237–252, 1998.
5. Ashish Patro, Srinivas Govindan, and Suman Banerjee. Observing home wireless experience through WiFi APs. In *Proceedings of ACM MobiCom*, September 30–October 4, 2013, Miami, FL.
6. Srikanth Sundaresan, Walter de Donato, Nick Feamster, Renata Teixeira, Sam Crawford, and Antonio Pescapè. Broadband Internet performance: A view from the gateway. In *Proceedings of ACM SIGCOMM*, August 15–19, 2011, Toronto, Ontario, Canada.
7. Richard Mortier, Tom Rodden, Peter Tolmie, Tom Lodge, Robert Spencer, Andy Crabtree, Joe Sventek, and Alexandros Koliousis. Homework: Putting interaction into the infrastructure. In *Proceedings of ACM UIST*, October 7–10, 2012, Cambridge, MA.
8. Jeonghwa Yang, W. Keith Edwards, and David Haslem. Eden: Supporting home network management through interactive visual tools. In *Proceedings of ACM UIST*, October 3–6, 2010, New York, NY.
9. Christos Gkantsidis, Thomas Karagiannis, Peter Key, Bozidar Radunovi, Elias Raftopoulos, and D. Manjunath. Traffic management and resource allocation in small wired/wireless networks. In *Proceedings of ACM CoNEXT*, December 1–4, 2009, Rome, Italy.
10. Claudio E. Palazzi, Matteo Brunati, and Marco Roccetti. An OpenWRT solution for future wireless homes. In *Proceedings of IEEE ICME*, July 19–23, 2010, Singapore.
11. J. Scott Miller, John R. Lange, and Peter A. Dinda. Emnet: Satisfying the individual user through empathic home networks. In *INFOCOM, 2010 Proceedings IEEE*, pages 1–9. IEEE, March 15–19, 2010, San Diego, CA.
12. Janne Seppanen and Marta Varela. Qoe-driven network management for real-time over-the-top multimedia services. In *Proceedings of IEEE WCNC*, pages 1621–1626. IEEE, April 7–10, 2013, Shanghai, China.
13. Marshini Chetty, Richard Banks, Richard Harper, Tim Regan, Abigail Sellen, Christos Gkantsidis, Thomas Karagiannis, and Peter Key. Who's hogging the bandwidth: The consequences of revealing the invisible in the home. In *Proceedings of the SIGCHI Conference on Human Factors in Computing Systems*, pages 659–668. ACM, April 10–15, 2010, Atlanta, GA.
14. Soumya Sen, Carlee Joe-Wong, Sangtae Ha, Jasika Bawa, and Mung Chiang. When the price is right: Enabling time-dependent pricing of broadband data. In *Proceedings of the SIGCHI Conference on Human Factors in Computing Systems*, pages 2477–2486. ACM, April 27–May 2, 2013, Paris, France.

15. Jeffrey K. MacKie-Mason, Liam Murphy, and John Murphy. Responsive pricing in the Internet. *Internet Economics*, The MIT Press, Cambridge, MA, pages 279–303, 1995.
16. Dongmyung Lee, Jeonghoon Mo, Jean Walrand, and Jinwoo Park. A token pricing scheme for internet services. In *Economics of Converged, Internet-Based Networks*, pages 26–37. Springer, 2011.
17. J. Scott Miller, Amit Mondal, Rahul Potharaju, Peter A. Dinda, and Aleksandar Kuzmanovic. Understanding end-user perception of network problems. In *Proceedings of the First ACM SIGCOMM Workshop on Measurements Up the Stack*, pages 43–48. ACM, August 19, 2011, Toronto, Ontario, Canada.
18. Cheng-Chun Tu, Kuan-Ta Chen, Yu-Chun Chang, and Chin-Laung Lei. Oneclick: A framework for capturing users network experiences. In *Proceedings of ACM SIGCOMM 2008 (poster)*, August 17–22, 2008, Seattle, WA.
19. Hari Balakrishnan, Hariharan S. Rahul, and Srinivasan Seshan. An integrated congestion management architecture for Internet hosts. In *Proceedings of ACM SIGCOMM*, August 31–September 3, 1999, Cambridge, MA.
20. Lampros Kalampoukas, Anujan Varma, and K. K. Ramakrishnan. Explicit window adaptation: A method to enhance TCP performance. In *Proceedings of IEEE INFOCOM*, March 29–April 2, 1998, San Francisco, CA.
21. Jeffrey Semke, Jamshid Mahdavi, and Matthew Mathis. Automatic TCP buffer tuning. In *Proceedings of ACM SIGCOMM*, September 2–4, 1998, Vancouver, British Columbia, Canada.
22. Peter Key, Laurent Massoulìè, and Bing Wang. Emulating low-priority transport at the application layer: A background transfer service. In *Proceedings of ACM SIGMETRICS/Performance*, June 10–14, 2004, New York, NY.
23. Youngbin Im, Carlee Joe-Wong, Sangtae Ha, Soumya Sen, Ted Taekyoung Kwon, and Mung Chiang. AMUSE: Empowering users for cost-aware offloading with throughput delay tradeoffs. In *Proceedings of IEEE INFOCOM*, April 14–19, 2013, Turin, Italy.
24. Neil T. Spring, Maureen Chesire, Mark Berryman, Vivek Sahasranaman, Thomas Anderson, and Brian Bershad. Receiver based management of low bandwidth access links. In *Proceedings of IEEE INFOCOM*, March 26–30, 2000, Tel Aviv, Israel.
25. Mike Fisk and Wu-Chun Feng. Dynamic right-sizing in TCP. In *Proceedings of LACSI Symposium*, October 15–18, 2001, Santa Fe, NM.
26. Marius A. Eriksen. Trickle: A userland bandwidth shaper for Unix-like systems. In *Proceedings of USENIX Annual Technical Conference*, April 10–15, 2005, Anaheim, CA.
27. Andrea Consiglio, Flavio Cocco, and Stavros A. Zenios. Scenario optimization asset and liability modelling for individual investors. *Annals of Operations Research*, 152(1):167–191, 2007.
28. Jae Yoon Chung, Yeongrak Choi, Byungchul Park, and James W.-K. Hong. Measurement analysis of mobile traffic in enterprise networks. In *Proceedings of APNOMS*, September 21–23, 2011, Taipei, Taiwan.
29. Felix Ming Fai Wong, Carlee Joe-Wong, Sangtae Ha, Zhenming Liu, and Mung Chiang. Improving user qoe for residential broadband: Adaptive traffic management at the network edge. In *Proceedings of IEEE/ACM IWQoS*, June 15 and 16, 2015, Portland, OR.

3 Socially-Aware Cooperative D2D and D4D Communications toward Fog Networking

XU CHEN,[1] JUNSHAN ZHANG,[1] and SATYAJAYANT MISRA[2]

[1]*School of ECEE, Arizona State University, Tempe, AZ, USA*
[2]*Department of Computer Science, New Mexico State University, Las Cruces, NM, USA*

3.1 INTRODUCTION

The past few years have witnessed the explosive growth of the mobile user population and the demands for bandwidth-eager multimedia content, which poses a significant challenge for wireless networks. The Cisco VNI report predicts that the number of mobile devices will grow from 4 billion in 2014 to 11 billion by 2019, and mobile data traffic is expected to reach 292 exabytes per month by 2019, up from 30 exabytes in 2012 [1]. In addition, according to the recent report from Juniper Research, the number of Internet of things (IoT) connected devices will reach 38.5 billion in 2020, and these devices will produce 10% of all the data generated worldwide. In short, large quantities of data will have to be communicated, stored, and analyzed.

One emerging cost-effective approach to address such challenges is fog networking and computing, which leverages a multitude of collaborative end-user clients or near-user edge devices to carry out a substantial amount of communication and computation to augment tasks and services in a network [2]. Motivated by the principle of fog networking, in this chapter we aim at boosting the network bandwidth by promoting collaborative communications between handheld devices at the network edge. Such device-to-device (D2D) and device-for-device (D4D) communications can offer a variety of advantages over traditional cellular communications, such as higher user throughput, improved spectral efficiency, and extended network coverage [3]. For example, a device can share the video content with neighboring devices who have similar interest, which can help to reduce the data traffic from the network operator.

Fog for 5G and IoT, First Edition. Edited by Mung Chiang, Bharath Balasubramanian, and Flavio Bonomi.

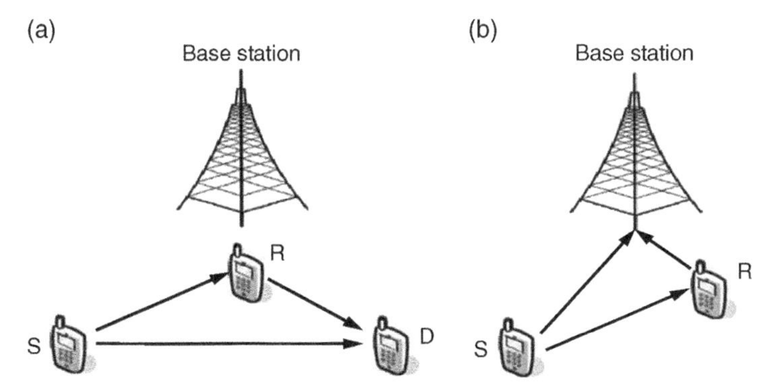

Figure 3.1 An illustration of cooperative D2D and D4D communication for cooperative networking. In sub-figure (a), device R serves as the relay for the D2D communication between devices S and D. In sub-figure (b), device R serves as the relay for the cellular communication between device S and the base station. In both cases, the D2D communication between devices S and R is part of cooperative networking.

Cooperative communication is an efficient D2D and D4D communication paradigm where devices can serve as relays for each other.[1] As illustrated in Figure 3.1, cooperative D2D and D4D communication can help to (i) improve the quality of D2D communication for direct data offloading between devices and (ii) enhance the performance of cellular communications between the base station and the devices as well. Moreover, cooperative D2D and D4D communication can achieve bandwidth boosting by exploiting different types of spectrum bands to support D2D communications [4]:

- *Inband D2D and D4D Communications.* The spectrum band that supports inband D2D communications is the same with the band for cellular communications. Therefore, D2D links share the same band with cellular links via the same air interface. In general, the control and management by the cellular network is required in order to efficiently use the cellular spectrum for both D2D and cellular links [5].
- *Outband D2D and D4D Communications.* The spectrum band that supports outband D2D communications is different from the band for cellular communications, which can be 2.4 GHz (e.g., Bluetooth), 5 GHz (e.g., Wi-Fi Direct) or 38 GHz bands (e.g., Millimeter Wave). Clearly, D2D links and cellular links can be active at the same time via two independent air interfaces. Outband D2D communication can be managed by the cellular network (i.e., controlled) or it can operate on its own (i.e., autonomous) [5].

Hence cooperative D2D and D4D communication can be a critical building block for efficient cooperative networking for future wireless networks, wherein individual

[1]There are many approaches for cooperative communications, and for ease of exposition, this study assumes cooperative relaying.

users cooperate to substantially boost the network capacity and cost-effectively provide rich multimedia services and applications, such as video conferencing and interactive media, anytime, anywhere.

3.1.1 From Social Trust and Social Reciprocity to D2D Cooperation

A key challenge for cooperative D2D and D4D communications-based fog networking is how to stimulate effective cooperation among devices. As different devices are usually owned by different individuals and they may pursue different interests, there is no good reason to assume that all devices would cooperate with each other. Since the handheld devices are carried by human beings, a natural question to ask is that "is it possible to leverage human social relationship to enhance D2D communications for cooperative networking?" Indeed, with the explosive growth of online social networks such as Facebook and Twitter, more and more people are actively involved in online social interactions, and social relationships among people are hence extensively broadened and significantly enhanced [4]. This has opened up a new avenue for cooperative D2D and D4D communication system design—we believe that it has potential to propel significant advances in mobile social networking.

One primary goal of this study is to establish a new D2D cooperation paradigm by leveraging two key social phenomena: social trust and social reciprocity. Social trust can be built up among humans, such as kinship, friendship, colleague relationship, and altruistic behaviors that are observed in many human activities [6], for example, when a device user is at home or work, typically family members, neighbors, colleagues, or friends are nearby. The device user can then exploit the social trust from these neighboring users to improve the quality of D2D communication, for example, by asking the best trustworthy device to serve as the relay. Another key social phenomenon, social reciprocity, is also widely observed in human society [7]. Social reciprocity is a powerful social paradigm to promote cooperation so that a group of individuals without social trust can exchange mutually beneficial actions, making all of them better off. For example, when a device user does not have any trusted friends in the vicinity, he or she may cooperate with the nearby strangers by providing relay assistance for each other to improve the quality of D2D communications. There are several ways to bootstrap this cooperation. One mechanism is through a symbiotic give-and-take barter relationship. That is, a node helps relay another node's data if the other node also relays its data. Another technique could also be based on assessing the number of hops the two nodes are apart in a trusted social network graph, such as Facebook or Twitter (can be obtained through a Facebook plug-in running on each node). This can be augmented through an idea similar to the *Web of Trust*, where the two nodes can create a trust chain [8]. This will help create a minimal infrastructure-based trust schema.

As illustrated in Figure 3.2, cooperative D2D communications based on social trust and social reciprocity can be projected onto two domains: the physical domain and the social domain. In the physical domain, different devices have different feasible relay selection relationships subject to the physical constraints. In the social domain, different devices have different assistance relationships based on social trust

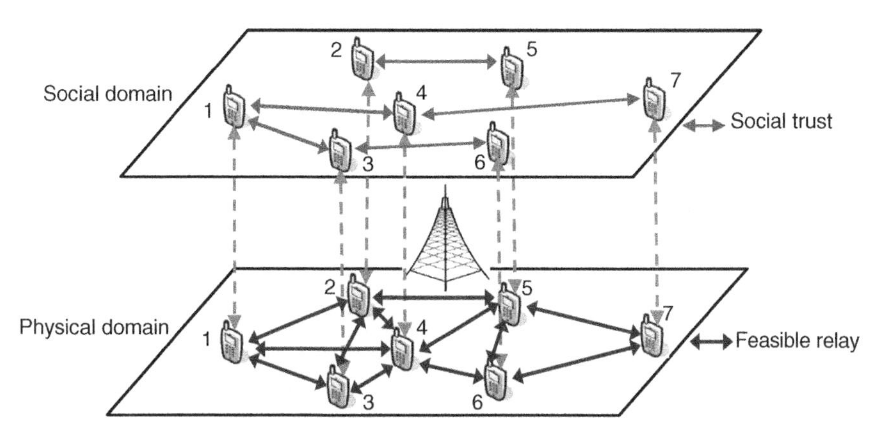

Figure 3.2 An illustration of the social trust model for cooperative D2D communications. In the physical domain, different devices have different feasible cooperation relationships subject to physical constraints. In the social domain, different devices have different assistance relationships based on social trust among the devices.

among the devices. In this case, each device has two options for relay selection: either seek relay assistance from another feasible device that has social trust toward him or her; or participate in a group formed based on social reciprocity by exchanging mutually beneficial relay assistance. The main thrust of this study is devoted to tackling two key challenges for the social trust- and social reciprocity-based approach. The first is which option a device should adopt for relay selection: social trust or social reciprocity. The second is how to efficiently form groups among the devices that adopt the social reciprocity-based relay selection. We will develop a coalitional game theoretic framework to address these challenges.

3.1.2 Smart Grid: An IoT Case for Socially-Aware Cooperative D2D and D4D Communications

The smart grid will be a fertile area for the implementation of the IoT technology and hence will leverage a significant amount of D2D and D4D communications. The smart devices, such as smartmeters, in-home smart devices, synchrophasors, substation sensors, and electric vehicles, will perform two-way communications in the grid, essentially forming a fog network. Further, the availability of an energy marketplace will allow energy consumers, producers, and prosumers (both consumer and producer of energy) to actively buy/sell energy among each other (as in the case of a microgrid [9]) and also transact with the grid. The way the smart grid networking and communication architecture is shaping, most of the communication will be of a distributed nature and leverage several communication technologies, such as wireless, wired, power line, and fiber [10, 11].

The proposed networking and communication paradigm for the smart grid will contain several cooperative D2D and D4D communication scenarios. For instance, grid monitoring and protection devices (e.g., synchrophasors, circuit monitors, and power quality monitors) will actively interact with each other at real time to identify and localize faults and reduce power quality variations [10, 12]. Smartmeters (or the consumer's smartphone) would be actively involved in energy transactions with smartmeters (or smartphones) representing other homes. Devices in a smart home will communicate with each other to identify optimal loads, schedules, and energy usage. A large proportion of these D2D and D4D communications will be done between independent agents that are not direct neighbors in the network.

For example, a PMU is connected to a local substation communicating with a PMU in another substation. Or, a smartmeter in a home negotiates an energy transaction with another smartmeter several homes away. All such communication would require the help of a cooperative intermediate relay(s). In turn, a relay incurs communication cost in receiving and forwarding the messages (in terms of bandwidth, interference, delay of its own communication, wireless transmission power, etc.), which will require incentives for cooperation. In all such D2D and D4D communications, social trust could be bootstrapped by the end users (neighbors trusting each other), or the trust could be built based on devices being part of a common trust framework. For example, all devices belonging to a particular grid entity, who can verify each other's credentials through certificates, trust each other. A trust broker can be created to verify credentials of nodes and give them tokens that they can use to bootstrap mutual trust.

The same way as users connected via their smartphones may have a social connection with each other (either based on social trust or social reciprocity), the smart grid IoT devices (operating under general qualitative guidelines) can be modeled as autonomous agents that operate based on trust and reciprocity. For example, the smartmeter of your house can "trust" the smartmeter of your neighbor to relay information, because you know your neighbor and you are best friends. The trust relationship may also be bootstrapped by trust brokers that form part of the grid and are maintained by the utilities. Your smartmeter may forward the data transmissions of another neighbor's smartmeter because that smartmeter helped deliver several messages in the past (reciprocity). The agents may also create trust relationships based on past interactions (e.g., two smartmeter agents were part of a prosumer bargaining cooperative; two PMUs have interacted with each other to report past faults correctly).

The IoT devices of the smart grid will be small embedded devices similar to the Raspberry Pis (smartmeters and synchrophasors) and low-power and low-computation-capable devices (smarthome power-outlet devices). The devices will be equipped with multiple communication interfaces with wired and wireless being the predominant ones. These devices will form a heterogeneous network with different unique characteristics and can cooperate with each other over several interfaces leveraging different communication technologies. We are using such devices in local testbed settings as shown in Figure 3.3. The figure shows a plug-in outlet (created in collaboration with power engineers) to which a load can be plugged in. The outlet has the capability to measure the characteristics of the electric signal (voltage, current, phase angle), and it interfaces with an ARM Cortex-based Texas Instrument board, which packetizes and transmits the information with an

(a)

(b)

Figure 3.3 Illustrative smart grid IoT devices. (a) Our outlet device with a Texas Instrument Microcontroller Board and XBee radio module; (b) Raspberry Pi to emulate a smartmeter or controller device.

off-the-shelf XBee radio module. This information is received by a Raspberry Pi, which acts as the smartmeter. We are building intelligence for the Raspberry Pi to issue commands to the outlet to reduce current and voltage to meet demand response needs, which it identifies through its interactions with the grid.

3.1.3 Summary of Main Results

The main contributions of this chapter are as follows:

- We propose a novel social trust- and social reciprocity-based framework to promote efficient cooperation among devices for cooperative D2D and D4D communications-based fog networking. By projecting D2D communications in a mobile social network onto both physical and social domains, we introduce the physical–social graphs to model the interplay therein while capturing the physical constraints for feasible D2D cooperation and the social relationships among devices for effective cooperation.

- We formulate the relay selection problem for social trust and social reciprocity-based cooperative D2D communications as a coalitional game. We show that the coalitional game admits the top-coalition property based on which we devise a core relay selection algorithm for computing the core solution to the game.
- We develop a network-assisted mechanism to implement the coalitional game-based solution. We show that the mechanism is immune to group deviations and individually rational, truthful, and computationally efficient. We further evaluate the performance of the mechanism by the real social data trace. Simulation results show that the proposed mechanism can achieve up to 122% performance gain over the case without D2D cooperation.

A primary goal of this chapter is to build a theoretically sound and practically relevant framework to understand social trust- and social reciprocity-based cooperative D2D and D4D communications. This framework highlights the interplay between potential physical network performance gain through efficient D2D cooperation and the exploitation of social relationships among device users to stimulate effective cooperation. Besides the cooperative D2D communication scenario where devices serve as relays for each other, the proposed social trust- and social reciprocity-based framework can also be applied to many other D2D cooperation scenarios, such as cooperative MIMO communications and mobile cloud computing. We believe that these initial steps presented here open a new avenue for mobile social networking and have great potential to enhance network capacity in future wireless networks.

3.2 RELATED WORK

D2D communications have recently drawn great attention from the wireless research community. Most existing literature has focused on the interference coordination issue between D2D communications and cellular communications. Authors in Refs. [13, 14] studied the power control problem for restricting cochannel interference from D2D communications to cellular communications. Janis *et al.* [15] utilized MIMO transmission schemes to mitigate interference from cellular downlink to D2D receivers sharing the same spectrum resources. Zulhasnine *et al.* [16] proposed to lessen interference to cellular communications by properly pairing the cellular and D2D users. Currently, more and more research efforts are devoted to cooperative D2D communications, which can significantly enhance the performance of D2D communications. Raghothaman *et al.* [17] proposed a system architecture that enables D2D communications with cooperative mobile relays. Ma *et al.* [18] developed a distributed relay selection algorithm for cooperative D2D communications.

Lee *et al.* [19] studied the multi-hop decode-and-forward (DF) relaying assisted cooperative D2D communications. The common assumption of these previous studies for cooperative D2D communications is that all the device users are cooperative and they are willing to help any other users. However, since each handheld device has limited battery and providing relaying assistance for cooperative D2D communications would incur significant energy consumption, there is no good reason to assume that all device users would cooperate with each other.

Much effort has been made in the literature to stimulate, via incentive mechanisms, cooperation in wireless networks. Payment-based mechanisms have been widely considered to incentivize cooperation for wireless ad hoc networks [20–22]. Another widely adopted approach for cooperation stimulation is reputation-based mechanisms, where a centralized authority or the whole user population collectively keeps records of the cooperative behaviors and punishes non-cooperating users [23–25]. However, it is yet clear whether these incentive mechanisms are feasible in practice since they require central authorities to monitor and regulate user behaviors and resolve disputes, which require extensive signaling overhead between users and central authorities, and can easily diminish the capacity gain of cooperative D2D communication. Moreover, incentive mechanisms typically assume that all users are fully rational and they act in a selfish manner. Such an assumption is not appropriate for D2D communications as handheld devices are carried by human beings and people typically act with bounded rationality and involve social interactions [26].

The social aspect is now becoming a new and important dimension for communication system design [4, 27]. As the development of online and mobile social networks such as Facebook and Twitter, more and more real-world data and traces of human social interactions are being generated. This enables researchers and engineers to observe, analyze, and incorporate the social factors into engineering system design in a way never previously possible [28]. Authors in Refs. [29, 30] exploited social structures such as social community to design efficient data forwarding and routing algorithms in delay tolerant networks. Chen *et al.* [31, 32] proposed a novel framework of social group utility maximization (SGUM) for cooperative networking design. Hui *et al.* [33] used the social betweenness and centrality as the forwarding metric. Costa *et al.* [34] proposed predictions based on metrics of social interaction to identify the best information carriers for content publish–subscribe. Authors in Refs. [35, 36] utilized the social influence phenomenon to devise efficient data dissemination mechanisms for mobile networks. The common assumption among these works, however, is that all users are always willing to help others, for example, for data forwarding and relaying. In this chapter we propose a novel framework to stimulate cooperation among device users while also taking the social aspect into account.

3.3 SYSTEM MODEL

In this section we present the system model of cooperative D2D and D4D communications based on social trust and social reciprocity—a new mobile social networking paradigm. As illustrated in Figure 3.2, cooperative D2D and D4D communications can be projected onto two domains: the physical domain and the social domain. In the physical domain, different devices have different feasible cooperation relationships for cooperative D2D communications subject to the physical constraints. In the social domain, different device users have different assistance relationships based on social relationships among them. We next discuss both physical and social domains in detail.

3.3.1 Physical (Communication) Graph Model

We consider a set of nodes $\mathcal{N} = \{1, 2, ..., N\}$ where N is the total number of nodes. Each node $n \in \mathcal{N}$ is a wireless device that would like to conduct D2D communication to transmit data packets to its corresponding destination d_n. Notice that a destination d_n may also be a transmit node in the set $\mathcal{N}$ of another D2D communication link, and hence a D2D traffic flow may traverse one hop or multiple hops among the devices. Similar to many previous studies in D2D communications [13–19], to enable tractable analysis, we consider a scenario where the locations of the nodes remain unchanged during a D2D communication scheduling period (e.g., several hundred milliseconds), while these may change across different periods due to users' mobility.[2]

The D2D communication is underlaid beneath a cellular infrastructure wherein there exists a base station controlling the uplink/downlink communications of the cellular devices. To avoid generating severe interference to the incumbent cellular devices, each node $n \in \mathcal{N}$ will first send a D2D communication establishment request message to the base station. The base station then computes the allowable transmission power level p_n for the D2D communication of node n based on the system parameters and the protection requirement of the neighboring cellular devices. For example, the proper transmission power p_n of the D2D communication can be computed according to the power control algorithms proposed in Refs. [13, 14]. Moreover, with the assistance by the base station, each node can detect a set of neighboring nodes, which can be potential relay candidates for cooperative D2D communications [3].

We consider a time division multiple access (TDMA) mechanism in which the transmission time is slotted and one node $n \in \mathcal{N}$ is scheduled to carry out its D2D communication in a time slot.[3] At the allotted time slot, node n can choose either to transmit to the destination node d_n directly or to use cooperative communication by asking another node m in its vicinity to serve as a relay.

Due to physical constraints such as signal attenuation, only a subset of nodes that are close enough (e.g., with a detectable signal strength) can be feasible relay candidates for the node n. To take such physical constraints into account, we introduce the physical graph[4] $\mathcal{G}^P \triangleq \{\mathcal{N}, \mathcal{E}^P\}$ where the set of nodes $\mathcal{N}$ is the vertex set and

[2]This assumption is valid for our case, since the proposed mechanism in Section 3.5 has a very low computational complexity, and hence the D2D communication scheduling can be carried out in a smaller time scale than that of users' mobility.
[3]Our methods are also applicable to other multiple access schemes.
[4]The graphs (e.g., physical graph and social graph) in this chapter can be directed.

$\mathcal{E}^P \triangleq \{(n,m): e^P_{nm} = 1, \forall n,m \in \mathcal{N}\}$ is the edge set where $e^P_{nm} = 1$ if and only if node m is a feasible relay for node n. An illustration of the physical graph is given in Figure 3.2. We also denote the set of nodes that can serve as a feasible relay of node n as $\mathcal{N}^P_n \triangleq \{m \in \mathcal{N} : e^P_{nm} = 1\}$. A recent work [37] shows that it is sufficient for a source node to choose the best relay node among multiple candidates to achieve full diversity. Specifically, an optimal power allocation procedure based on user's channel side information (CSI) is carried out prior to the relay selection. For ease of exposition, we hence consider the single relay selection scheme such that each node n selects at most one neighboring node $m \in \mathcal{N}^P_n$ as the relay. Moreover, since multiple relay selection scheme typically requires the synchronization among the relays, the single relay selection scheme demands less signaling overhead and is easier to be implemented in practice.

For ease of exposition, we consider the full duplex DF relaying scheme [38] for the cooperative D2D communication. Let $r_n \in \mathcal{N}^P_n$ denote the relay node chosen by node $n \in \mathcal{N}$ for cooperative communication. The data rate achieved by node n is then given as [38]

$$Z^{\mathrm{DF}}_{n,r_n} = \frac{W}{N}\min\{\log(1+\mu_{nr_n}), \log(1+\mu_{nd_n}+\mu_{r_nd_n})\},$$

where W denotes the channel bandwidth and μ_{ij} denotes the signal-to-noise ratio (SNR) at device j when device i transmits a signal to device j. As an alternative, the node n can also choose to transmit directly without any relay assistance and achieve a data rate of

$$Z^{\mathrm{Dir}}_n = \frac{W}{N}\log(1+\mu_{nd_n}).$$

For simplicity, we define the data rate function of node n as $R_n : \mathcal{N}^P_n \cup \{n\} \rightarrow \mathbb{R}_+$, which is given by

$$R_n(r_n) = \begin{cases} Z^{\mathrm{DF}}_{n,r_n}, & \text{if } r_n \neq n, \\ Z^{\mathrm{Dir}}_n, & \text{if } r_n = n. \end{cases} \tag{3.1}$$

We will use the terminology that node n chooses itself as the relay for the situation in which node n transmits directly to its destination d_n.

3.3.2 Social Graph Model

We next introduce the social trust model for cooperative D2D communications. The underlying rationale of using social trust is that the handheld devices are carried by human beings and the knowledge of human social ties can be utilized to achieve effective and trustworthy relay assistance for cooperative D2D communications.

More specifically, we introduce the social graph $\mathcal{G}^S = \{\mathcal{N}, \mathcal{E}^S\}$ to model the social trust among the nodes. Here the vertex set is the same as the node set $\mathcal{N}$, and the edge set is given as $\mathcal{E}^S = \{(n,m): e^S_{nm} = 1, \forall n,m \in \mathcal{N}\}$ where $e^S_{nm} = 1$ if and only if nodes n and m have social trust toward each other, which can be kinship, friendship, or colleague relationship between two nodes. We denote the set of nodes that have social trust toward node n as $\mathcal{N}_n^S = \{m : e_{nm}^S = 1, \forall m \in \mathcal{N}\}$, and we assume that the nodes in $\mathcal{N}_n^S$ are willing to serve as the relay of node n for cooperative communication.

One critical task here is to identify the social relationships among device users. To this end, we can adopt a network-assisted approach such that two device users carry out the identification process through the cellular communications. Two device users can detect their social relationship by carrying out the "matching" process to identify the common social features among them. For example, two users can match their mobile phones' contact books. If they have the phone numbers of each other or many of their phone numbers are the same, then it is very likely that they know each other. As another example, two device users can match their home and working addresses and identify whether they are neighbors or colleagues. Furthermore, two device users can detect the social relationship among them by accessing to the online social networks such as Facebook and Twitter. For example, Facebook has exposed access to their social graph including the objects of friends, events, groups, profile information, and photos. Any authenticated Facebook user can have access to these information through the Open Graph API [39]. To preserve the privacy of the device users, the private set intersection technique in Refs. [40–44] can be adopted to design a privacy-preserving social relationship identification mechanism such that the intersection of private social information of two device users can be obtained without leaking any additional private information. Interested readers can refer to Refs. [40–44] for the detailed discussion of the privacy-preserving social relationship identification mechanism design. To further protect device user's personal information such as identity and visited locations, we can adopt the privacy-preserving scheme in Ref. [45].

Based on the physical graph $\mathcal{G}^P$ and social graph $\mathcal{G}^S$ earlier, each node $n \in \mathcal{N}$ can classify the set of feasible relay nodes in $\mathcal{N}_n^P$ into two types: nodes with social trust and nodes without social trust. A node n then has two options for relay selection. On the one hand, the node n can choose to seek relay assistance from another feasible device that has social trust toward him or her. On the other hand, the node n can choose to participate in a group formed based on social reciprocity by exchanging mutually beneficial relay assistance. In the following, we will study (i) how to choose between social trust- and social reciprocity-based relay selections for each node and (ii) how to efficiently form reciprocal groups among the nodes without social trust.

3.4 SOCIALLY-AWARE COOPERATIVE D2D AND D4D COMMUNICATIONS TOWARD FOG NETWORKING

In this section, we study the cooperative D2D communications based on social trust and social reciprocity. As mentioned, each node $n \in \mathcal{N}$ has two options for relay selection: social trust and social reciprocity. We next address the issues of choosing between social trust- and social reciprocity-based relay selections for each node and the reciprocal group forming among the nodes without social trust.

3.4.1 Social Trust-Based Relay Selection

We first consider social trust-based relay selection for D2D cooperation. The key motivation for using social trust is to utilize the knowledge of human social ties to achieve effective and trustworthy relay assistance among the devices for cooperative D2D communications. For example, when a device user is at home or working place, he or she typically has family members, neighbors, colleagues, or friends in the vicinity. The device user can then exploit the social trust from neighboring users to improve the quality of D2D communication by asking the best trustworthy device to serve as the relay.

To take both the physical and social constraints into account, we define the *physical–social graph* $\mathcal{G}^{PS} \triangleq \{\mathcal{N}, \mathcal{E}^{PS}\}$ where the vertex set is the node set $\mathcal{N}$ and the edge set $\mathcal{E}^{PS} = \{(n,m) : e_{nm}^{PS} \triangleq e_{nm}^{P} \cdot e_{nm}^{S} = 1, \forall n, m \in \mathcal{N}\}$ where $e_{nm}^{PS} = 1$ if and only if node m is a feasible relay (i.e., $e_{nm}^{P} = 1$) and has social trust toward node n (i.e., $e_{nm}^{S} = 1$). An illustration of the physical–social graph is given in Figure 3.4. We also denote the set of nodes that have social trust toward node n and are also feasible relay candidates for node n as $\mathcal{N}_n^{PS} = \{m : e_{nm}^{PS} = 1, \forall m \in \mathcal{N}\}$.

For cooperative D2D communications based on social trust, each node $n \in \mathcal{N}$ can choose the best relay $r_n^S = \arg\max_{r_n \in \mathcal{N}_n^{PS} \cup \{n\}} R_n(r_n)$ to maximize its data rate subject to both physical and social constraints.

3.4.2 Social Reciprocity-Based Relay Selection

Next, we study the social reciprocity-based relay selection. Different from D2D cooperation based on social trust that requires strong social ties among device users, social reciprocity is a powerful mechanism for promoting mutual beneficial cooperation among the nodes in the absence of social trust. For example, when a device user does not have any friends in the vicinity, he or she may cooperate with the nearby strangers by providing relay assistance for each other to improve the quality of D2D communications. In general, there are two types of social reciprocity: direct reciprocity and indirect reciprocity[5] (see Figure 3.5 for an illustration). Direct reciprocity is captured in the principle of "you help me, and I will help you." That is,

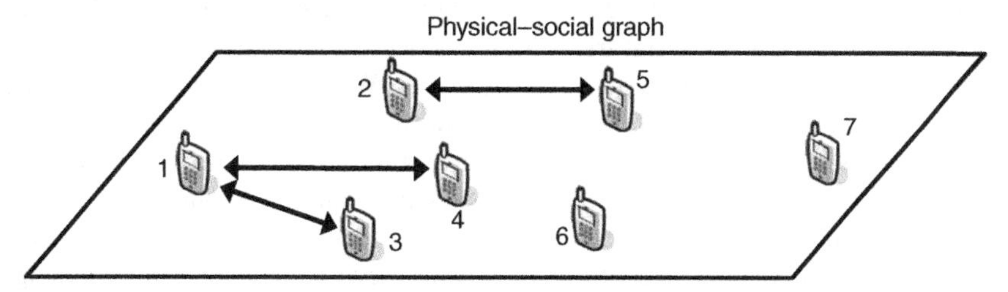

Figure 3.4 The physical–social graph based on the physical graph and social graph in Figure 3.2. For example, there exists an edge between nodes 1 and 3 in the physical–social graph since they can serve as the feasible relay for each other and also have social trust toward each other.

[5]Reciprocity in this study refers to social reciprocity.

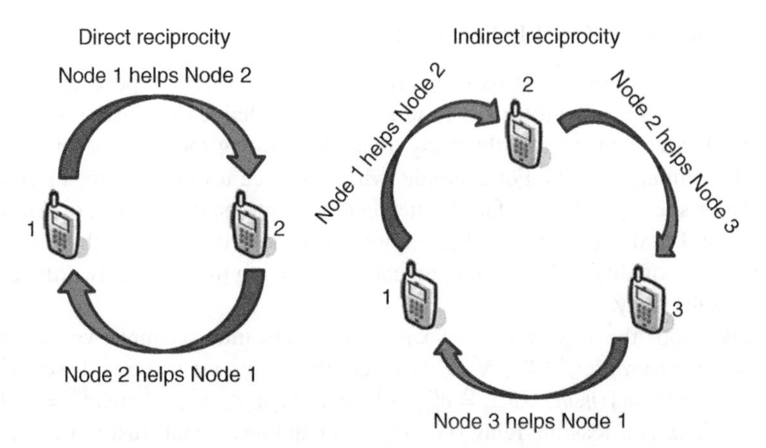

Figure 3.5 An illustration of direct and indirect reciprocity.

two individuals exchange altruistic actions so that both obtain a net benefit. Indirect reciprocity is essentially the concept of "I help you, and someone else will help me." That is, a group of individuals exchange altruistic actions so that all of them can be better off.

Note that in this chapter, we consider that the objective of each device user is to increase the throughput of its D2D communication, and hence a user is willing to participate in a reciprocal group if its communication performance can be improved. Our result can be extended to the case when the cost (e.g., energy consumption) of serving as a relay for other users is taken into account. In this case, each user will make the decision of participating a reciprocal group based on its net utility (i.e., the achieved throughput of getting relaying assistance minus the cost of serving as a relay for others). If the cost of a user is too high, then the user would not join any reciprocal relay groups and choose the direct communication without any relay.

To better describe the possible cooperation relationships among the set of nodes without social trust, we introduce the *physical-coalitional graph* $\mathcal{G}^{PC} = \{\mathcal{N}, \mathcal{E}^{PC}\}$. Here the vertex set is the node set $\mathcal{N}$ and the edge set $\mathcal{E}^{PC} = \{(n,m) : e_{nm}^{PC} \triangleq e_{nm}^{P} \cdot (1-e_{nm}^{S}) = 1, \forall n,m \in \mathcal{N}\}$ where $e_{nm}^{PC} = 1$ if and only if node m is a feasible relay (i.e., $e_{nm}^{P} = 1$) and has no social trust toward node n (i.e., $e_{nm}^{S} = 0$). An illustration of physical-coalitional graph is given in Figure 3.6. We also denote the set of nodes that have no social trust toward user n but are feasible relay candidates of node n as $\mathcal{N}_n^{PC} \triangleq \{m : e_{nm}^{PC} = 1, \forall m \in \mathcal{N}\}$. For social reciprocity-based relay selection, a key challenge is how to efficiently divide the nodes into multiple groups such that the nodes can significantly improve their data rates by the reciprocal cooperation within the groups. We will propose a coalitional game framework to address this challenge.

3.4.2.1 Introduction to Coalitional Game For the sake of completeness, we first give a brief introduction to the coalitional game [46]. Formally, a coalitional game consists of a tuple $\Omega = (\mathcal{N}, \mathcal{X}_{\mathcal{N}}, V, (\succ_n)_{n\in\mathcal{N}})$, where:

- $\mathcal{N}$ is a finite set of players.
- $\mathcal{X}_{\mathcal{N}}$ is the space of feasible cooperation strategies of all players.

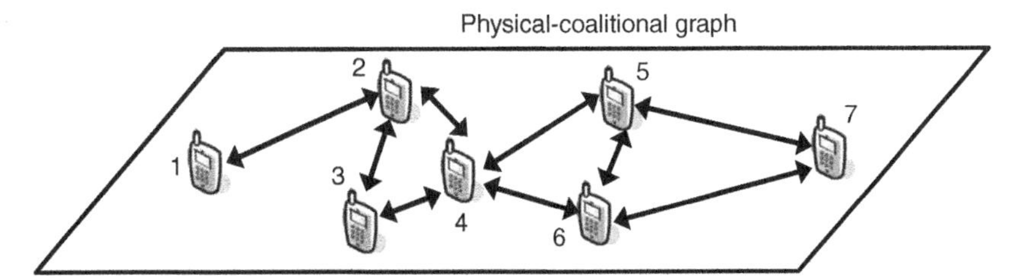

Figure 3.6 The physical-coalitional graph based on the physical graph and social graph in Figure 3.7. For example, there exists an edge between nodes 1 and 2 in the physical-coalitional graph since they can serve as the feasible relay for each other and have no social trust toward each other.

- V is a characteristic function that maps from every nonempty subset of players $\mathcal{S} \subseteq \mathcal{N}$ (a coalition) to a subset of feasible cooperation strategies $V(\mathcal{S}) \subseteq \mathcal{X}_{\mathcal{N}}$. This represents the possible cooperation strategies among the players in the coalition $\mathcal{S}$ given that other players out of the coalition $\mathcal{S}$ do not participate in any cooperation.
- $\succ_n$ is a preference order (reflexive, complete, and transitive binary relation) on $\mathcal{X}_{\mathcal{N}}$ for each player $n \in \mathcal{N}$. This captures the idea that different players may have different preferences over different cooperation strategies.

In the same spirit as Nash equilibrium in a non-cooperative game, the "core" plays a critical role in the coalitional game.

Definition 3.1 *The core is the set of $\mathbf{x} \in V(\mathcal{N})$ for which there does not exist a coalition $\mathcal{S}$ and $\mathbf{y} \in V(\mathcal{S})$ such that $\mathbf{y} \succ_n \mathbf{x}$ for all $n \in \mathcal{S}$.*

Intuitively, the core is a set of cooperation strategies such that no coalition can deviate and improve for all its members by cooperation within the coalition [46].

3.4.2.2 Coalitional Game Formulation We then cast the social reciprocity-based relay selection problem as a coalitional game $\Omega = (\mathcal{N}, \mathcal{X}_{\mathcal{N}}, V, (\succ_n)_{n\in\mathcal{N}})$ as follows:

- The set of players $\mathcal{N}$ is the set of nodes.
- The set of cooperation strategies $\mathcal{X}_{\mathcal{N}} = \{(r_n)_{n\in\mathcal{N}} : r_n \in \mathcal{N}_n^{PC} \cup \{n\}, \forall n \in \mathcal{N}\}$, which describes the set of possible relay selections for all nodes based on the physical-coalitional graph $\mathcal{G}^{PC}$.
- The characteristic function $V(\mathcal{S}) = \{(r_n)_{n\in\mathcal{N}} \in \mathcal{X}_{\mathcal{N}} : \{r_n\}_{n\in\mathcal{S}} = \{n\}_{n\in\mathcal{S}}$ and $r_m = m, \forall m \in \mathcal{N}\backslash\mathcal{S}\}$ for each coalition $\mathcal{S} \subseteq \mathcal{N}$. Here the condition "$\{r_n\}_{n\in\mathcal{S}} = \{n\}_{n\in\mathcal{S}}$" represents the possible relay assistance exchange among the nodes in the coalition $\mathcal{S}$. The condition "$r_m = m, \forall m \in \mathcal{N}\backslash\mathcal{S}$" states that the nodes out of the coalition $\mathcal{S}$ will not participate in any cooperation and choose to transmit directly. For example, in Figure 3.5, the coalition

$\mathcal{S} = \{1, 2\}$ in the direct reciprocity case adopts the cooperation strategy $r_1 = 2$ and $r_2 = 1$, and the coalition $\mathcal{S} = \{1, 2, 3\}$ in the indirect reciprocity case adopts the cooperation strategy $r_1 = 3$, $r_2 = 1$ and $r_3 = 2$.

- The preference order $\succ_n$ is defined as $(r_m)_{m\in\mathcal{N}} \succ_n (r'_m)_{m\in\mathcal{N}}$ if and only if $r_n \succ_n r'_n$. That is, node n prefers the relay selection $(r_m)_{m\in\mathcal{N}}$ to another selection $(r'_m)_{m\in\mathcal{N}}$ if and only if its assigned relay r_n in the former selection $(r_m)_{m\in\mathcal{N}}$ is better than the assigned relay r'_n in the latter selection $(r'_m)_{m\in\mathcal{N}}$. In the following, we define that $r_n \succ_n r'_n$ when $R_n(r_n) > R_n(r'_n)$, and if $R_n(r_n) = R_n(r'_n)$, then ties are broken arbitrarily.

The core of this coalitional game is a set of $(r^*_n)_{n\in\mathcal{N}} \in V(\mathcal{N})$ for which there does not exist a coalition $\mathcal{S}$ and $(r_n)_{n\in\mathcal{N}} \in V(\mathcal{S})$ such that $(r_n)_{n\in\mathcal{N}} \succ_n (r^*_n)_{n\in\mathcal{N}}$ for all $n \in \mathcal{S}$. In other words, no coalition of nodes can deviate and improve their relay selection by cooperation in the coalition. We will refer the solution $(r^*_n)_{n\in\mathcal{N}}$ as the core relay selection in the sequel.

3.4.2.3 Core Relay Selection We now study the existence of the core relay selection. To proceed, we first introduce the following key concepts of coalitional game.

Definition 3.2 *Given a coalitional game $\Omega = (\mathcal{N}, \mathcal{X}_\mathcal{N}, V, (\succ_n)_{n\in\mathcal{N}})$, we call a coalitional game $\Phi = (\mathcal{M}, \mathcal{X}_\mathcal{M}, V, (\succ_m)_{m\in\mathcal{M}})$ a coalitional sub-game of the game Ω if and only if $\mathcal{M} \subseteq \mathcal{N}$ and $\mathcal{M} \neq \emptyset$.*

In other words, a coalitional sub-game Φ is a coalitional game defined on a subset of the players of the original coalitional game Ω.

Definition 3.3 *Given a coalitional sub-game $\Phi = (\mathcal{M}, \mathcal{X}_\mathcal{M}, V, (\succ_m)_{m\in\mathcal{M}})$, a non-empty subset $\mathcal{S} \subseteq \mathcal{M}$ is a top-coalition of the game Φ if and only if there exists a cooperation strategy $(\tilde{r}_m)_{m\in\mathcal{M}} \in V(\mathcal{S})$ such that for any $\mathcal{K} \subseteq \mathcal{M}$ and any cooperation strategy $(r_m)_{m\in\mathcal{M}} \in V(\mathcal{K})$ satisfying $\tilde{r}_m \neq r_m$ for any $m \in \mathcal{S}$, we have $\tilde{r}_m \succ_m r_m$ for any $m \in \mathcal{S}$.*

That is, by adopting the cooperation strategy $(\tilde{r}_m)_{m\in\mathcal{S}}$, the coalition $\mathcal{S}$ is a group that is mutually best for all its members [47].

Definition 3.4 *A coalitional game $\Omega = (\mathcal{N}, \mathcal{X}_\mathcal{N}, V, (\succ_n)_{n\in\mathcal{N}})$ satisfies the top-coalition property if and only if there exists a top-coalition for any of its coalitional sub-game Φ.*

We then show that the proposed coalitional game for social reciprocity-based relay selection satisfies the top-coalition property. For simplicity, we first denote $\tilde{\mathcal{N}}^{PC}_n \triangleq \mathcal{N}^{PC}_n \cup \{n\}$. For a coalitional sub-game $\Phi = (\mathcal{M}, \mathcal{X}_\mathcal{M}, V, (\succ_m)_{m\in\mathcal{M}})$, we denote the mapping $\gamma(n, \mathcal{M})$ as the most preferable relay of node $n \in \mathcal{M}$ in the set of nodes $\mathcal{M} \cap \tilde{\mathcal{N}}^{PC}_n$, that is, $\gamma(n, \mathcal{M}) \succ_n i$ for any $i \neq \gamma(n, \mathcal{M})$ and $i \in \mathcal{M} \cap \tilde{\mathcal{N}}^P_n$. Based on the mapping γ, we can define the concept of reciprocal relay selection cycle as follows.

Definition 3.5 *Given a coalitional sub-game* $\Phi = (\mathcal{M}, \mathcal{X}_{\mathcal{M}}, V, (\succ_m)_{m \in \mathcal{M}})$*, a node sequence* $(n_1, ..., n_L)$ *is called a reciprocal relay selection cycle of length L if and only if* $\gamma(n_l, \mathcal{M}) = n_{l+1}$ *for* $l = 1, ..., L-1$ *and* $\gamma(n_L, \mathcal{M}) = n_1$.

Notice that when $L = 1$ (i.e., $\gamma(n, \mathcal{M}) = n$), the most preferable choice of node n is to choose to transmit directly; when $L = 2$, this corresponds to the direct reciprocity case; when $L \geq 3$, this corresponds to the indirect reciprocity case. Since the number of nodes (i.e., $|\mathcal{M}|$) is finite, there hence must exist at least one reciprocal relay selection cycle for the coalitional sub-game Φ. This leads to the following result.

Lemma 3.1 *Given a coalitional sub-game* Φ*, there exists at least one reciprocal relay selection cycle. Any reciprocal relay selection cycle is a top-coalition of the coalitional sub-game* Φ.

Proof: For the first part of the lemma, we can choose any node $n \in \mathcal{M}$ as the starting node n_1. Then we can find the second node $n_2 = \gamma(n_1, \mathcal{M})$ and continue in this manner. If no cycle exists, the node sequence $(n_1, n_2, ...)$ can grow infinitely long, and any two nodes in the sequence are different. This obviously contradicts with the fact that the set of nodes $\mathcal{M}$ is finite.

For the second part of the lemma, given a reciprocal relay selection cycle $(n_1, ..., n_L)$, we denote the set of nodes in the cycle as C. We can then adopt the cooperation strategy for the nodes in the cycle as $\tilde{r}_{n_l} = n_{l+1}$ for $l = 1, ..., L = 1$ and $\tilde{r}_{n_L} = n_1$. According to Definition 3.5, each node $n \in C$ is allocated with its most preferable relay in the coalitional sub-game Φ. Thus for any other relays $r_n \neq \tilde{r}_n$, we have that $\tilde{r}_n \succ_n r_n$ for any $n \in C$. ∎

According to Lemma 3.1, we have the following result.

Lemma 3.2 *The coalitional game* Ω *for cooperative D2D communications satisfies the top-coalition property.*

Similar to the top trading cycle scheme for the housing market [48], based on the top-coalition property [47], we can then construct the core relay selection in an iterative manner. Let $\mathcal{M}_t$ denote the set of nodes of the coalitional sub-game $\Phi_t = (\mathcal{M}_t, \mathcal{X}_{\mathcal{M}_t}, V, (\succ_m)_{m \in \mathcal{M}_t})$ in the t-th iteration. Based on the mapping γ and the given set of nodes $\mathcal{M}_t$, we can then find all the reciprocal relay selection cycles as $C_1^t, ..., C_{Z_t}^t$ where each cycle $C_z^t = (n_1^t, ..., n_{|C_z^t|}^t)$ is a node sequence and Z_t denotes the number of cycles at the t-th iteration. Abusing notation, we will also use C_z^t to denote the set of nodes in the cycle C_z^t. We can then construct the core relay selection as follows. For the first iteration $t = 1$, we set $\mathcal{M}_1 = \mathcal{N}$ and find the reciprocal relay selection cycles as $C_1^1, ..., C_{Z_1}^1$ based on the set of nodes $\mathcal{M}_1$. For the second iteration $t = 2$, we can then set that $\mathcal{M}_2 = \mathcal{M}_1 \setminus \cup_{i=1}^{Z_1} C_i^1$ (i.e., remove the nodes in the cycles in the previous iteration) and find the new reciprocal relay selection cycles as $C_1^2, ..., C_{Z_2}^2$ based on the set of nodes $\mathcal{M}_2$. This procedure repeats until the set of nodes $\mathcal{M}_t = \emptyset$ (i.e., no operation can be further carried out). We summarize the aforementioned procedure for constructing the core relay selection in Algorithm 3.1

Algorithm 3.1 Core Relay Selection Algorithm

1: **initialization:**
2: **set** initial set of nodes $\mathcal{M}_1 = \mathcal{N}$.
3: **set** iteration index $t = 1$.
4: **end initialization**

5: **loop** until $\mathcal{M}_t = \emptyset$:
6: **find** all the reciprocal relay selection cycles $C_1^t, ..., C_{Z_t}^t$.
7: **remove** the set of nodes in the cycles from the current set of nodes $\mathcal{M}_t$, i.e., $\mathcal{M}_{t+1} = \mathcal{M}_t \setminus \cup_{i=1}^{Z_t} C_i^t$.
8: **set** $t = t + 1$.
9: **end loop**

Suppose that the algorithm takes T iterations to converge. We can obtain the set of reciprocal relay selection cycles in all T iterations as $\{C_i^t : \forall i = 1, ..., Z_t \text{ and } t = 1, ..., T\}$. Since the mapping $\gamma(n, \mathcal{M}_t)$ is unique for each node $n \in \mathcal{M}_t$, we must have that $\cup_{i=1,...,Z_t}^{t=1,...,T} C_i^t = \mathcal{N}$ (i.e., all the nodes are in the cycles) and $C_i^t \cap C_j^{t'} = \emptyset$ for any $i \neq j$ and $t, t' = 1, ..., T$ (i.e., there do not exist any intersecting cycles). For each cycle $C_i^t = (n_1^t, ..., n_{|C_i^t|}^t)$, we can then define the relay selection as $r_{n_l^t}^* = n_{l+1}^t$ for any $l = 1, 2..., |C_i^t| - 1$ and $r_{n_{|C_i^t|}^t}^* = n_1^t$. We show that $(r_n^*)_{n\in\mathcal{N}}$ is a core relay selection of the coalitional game Ω for the social reciprocity-based relay selection.

Theorem 3.1 *The relay selection $(r_n^*)_{n\in\mathcal{N}}$ is a core solution to the coalitional game Ω for the social reciprocity-based relay selection.*

Proof: We prove the result by contradiction. We assume that there exists a nonempty coalition $S \subseteq \mathcal{N}$ with another relay selection $(r_m)_{m\in\mathcal{N}} \in V(S)$ satisfying $(r_m)_{m\in\mathcal{N}} \succ_n (r_m^*)_{m\in\mathcal{N}}$ for any $n \in S$. Let $C^t = \cup_{i=1}^{Z_t} C_i^t$ be the set of nodes in the reciprocal relay selection cycles obtained in the t-th iteration. According to Lemma 3.1, we know that each cycle C_i^1 is a top-coalition given the set of nodes $\mathcal{M}_1 = \mathcal{N}$. By the definition of top-coalition, we must have that $S \cap C^1 = \emptyset$. In this case, we have that $S \subseteq \mathcal{M}_2 \triangleq \mathcal{M}_1 \setminus C^1$. Similarly, each cycle C_i^2 is a top-coalition given the set of nodes $\mathcal{M}_2$. We thus also have that $S \cap C^2 = \emptyset$. Repeating this argument, we can find that $S \cap C^t = \emptyset$ for any $t = 1, ..., T$. Since $\mathcal{N} = \cup_{t=1}^T C^t$, we must have that $S \cap \mathcal{N} = \emptyset$, which contradicts with the hypothesis that $S \subseteq \mathcal{N}$ and $S \neq \emptyset$. This completes the proof. ■

3.4.3 Social Trust and Social Reciprocity-Based Relay Selection

According to the principles of social trust and social reciprocity earlier, each node $n \in \mathcal{M}$ has two options for relay selection. The first option is that node n can choose the best relay $r_n^S = \arg\max_{r_n \in \mathcal{N}_n^{PS} \cup \{n\}} R_n(r_n)$ from the set of nodes with social trust

$\mathcal{N}_n^{PS}$. Alternatively, node n can choose a relay $r_n \in \mathcal{N}_n^{PC}$ from the set of nodes without social trust by participating in a directly or indirectly reciprocal cooperation group.

We next address the issue of choosing between social trust and social reciprocity-based relay selections for each node, by generalizing the core relay selection $(r_n^*)_{n\in\mathcal{N}}$ in Section 3.4.2.3. The key idea is to adopt the social trust-based relay selection r_n^S as the benchmark for participating in the social reciprocity-based relay selection. That is, a node n prefers social reciprocity-based relay selection to social trust-based relay selection if the social reciprocity-based relay selection offers better performance. More specifically, we define that $r_n \succ_n n$ if and only if $r_n \succ_n r_n^S$, and the selection "$r_n = n$" represents that node n will select the relay r_n^S based on social trust. Based on this, we can then compute the core relay selection $(r_n^*)_{n\in\mathcal{N}}$ according to Algorithm 3.1. In this case, if we have $r_m^* = m$ in the core relay selection $(r_n^*)_{n\in\mathcal{N}}$, then node m will select the relay r_n^S based on social trust. If we have $r_m^* \neq m$ in the core relay selection $(r_n^*)_{n\in\mathcal{N}}$, then node m will select the relay based on social reciprocity.

In a nutshell, we have studied the cooperative D2D communications based on social trust and social reciprocity. We have developed a coalitional game approach for efficiently forming the reciprocal groups among the nodes and also addressed the issue of choosing between social trust and social reciprocity-based relay selections for each node.

3.5 NETWORK ASSISTED RELAY SELECTION MECHANISM

In this section, we turn our attention to the implementation of the core relay selection for social trust- and social reciprocity-based cooperative D2D communications. A key challenge here is how to find the reciprocal relay selection cycles in the proposed core relay selection algorithm (see Algorithm 3.1). In the following, we will first propose a reciprocal relay selection cycle-finding algorithm to address this issue and then develop a network-assisted mechanism to implement the core relay selection solution in practical D2D communication systems.

3.5.1 Reciprocal Relay Selection Cycle Finding

We first consider the issue of reciprocal relay selection cycle finding in the core relay selection algorithm. We introduce a graphical approach to address this issue. More specifically, given the set of nodes $\mathcal{M}_t$ and the mapping γ, we can construct a graph $\mathcal{G}^{\mathcal{M}_t} = \{\mathcal{M}_t, \mathcal{E}^{\mathcal{M}_t}\}$. Here the set of vertices is $\mathcal{M}_t$ and the set of edges $\mathcal{E}^{\mathcal{M}_t} = \{(nm) : e_{nm}^{\mathcal{M}_t} = 1, \forall n, m \in \mathcal{M}_t\}$ where there is an edge directed from node n to m (i.e., $e_{nm}^{\mathcal{M}_t} = 1$) if and only if $\gamma(n, \mathcal{M}_t) = m$. For the graph $\mathcal{G}^{\mathcal{M}_t}$, we have the following key observations.

Lemma 3.3 *The out-degree of each node in the graph $\mathcal{G}^{\mathcal{M}_t}$ is one.*

This is due to the fact that the node m generated by the mapping $\gamma(n, \mathcal{M}_t)$ is unique.

We next introduce the concept of path in graph theory. A path of length I on a graph is a sequence of nodes $(n_1, n_2, ..., n_I)$ where there is an edge directed from node n_i to n_{i+1} on the graph for any $i = 1, ..., I-1$. A cycle of the graph is a path in which the first and last nodes are identical. A reciprocal relay selection cycle of the coalitional game then corresponds to a cycle of the graph $\mathcal{G}^{\mathcal{M}_t}$. When $\gamma(n, \mathcal{M}_t) = n$, the cycle degenerates to a self-loop of node n. In the following section, we say a path $(n_1, n_2, ..., n_I)$ induces a cycle if there exists a path beginning from node n_I that is a cycle. If two cycles are a cyclic permutation of each other, we will regard them as one cycle.

Lemma 3.4 *Any sufficiently long path beginning from any node on the graph $\mathcal{G}^{\mathcal{M}_t}$ induces one and only one cycle.*

Proof: We first show that a path beginning from a node n_1 induces a cycle. Since each node has an out-degree of one, this implies we can construct a path $(n_1, n_2, ...)$ of an infinitely large length if the path does not induce a cycle. This contradicts with the fact that the number of nodes on the graph is finite.

On the other hand, if the path induces multiple distinct cycles, there must exist a node with more than one outward directed edge. This contradicts Lemma 3.3. ■

Based on Lemmas 3.3 and 3.4, we propose an algorithm to find the reciprocal relay selection cycles in Algorithm 3.2. The key idea of the algorithm is to explore the paths beginning from each node. More specifically, if a path beginning from a node induces an unfound cycle, then we find a new cycle. We will set the nodes in both the path and cycle as visited nodes since any path beginning from these nodes would induce the same cycle. If a path beginning from a node leads to a visited node, the path would induce a cycle that has already been found if we continue to construct the path on the visited nodes. We will also set the nodes in the path as visited nodes. Since each node will be visited once in the algorithm, the computational complexity of the reciprocal relay selection cycles finding algorithm is $\mathcal{O}(|\mathcal{M}_t|)$.

3.5.2 NARS Mechanism

We now propose a network-assisted relay selection (NARS) mechanism to implement the core relay selection, which works as follows:

- Each node $n \in \mathcal{N}$ first determines its preference list $\mathcal{L}_n^{\mathrm{P}}$ for the set of feasible relay selections $\tilde{\mathcal{N}}_n^{\mathrm{P}} \triangleq \mathcal{N}_n^{\mathrm{P}} \cup \{n\}$ based on the physical graph $\mathcal{G}^{\mathrm{P}}$. Here $\mathcal{L}_n = (r_n^1, ..., r_n^{|\tilde{\mathcal{N}}_n^{\mathrm{P}}|})$ is a permutation of all the feasible relays in $\tilde{\mathcal{N}}_n^{\mathrm{P}}$ satisfying that $r_n^i \succ_n r_n^{i+1}$ for any $i = 1, ..., |\tilde{\mathcal{N}}_n^{\mathrm{P}}| - 1$. This step can be done through the channel probing procedure to measure the achieved data rate resulting from choosing with different relays.
- Each node $n \in \mathcal{N}$ then computes the best social trust-based relay selection $r_n^{\mathrm{S}} = \arg\max_{r_n \in \mathcal{N}_n^{\mathrm{PS}} \cup \{n\}} R_n(r_n)$ based on the physical–social graph $\mathcal{G}^{\mathrm{PS}}$ and the preference list $\mathcal{L}_n^{\mathrm{P}}$.

Algorithm 3.2 Reciprocal Relay Selection Cycle Finding Algorithm

1: **initialization:**
2: **construct** the graph $\mathcal{G}^{\mathcal{M}_t}$ based on the set of nodes $\mathcal{M}_t$ and the mappings $\{\gamma(n, \mathcal{M}_t)\}_{n\in\mathcal{M}_t}$.
3: **set** the set of visited nodes $\mathcal{V} = \emptyset$ and the set of unvisited nodes $\mathcal{U} = \mathcal{M}_t \backslash \mathcal{V}$.
4: **set** the set of identified cycles $\triangle = \emptyset$.
5: **end initialization**

6: **loop** until $\mathcal{U} = \emptyset$:
7: **select** one node $n_a \in \mathcal{U}$ randomly.
8: **set** the set of visited nodes in the current path $\mathcal{H} = \{n_a\}$.
9: **set** the flag $F = 0$.
10: **loop** until $F = 1$:
11: **generate** the next node $n_b = \gamma(n_a, \mathcal{M}_t)$.
12: **if** $n_b \in \mathcal{V}$ **then**
13: **set** $\mathcal{V} = \mathcal{V} \cup \mathcal{H}$ and $\mathcal{U} = \mathcal{M}_t \backslash \mathcal{V}$.
14: **set** $F = 1$.
15: **else if** $n_b \in \mathcal{H}$ **then**
16: **set** the identified cycle as $C = (n_1 = n_b, ..., n_i = \gamma(n_{i-1}, \mathcal{M}_t), ..., n_I = n_a)$.
17: **set** the set of identified cycles $\triangle = \triangle \cup \{C\}$.
18: **set** $\mathcal{V} = \mathcal{V} \cup \mathcal{H}$ and $\mathcal{U} = \mathcal{M}_t \backslash \mathcal{V}$.
19: **set** $F = 1$.
20: **else**
21: **set** $\mathcal{H} = \mathcal{H} \cup \{n_b\}$.
22: **set** $n_a = n_b$.
23: **end if**
24: **end loop**
25: **end loop**

- Each node $n \in \mathcal{N}$ next determines its preference list $\mathcal{L}_n^{\text{PC}}$ for the set of relay selections $\mathcal{N}_n^{\text{PC}} \cup \{n\}$ based on the physical-coalitional graph $\mathcal{G}^{\text{PC}}$. Notice that we have that $r_n \succ_n n$ in the preference list $\mathcal{L}_n^{\text{PC}}$ if and only if $r_n \succ_n r_n^{\text{S}}$ in the preference list $\mathcal{L}_n^{\text{P}}$.
- Each node $n \in \mathcal{N}$ then reports its preference list $\mathcal{L}_n^{\text{PC}}$ to the base station.
- Based on the preference lists $\mathcal{L}_n^{\text{PC}}$ of all nodes, the base station computes the core relay selection $(r_n^*)_{n\in\mathcal{N}}$ according to Algorithms 3.1 and 3.2 and broadcasts the relay selection $(r_n^*)_{n\in\mathcal{N}}$ to all nodes.

As mentioned in Section 3.4.3, if $r_m^* = m$ in the core relay selection $(r_n^*)_{n\in\mathcal{N}}$, then node m will select the relay r_n^{S} based on social trust. If $r_m^* \neq m$ in the core relay selection $(r_n^*)_{n\in\mathcal{N}}$, then node m will select the relay based on social reciprocity.

TABLE 3.1 The Preference Lists of $N = 7$ Nodes Based on the Physical Graph $\mathcal{G}^{\mathrm{P}}$ and Social Graph $\mathcal{G}^{\mathrm{S}}$ in Figure 3.2

Node n	Preference List $\mathcal{L}_n^{\mathrm{P}}$	Relay r_n^{S}	Preference List $\mathcal{L}_n^{\mathrm{PC}}$
1	(1,2,3,4)	1	(1,2)
2	(1,3,2,4,5)	2	(1,3,2,4)
3	(2,3,4,1)	3	(2,3,4)
4	(2,1,4,3,5,6)	1	(2,4,3,5,6)
5	(4,6,7,5,2)	5	(4,6,7,5)
6	(7,5,4,6)	6	(7,5,4,6)
7	(5,6,7)	7	(5,6,7)

We now use an example to illustrate how the NARS mechanism works. We consider the network of $N = 7$ nodes based on the physical graph $\mathcal{G}^{\mathrm{P}}$ and the social graph $\mathcal{G}^{\mathrm{S}}$ in Figure 3.2. According to NARS mechanism, each node n first determines its preference list $\mathcal{L}_n$ for the set of feasible relay selections $\mathcal{N}_n^{\mathrm{P}} \cup \{n\}$. We will use the preference lists $\mathcal{L}_n^{\mathrm{P}}$ in Table 3.1. For example, in the table the feasible relays for node 7 on the physical graph $\mathcal{G}^{\mathrm{P}}$ are $\{5, 6, 7\}$. The preference list $(5, 6, 7)$ represents that $5 \succ_7 6 \succ_7 \succ 7$, that is, node 7 prefers choosing node 5 as the relay to choosing node 6, and transmitting directly offers the worst performance. Then based on the physical–social graph $\mathcal{G}^{\mathrm{PS}}$ in Figure 3.4 and the preference list $\mathcal{L}_n^{\mathrm{P}}$, each node n computes the best social trust-based relay selection r_n^{S}, for example, node 4's best social trust-based relay selection $r_n^{\mathrm{S}} = 1$ (i.e., node 1). Each node n next determines the preference list $\mathcal{L}_n^{\mathrm{PC}}$ based on the physical–social graph $\mathcal{G}^{\mathrm{PS}}$ in Figure 3.6.

All the nodes then report the preference lists $\mathcal{L}_n^{\mathrm{PC}}$ to the base station. Based on the preference lists, the base station will compute the core relay selection $(r_n^*)_{n\in\mathcal{N}}$ according to the core relay selection algorithm in Algorithm 3.1. We illustrate the iterative procedure of the core relay selection algorithm in Figure 3.7 by adopting the graphical representation $\mathcal{G}^{\mathcal{M}_t}$ introduced in Section 3.5.1. Recall that there is an edge directed from node n to node m on graph $\mathcal{G}^{\mathcal{M}_t}$ if node m is the most preferable relay of node n given the set of nodes $\mathcal{M}_t$. At iteration $t = 1$, given that $\mathcal{M}_1 = \mathcal{N}$, the base station identifies one cycle, that is, a self-loop formed by node 1. At iteration $t = 2$, given that $\mathcal{M}_2 = \mathcal{M}_1 \backslash \{1\}$, the base station then identifies one cycle formed by nodes 2 and 3. Notice that graph $\mathcal{G}^{\mathcal{M}_2}$ can be derived from graph $\mathcal{G}^{\mathcal{M}_1}$ by removing node 1 and any edges directed to node 1. For each node (e.g., node 2) from which there is a removed edge directed to node 1, we add a new edge directed from the node to its most preferable node among the set of nodes $\mathcal{M}_2$ (e.g., the edge $2 \rightarrow 3$). We continue in this manner until all the nodes have been removed from the graph. Figure 3.8 shows all the reciprocal relay selection cycles identified by the core relay selection algorithm in Figure 3.7. In this case, the core relay selection is as follows: (i) since $r_1^S = 1$, node 1 transmits directly; (ii) nodes 2 and 3 serve as the relay of each other (i.e., direct reciprocity-based relay selection); (iii) since $r_4^S = 1$, node 4 seeks relay assistance from node 1 (i.e., social trust-based relay selection); (iv) node 5 serves as the relay of node 7, which in turn serves as the relay of node 6; and node 6 in turn is the relay of node 5 (i.e., indirect reciprocity-based relay selection).

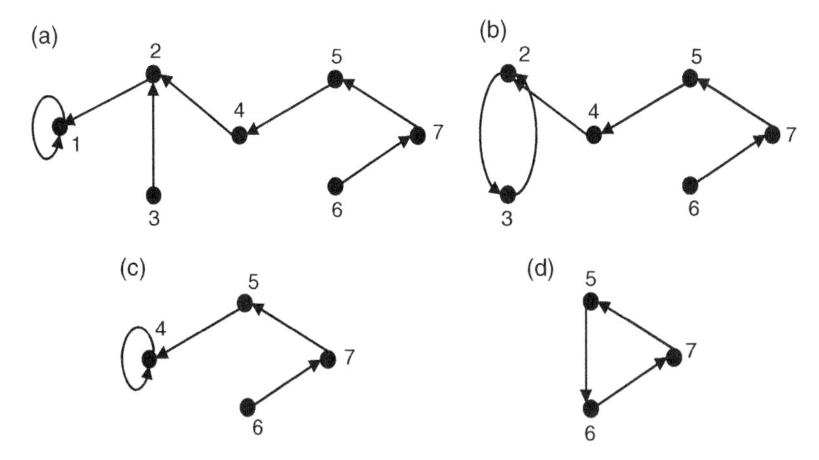

Figure 3.7 An illustration of the resulting graphs $\mathcal{G}^{\mathcal{M}_t}$ at each iteration t of the core relay selection algorithm. (a) $t = 1$, (b) $t = 2$, (c) $t = 3$, and $t = 4$.

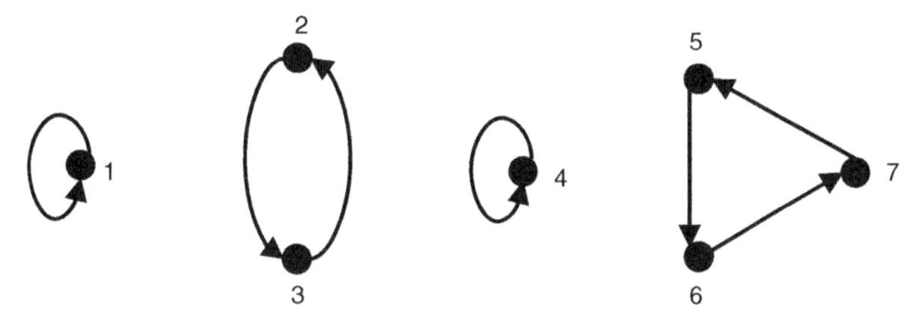

Figure 3.8 The reciprocal relay selection cycles identified by the core relay selection algorithm in Figure 3.7.

3.5.3 Properties of NARS Mechanism

We next study the properties of the proposed NARS mechanism. First of all, according to the definition of the core solution of coalitional game, we know the following.

Lemma 3.5 *The core relay selection $(r_n^*)_{n \in \mathcal{N}}$ by NARS mechanism is immune to group deviations, that is, no group of nodes can deviate and improve by cooperation within the group.*

We can then show that the mechanism guarantees individual rationality, which means that each participating node will not achieve a lower data rate than that when the node does not participate (i.e., in this case the node will transmit directly).

Lemma 3.6 *The core relay selection $(r_n^*)_{n \in \mathcal{N}}$ by NARS mechanism is individually rational, that is, each node $n \in \mathcal{N}$ will be assigned a relay r_n^* which satisfies either $r_n^* \succ_n n$ or $r_n^* = n$.*

Proof: If the assigned relay $r_n^* <_n n$ for some node $n \in \mathcal{N}$, then the node n can deviate from the current coalition and improve its data rate by transmitting directly (i.e., $r_n^* = n$). This contradicts with the fact that $(r_n^*)_{n\in\mathcal{N}}$ is a core relay selection. ■

We next explore the truthfulness of NARS mechanism. A mechanism is truthful if no node can improve by reporting a preference list different from its true preference list, given that other nodes report truthfully.

Lemma 3.7 *NARS mechanism is individually truthful.*

Proof: Let C^t be the set of nodes in the reciprocal relay selection cycles obtained in the t-th iteration of core relay selection algorithm. Suppose that the node m reports another preference list that is different from its true preference list. Let τ be the index such that $m \in C^\tau$. Given that the nodes in the set $\cup_{t=1}^{\tau-1} C^t$ truthfully report, they will be assigned the relays in the core relay selection regardless of what the nodes out of the set $\cup_{t=1}^{\tau-1} C^t$ report. In this case, given the set of remaining nodes $\mathcal{M}_\tau = \mathcal{N} \setminus \cup_{t=1}^{\tau-1} C^t$, the most preferable relay of node m is the relay r_m^* in the core relay selection. This is exactly what the node m achieves by reporting truthfully. Thus, the node m cannot improve by reporting another preference list. ■

We further show a stronger result of collective truthfulness. A mechanism is collectively truthful if no group of nodes can improve by joint reporting their preference lists different from their true preference lists, given that other nodes report truthfully.

Lemma 3.8 *NARS mechanism is collectively truthful.*

Proof: Suppose that a group of nodes $\mathcal{S}$ report other preference lists that are different from their true preference lists. Let τ be the smallest index such that $\mathcal{S} \cap C^\tau \neq \emptyset$. Given that the nodes in the set $\cup_{t=1}^{\tau-1} C^t$ truthfully report, they will be assigned the relays in the core relay selection regardless of what the nodes out of the set $\cup_{t=1}^{\tau-1} C^t$ report. Furthermore, given that nodes in the set $\mathcal{N} \setminus \mathcal{S}$ report truthfully, for any node $m \in \mathcal{S} \cap C^\tau$, the most preferable relay of node m among the remaining nodes $\mathcal{M}_\tau = \mathcal{N} \setminus \cup_{t=1}^{\tau-1} C^t$ is the relay r_m^* in the core relay selection. This is exactly what the node m achieves by reporting truthfully. Thus, a node $m \in \mathcal{S} \cap C^\tau$ cannot improve by reporting another preference list. Similarly, we can show that for a node m in the set $\mathcal{S} \cap C^{\tau+1}$, the most preferable relay of node m among the remaining nodes $\mathcal{M}_{\tau+1} = \mathcal{N} \setminus \cup_{t=1}^{\tau} C^t$ is the relay r_m^* in the core relay selection. We can repeat the same augment for k times until that $\mathcal{S} \cap C^{\tau+k} = \emptyset$, which completes the proof. ■

We finally consider the computational complexity of NARS mechanism. We say the mechanism is computationally efficient if the solution can be computed in polynomial time.

Lemma 3.9 *NARS mechanism is computationally efficient.*

Proof: Recall that the reciprocal relay selection cycle finding algorithm in Algorithm 3.2 has a complexity of $\mathcal{O}(|\mathcal{M}_t|)$. Since the reciprocal relay selection cycle finding algorithm is the dominating step in each iteration, the core relay selection algorithm hence has a complexity of $\mathcal{O}(\sum_{t=1}^{T}|\mathcal{M}_t|)$. As $\sum_{t=1}^{T}|\mathcal{M}_t| = N + \sum_{t=2}^{T}(N - \sum_{\tau=1}^{t-1}|C^{\tau}|)$ and $\sum_{t=1}^{T}|C^{\tau}| = N$, by setting $|C^{\tau}| = 1$ for $\tau = 1, ..., T$, we have the worst case that $\sum_{t=1}^{T}|\mathcal{M}_t| = \sum_{i=1}^{N} i = \frac{N(N+1)}{2}$. Thus, the mechanism has a complexity of at most $\mathcal{O}(N^2)$. ■

The aforementioned five Lemmas together prove the following theorem.

Theorem 3.2 *NARS mechanism is immune to group deviations and individually rational, individually and collectively truthful, and computationally efficient.*

To summarize, in this section we have developed a graphical-based algorithm for finding the reciprocal relay selection cycles and have further devised an efficient NARS mechanism with nice property guarantee for implementing the social trust- and social reciprocity-based relay selection solution in practical D2D communication systems.

3.6 SIMULATIONS

In this section we evaluate the performance of the proposed social trust- and social reciprocity-based relay selection for cooperative D2D communications through simulations. For the purpose of illustration, here we use the outband D2D communications as a study case. Nevertheless our mechanisms can be also implemented in the scenario of inband D2D communications given the spectrum sharing among D2D and cellular links is controlled by the base station.

We consider that multiple nodes are randomly scattered across a square area with a side length of 1000 m. Two nodes within a distance of 250 m are randomly matched into a source–destination D2D communication link. The motivation of randomly matching source–destination pairs is as follows: (i) due to the mobility, a user may have opportunities to conduct D2D communications with different users at different time periods and different locations; (ii) a user may have diverse interest to carry out D2D communications with different users for sharing different content. We compute the SNR value μ_{ij} according to the physical interference model, that is, $\mu_{ij} = \frac{p_i}{\omega_0 \cdot ||i,j||^{\alpha}}$ with the transmission power $p_i = 1$ W, the background noise $\omega_0 = 10^{-10}$ W, and the path loss factor $\alpha = 4$ [49]. Based on the SNR μ_{ij}, we set the bandwidth $W = 10$ MHz and then compute the data rate achieved by using different relays according to Equation (3.1). We construct the physical graph $\mathcal{G}^{\mathrm{P}}$ by setting $e_{nm}^{\mathrm{P}} = 1$ (i.e., node m can be a relay candidate of node n) if and only if the distance between nodes n and m is not greater than a threshold $\delta = 500$ m (i.e., $||n,m|| \leq \delta$). We set a relatively large distance threshold due to the fact that in the D2D communication, the detection of neighboring relay nodes can be significantly enhanced with the assistance by the base–station [3]. For the social trust model, we will consider two types of social graphs: Erdos–Renyi (ER) social graph and real data trace-based social graph.

3.6.1 Erdos–Renyi Social Graph

We first consider $N = 100$ nodes with the social graph $\mathcal{G}^{\mathrm{S}}$ represented by the ER graph model [50] where a social link exists between any two nodes with a probability of P_{L}. To evaluate the impact of social link density of the social graph, we implement the simulations with different social link probabilities $P_{\mathrm{L}} = 0, 0.05, 0.1, ..., 1.0$, respectively. For each given P_{L}, we average over 1000 runs. As the benchmark, we also implement the solution that each node transmits directly and selects the relay based on social trust only (i.e., $r_n = r_n^{\mathrm{S}}$) and on social reciprocity only by assuming that there is no social trust among the nodes. Furthermore, we also compute the throughput upper bound by letting each node select the best relay $\overline{r}_n = \arg\max_{r_n \in \mathcal{N}_n^{\mathrm{P}} \cup \{n\}} R_n(r_n)$ among all its feasible relays. Notice that the throughput upper bound can only be achieved when all the nodes are willing to help each other (i.e., all the nodes are cooperative).

We show the average system throughput in Figure 3.9. We see that the performance of the social trust and social reciprocity-based relay selection dominates that of social trust-only-based relay selection and social reciprocity-only-based relay selection. When the social link probability P_{L} is small, the social trust and social reciprocity-based relay selection achieves up to 64.5% performance gain over the social trust-only-based relay selection. When the social link probability P_{L} is large, the social trust- and social reciprocity-based relay selection achieves up to 24%

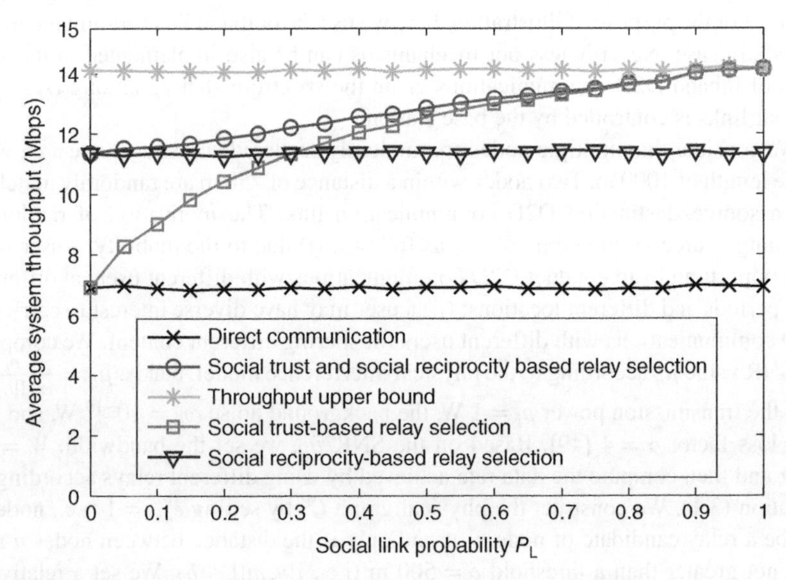

Figure 3.9 System throughput with the number of nodes $N = 100$ and different social network density.

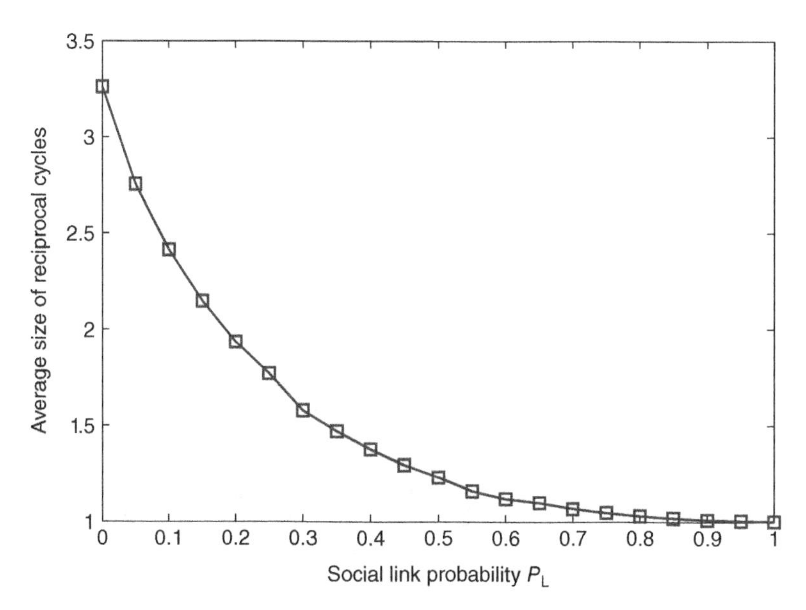

Figure 3.10 Average size of the reciprocal relay selection cycles in the social trust and social reciprocity-based relay selection with $N = 100$ and different social network density.

performance gain over the social reciprocity-only-based relay selection. We also observe that the social trust- and social reciprocity-based relay selection achieves up to 100.4% performance gain over the case that all the nodes transmit directly. Compared with the throughput upper bound, the performance loss of the social trust- and social reciprocity-based relay selection is at most 24%. As the social link probability P_L increases, the social trust- and social reciprocity-based relay selection improves and approaches the throughput upper bound. This is due to the fact that when the social link probability P_L is large, each node will have a high probability of having social trust from any other node, and hence each node is likely to have social trust from its best relay node. This can be illustrated by Figure 3.10, which shows the average size of the reciprocal relay selection cycles in the social trust- and social reciprocity-based relay selection. We observe that as the social link probability P_L increases, the average size of the reciprocal relay selection cycles decreases. This is because as the social link probability P_L increases, more nodes are able to select their best relay nodes based on social trust. As a result, less nodes would select relay nodes based on social reciprocity, and hence the average size of the reciprocal relay selection cycles decreases.

To investigate the impact of the distance threshold δ for relay detection, we implement the simulations with the number of nodes $N = 100$, the social link probability $P_L = 0.2$, and the distance threshold $\delta = 50, 100, ..., 600$ m, respectively. We see from Figure 3.11 that initially the system performance of social trust- and social reciprocity-based relay selection improves as the distance threshold δ increases. When the distance threshold δ is large, however, the performance of social

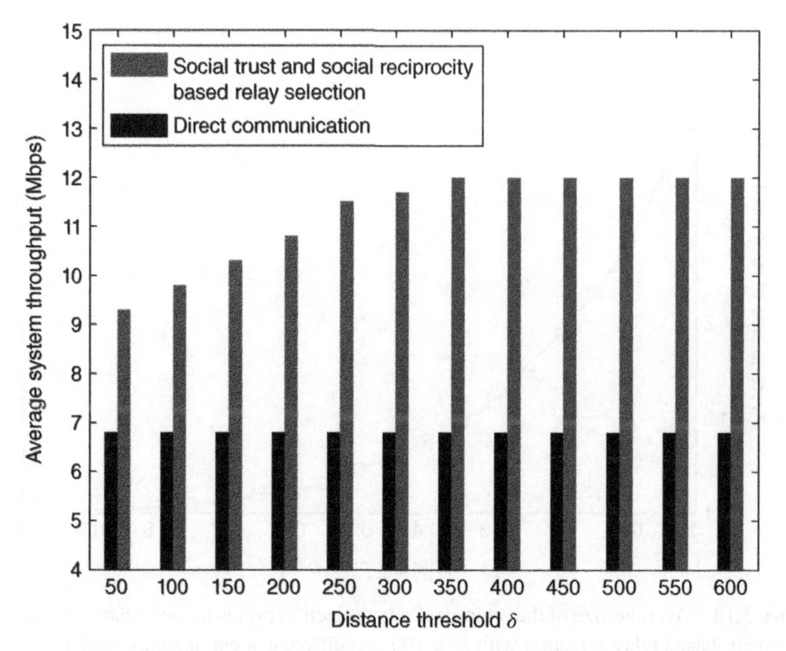

Figure 3.11 System throughput of nodes $N = 100$ and different distance threshold δ for relay detection.

trust- and social reciprocity-based relay selection levels off. This is because initially as the distance threshold δ increases, more and more good relay nodes are available. Once the distance threshold δ is large enough, only those nodes that are within a relatively short distance can be good relays for cooperative D2D communications. For those nodes that have a long distance, they will not be chosen as relays since they would offer worse performance than that of the direct communication case.

3.6.2 Real Trace Based Social Graph

We then evaluate the proposed social trust- and social reciprocity-based relay selection using the real data trace Brightkite [51]. Brightkite is a data trace collected from a location-based social networking service platform where users share their location check-ins. Brightkite contains an explicit friendship network among the users. Different from the ER social graph, the friendship network of Brightkite is scale-free such that the node degree distribution follows a power law [52]. We implement simulations as the number of nodes $N = 250, 500, ..., 1500$, respectively. We randomly select N nodes from Brightkite and construct the social graph based on the friendship relationship among these N nodes in the friendship network of Brightkite. For each given N, we average over 1000 runs. Figure 3.12 shows the average number of social links among these nodes of the social graphs when using the real data trace Brightkite.

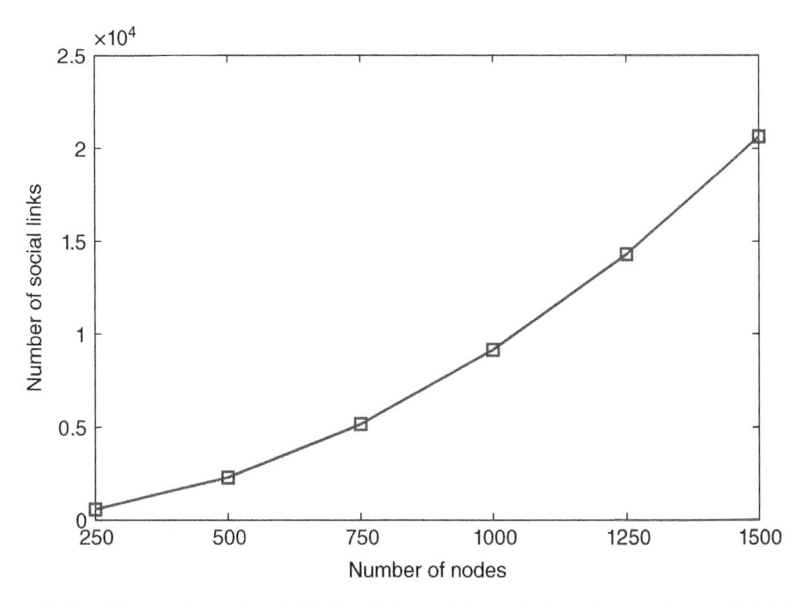

Figure 3.12 The number of social links of the social graphs based on real trace Brightkite.

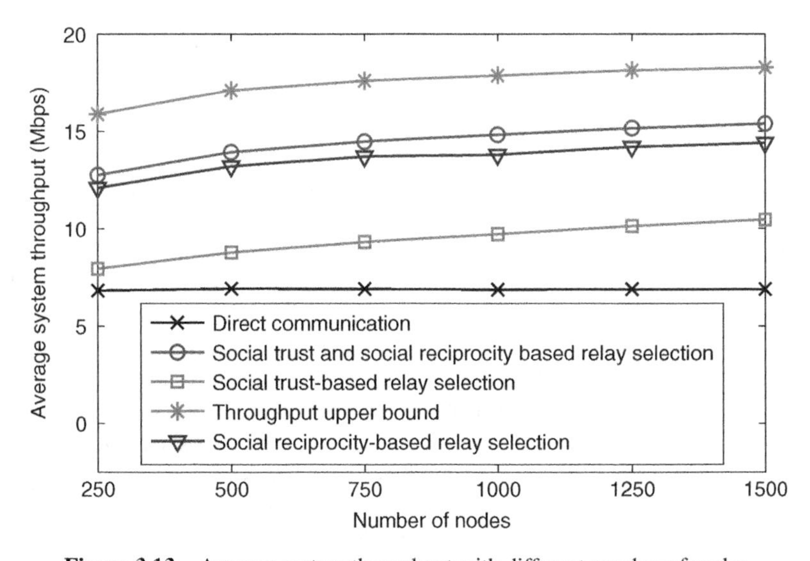

Figure 3.13 Average system throughput with different number of nodes.

We show the average system throughput in Figure 3.13. We see that the system throughput of the social trust- and social reciprocity-based relay selection increases as the number of users N increases. This is because more cooperation opportunities among the nodes are present when the number of users N increases. Moreover, the social trust- and social reciprocity-based relay selection achieves up to 122%

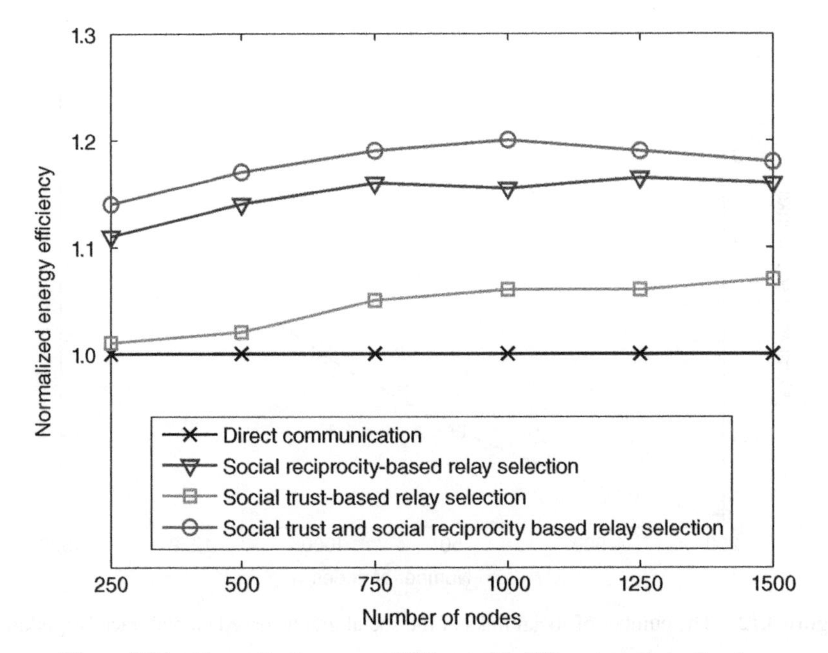

Figure 3.14 Normalized energy efficiency with different number of nodes.

performance gain over the solution that all users transmit directly. Compared with the throughput upper bound, the performance loss by the social trust- and social reciprocity-based relay selection is at most 21%.

We then evaluate the energy efficiency of the proposed NARS mechanism. We adopt the common practice in literature [53] and compute the energy efficiency as the ratio between the system-wide throughput and the system-wide energy consumption. Figure 3.14 shows the normalized energy efficiency of different relay selection schemes with respect to that of direct communication. It demonstrates that the proposed social trust- and social reciprocity-based relay selection scheme is energy efficient and achieves the highest energy efficiency among all the schemes.

We next show the computational complexity of the NARS mechanism for computing the social trust- and social reciprocity-based relay selection solution in Figure 3.15. We see that the average number of iterations of the mechanism grows linearly as the number of nodes N increases. We also measure the running time of the NARS mechanism on a 64-bit Windows PC with 2.5GHz quad-core CPU and 16GB memory in Figure 3.16. We observe that the running time of the mechanism increases linearly as the number of nodes N increases and the running time is less than 1 second in all cases. Notice that when the NARS mechanism is implemented in practical D2D systems, the base station typically has a much stronger computational capability than a PC, and the running time of the NARS mechanism can be further significantly reduced. This demonstrates that the proposed NARS mechanism is computationally efficient.

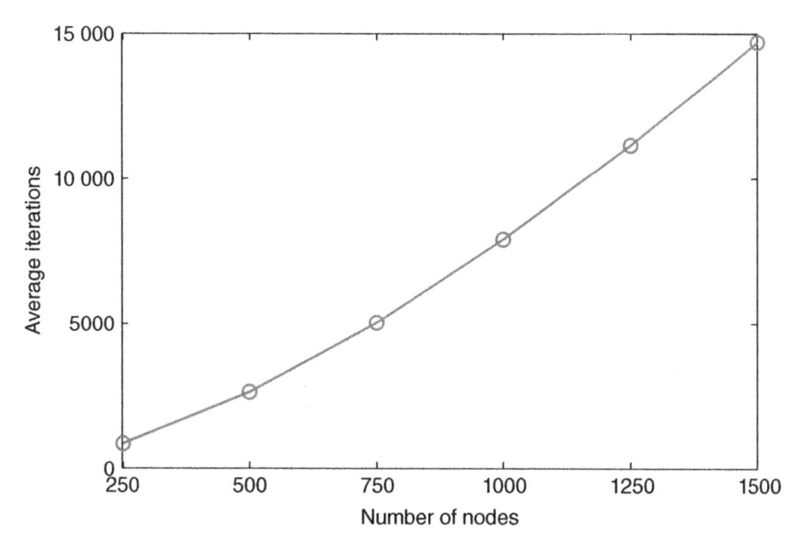

Figure 3.15 Average number of iterations of the NARS mechanism.

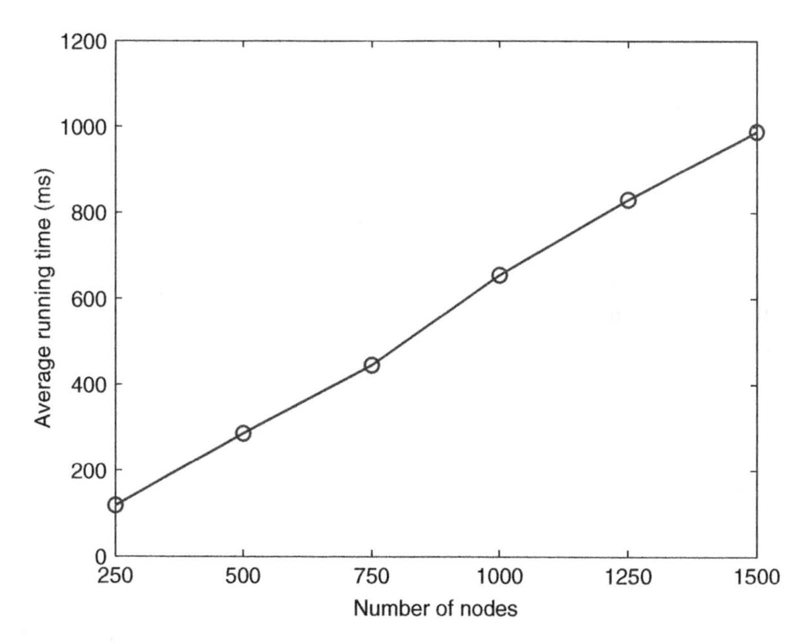

Figure 3.16 Average running time of the NARS mechanism.

3.7 CONCLUSION

In this chapter, we studied socially aware cooperative D2D- and D4D-based fog networking for bandwidth boosting. We introduced the physical–social graphs to capture the physical constraints for feasible D2D cooperation and the social relationships among devices for effective cooperation. We proposed a coalitional game theoretic approach to find the efficient D2D cooperation strategy and developed a NARS mechanism for implementing the coalitional game solution. We showed that the devised mechanism is immune to group deviations and individually rational, truthful, and computationally efficient. We further evaluated the performance of the mechanism based on ER social graphs and real data trace-based social graphs. Simulation results show that the proposed mechanism can achieve up to 122% performance gain over the case without D2D cooperation.

We focused on the case where the cooperative D2D communications between the relay node and the destination node use in-band communication (i.e., using cellular spectrum). To achieve better network connectivity and enhance the communication performance, both the in-band and out-band (i.e., using Wi-Fi spectrum) D2D communications can be utilized. For instance, two users can adopt the Wi-Fi Direct to conduct out-band D2D communication; alternatively, the users can conduct the in-band D2D communication by using the cellular spectrum. To this end, we have built a prototype system on cooperative D2D communications using out-band (Wi-Fi Direct) communications.

We are currently developing a framework for socially aware cooperative D2D- and D4D-based fog networking and computing. A key observation is that mobile devices and IoT devices have become next of kin (NOK), in the sense that there is a human being associated with each of such devices and these connected devices have kinship. Since these devices are connected (e.g., via D2D and D4D) and they have capabilities (bandwidth, storage, computing, sharing) that are often unused (idle most of the time), a natural question is that "Why cannot these devices in the kinship help each other just like families, relatives, and friends do?" Clearly the answer is yes—this leads to our vision of "Internet of NOK" where these connected devices form a fog network on the fly to help each other. Our study on the theoretic foundation and prototyping for "Internet of NOK" is underway.

ACKNOWLEDGMENTS

Junshan Zhang would like to thank Kaushik Pillalamarri and Mung Chiang for stimulating discussions on the concept of "Internet of NOK." This research was supported in part by the US National Science Foundation under Grants CNS-1218484, CNS-1422277, CNS-1248109, and HRD-1345232 and in part by the Defense Threat Reduction Agency under Grant HDTRA1-13-1-0029 and the US Army Research Office W911NF-15-1-0393.

REFERENCES

1. Cisco (2012) Global mobile data traffic data forecast update, 2011–2016, Cisco white paper.
2. Cisco (2014) Cisco delivers vision of fog computing to accelerate value from billions of connected devices, Cisco Tech. Rep. 1334100.
3. Fodor, G., Dahlman, E., Mildh, G., Parkvall, S., Reider, N., Miklós, G., and Turányi, Z. (2012) Design aspects of network-assisted device-to-device communications. *IEEE Communications Magazine*, **50** (3), 170–177.
4. Kayastha, N., Niyato, D., Wang, P., and Hossain, E. (2011) Applications, architectures, and protocol design issues for mobile social networks: A survey. *Proceedings of the IEEE*, **99** (12), 2130–2158.
5. Asadi, A., Wang, Q., and Mancuso, V. (2014) A survey on device-to-device communication in cellular networks. *IEEE Communications Surveys & Tutorials*, **16** (4), 1801–1819.
6. Govier, T. (1997) *Social trust and human communities*, McGill-Queen's University Press Montreal.
7. Gintis, H. (2000) Strong reciprocity and human sociality. *Journal of Theoretical Biology*, **206** (2), 169–179.
8. Schneier, B. and Ferguson, N. (2003) *Practical cryptography*, Wiley, New York, 1st edn.
9. Lasseter, R. and Paigi, P. (2004) Microgrid: A conceptual solution, in *IEEE Power Electronics Specialists Conference*, vol. 6, June 20–26, Aachen, Germany, pp. 4285–4290.
10. Fang, X., Misra, S., Xue, G., and Yang, D. (2012) Smart grid–the new and improved power grid: A survey. *IEEE Communications Surveys and Tutorials (CST)*, **14** (4), 944–980.
11. Fang, X., Misra, S., Xue, G., and Yang, D. (2012) Managing smart grid information in the cloud: Opportunities, model, and applications. *IEEE Network*, **26** (4), 32–38.
12. Fang, X., Misra, S., Xue, G., and Yang, D. (2013) How smart devices, online social networks and the cloud will affect the smart grid's evolution. *IEEE Smart Grid Newsletter* URL http://smartgrid.ieee.org/newsletters/january-2013/how-smart-devices-online-social-networks-and-the-cloud-will-affect-the-smart-grid-s-evolution? (accessed on October 26, 2016).
13. Yu, C.H., Tirkkonen, O., Doppler, K., and Ribeiro, C. (2009) On the performance of device-to-device underlay communication with simple power control, in *IEEE 69th Vehicular Technology Conference (VTC Spring)*, IEEE, pp. 1–5.
14. Yu, C.H., Tirkkonen, O., Doppler, K., and Ribeiro, C. (2009) Power optimization of device-to-device communication underlaying cellular communication, in *IEEE International Conference on Communications (ICC)*, IEEE, pp. 1–5.
15. Janis, P., Koivunen, V., Ribeiro, C.B., Doppler, K., and Hugl, K. (2009) Interference-avoiding mimo schemes for device-to-device radio underlaying cellular networks, in *IEEE 20th International Symposium on Personal, Indoor and Mobile Radio Communications*, IEEE, pp. 2385–2389.
16. Zulhasnine, M., Huang, C., and Srinivasan, A. (2010) Efficient resource allocation for device-to-device communication underlaying LTE network, in *IEEE Sixth International Conference on Wireless and Mobile Computing, Networking and Communications*, IEEE, pp. 368–375.
17. Raghothaman, B., Sternberg, G., Kaur, S., Pragada, R., Deng, T., and Vanganuru, K. (2011) System architecture for a cellular network with cooperative mobile relay, in *IEEE Vehicular Technology Conference (VTC Fall)*, IEEE, pp. 1–5.

18. Ma, X., Yin, R., Yu, G., and Zhang, Z. (2012) A distributed relay selection method for relay-assisted device-to-device communication system, in *IEEE 23rd International Symposium on Personal Indoor and Mobile Radio Communications (PIMRC)*, IEEE, pp. 1020–1024.
19. Lee, D., Kim, S.I., Lee, J., and Heo, J. (2012) Performance of multihop decode-and-forward relaying-assisted device-to-device communication underlaying cellular networks, in *2012 International Symposium on Information Theory and Its Applications (ISITA)*, IEEE, pp. 455–459.
20. Zhong, S., Chen, J., and Yang, Y. (2003) Sprite: A simple, cheat-proof, credit-based system for mobile ad-hoc networks, in *IEEE INFOCOM* March 2003, San Francisco, CA, pp. 1987–1997.
21. Marbach, P. and Qiu, Y. (2005) Cooperation in wireless ad hoc networks: A market-based approach. *IEEE/ACM ToN*, **13** (6), 1325–1338.
22. Neely, M. (2009) Optimal pricing in a free market wireless network. *Wireless Networks*, **15** (7), 901–915.
23. Molva, R. and Michiardi, P. (2001) Core: A collaborative reputation mechanism to enforce node cooperation in mobile ad hoc networks. *Institute Eurecom Research Report RR-02-062* EURECOM, Biot.
24. Gao, Y., Chen, Y., and Liu, K. (2011) Cooperation stimulation in cooperative communications: An indirect reciprocity game, in *IEEE International Conference on Communications (ICC)*, Ottawa June 2012, pp. 5163–5167.
25. Milan, F., Jaramillo, J., and Srikant, R. (2006) Achieving cooperation in multihop wireless networks of selfish nodes, in *GameNets*, p. 3.
26. Maddux, J.E. and Snyder, C. (1997) *Social cognitive psychology: History and current domains*, Springer New York.
27. Chen, X. and Huang, J. (2015) Imitation-based social spectrum sharing. *IEEE Transactions on Mobile Computing*, **14** (6), 1189–1202.
28. Kleinberg, J. (2008) The convergence of social and technological networks. *Communications of the ACM*, **51** (11), 66–72.
29. Gao, W., Li, Q., Zhao, B., and Cao, G. (2009) Multicasting in delay tolerant networks: A social network perspective, in *Proceedings of the 10th ACM International Symposium on Mobile Ad Hoc Networking and Computing*, ACM, New York, pp. 299–308.
30. Cabaniss, R., Madria, S., Rush, G., Trotta, A., and Vulli, S. (2010) Dynamic social grouping based routing in a mobile ad-hoc network, in IEEE International Conference on High Performance Computing, New Orleans, LA, December 2010, pp. 1–8.
31. Chen, X., Gong, X., Yang, L., and Zhang, J. (2014) A social group utility maximization framework with applications in database assisted spectrum access, in *Proceedings of IEEE INFOCOM*, IEEE, pp. 1959–1967.
32. Chen, X., Gong, X., Yang, L., and Zhang, J. (2016) Exploiting social tie structure for cooperative wireless networking: A social group utility maximization framework. *ACM/IEEE Transactions on Networking*, DOI: 10.1109/TNET.2016.2530070.
33. Hui, P., Crowcroft, J., and Yoneki, E. (2011) Bubble rap: Social-based forwarding in delay-tolerant networks. *IEEE TMC*, **10** (11), 1576–1589.
34. Costa, P., Mascolo, C., Musolesi, M., and Picco, G. (2008) Socially-aware routing for publish-subscribe in delay-tolerant mobile ad hoc networks. *IEEE JSAC*, **26** (5), 748–760.

35. Boldrini, C., Conti, M., and Passarella, A. (2008) Contentplace: Social-aware data dissemination in opportunistic networks, in *ACM MSWiM* Pisa, Italy, October 2008, pp. 203–210.
36. Han, B., Hui, P., Kumar, V., Marathe, M., Shao, J., and Srinivasan, A. (2012) Mobile data offloading through opportunistic communications and social participation. *IEEE TMC*, **11** (5), 821–834.
37. Zhao, Y., Adve, R., and Lim, T. (2006) Improving amplify-and-forward relay networks: Optimal power allocation versus selection, in *IEEE ISIT* Seattle, WA, July 2006, pp. 1234–1238.
38. Host-Madsen, A. and Zhang, J. (2005) Capacity bounds and power allocation for wireless relay channels. *IEEE TIT*, **51** (6), 2020–2040.
39. Ko, M.N., Cheek, G.P., Shehab, M., and Sandhu, R. (2010) Social-networks connect services. *Computer*, **43** (8), 37–43.
40. Kissner, L. and Song, D. (2005) Privacy-preserving set operations, in *Advances in cryptology–CRYPTO 2005*, Springer, Berlin, pp. 241–257.
41. Zhang, R., Zhang, Y., Sun, J., and Yan, G. (2012) Fine-grained private matching for proximity-based mobile social networking, in *IEEE INFOCOM*, IEEE, pp. 1969–1977.
42. Von Arb, M., Bader, M., Kuhn, M., and Wattenhofer, R. (2008) Veneta: Serverless friend-of-friend detection in mobile social networking, in *IEEE International Conference on Wireless and Mobile Computing, Networking and Communications*, IEEE, pp. 184–189.
43. Li, M., Cao, N., Yu, S., and Lou, W. (2011) Findu: Privacy-preserving personal profile matching in mobile social networks, in *International Conference on Computer Communications (INFOCOM)*, IEEE, pp. 2435–2443.
44. Zhang, R., Zhang, J., Zhang, Y., Sun, J., and Yan, G. (2013) Privacy-preserving profile matching for proximity-based mobile social networking. *IEEE Journal on Selected Areas in Communications*, **31** (9), 656–668.
45. Liang, X., Li, X., Luan, T.H., Lu, R., Lin, X., and Shen, X. (2012) Morality-driven data forwarding with privacy preservation in mobile social networks. *IEEE Transactions on Vehicular Technology*, **61** (7), 3209–3222.
46. Myerson, R. (1997) *Game theory: Analysis of conflict*, Harvard University Press Cambridge, MA.
47. Banerjee, S., Konishi, H., and Sönmez, T. (2001) Core in a simple coalition formation game. *Social Choice and Welfare*, **18** (1), 135–153.
48. Shapley, L. and Scarf, H. (1974) On cores and indivisibility. *Journal of Mathematical Economics*, **1** (1), 23–37.
49. Rappaport, T.S. (1996) *Wireless communications: Principles and practice*, vol. 2, Prentice Hall PTR, NJ.
50. Newman, M., Watts, D., and Strogatz, S. (2002) Random graph models of social networks. *PANS*, **99** (1), 2566–2572.
51. Leskovec, J. (2012) Brightkite dataset. URL http://snap.stanford.edu/data (accessed September 12, 2016), Stanford University.
52. Kunegis, J. (2013) Network analysis of brightkite, in *KONECT - The Koblenz Network Collection*. URL http://konect.uni-koblenz.de (accessed September 12, 2016).
53. Li, G.Y., Xu, Z., Xiong, C., Yang, C., Zhang, S., Chen, Y., and Xu, S. (2011) Energy-efficient wireless communications: Tutorial, survey, and open issues. *Wireless Communications, IEEE*, **18** (6), 28–35.

4 You Deserve Better Properties (From Your Smart Devices)

STEVEN Y. KO

University at Buffalo, The State University of New York, Buffalo, NY, USA

This chapter offers a viewpoint for operating systems in fog networking. The central claim is the following—*we need to provide better properties for the operating systems in fog networking*. This means that the current operating systems do not provide necessary properties for fog networking, and the needs for new and better properties are arising. To support this claim in a comprehensive manner, we answer four questions. The first is *why*—why we need to provide better properties. The answer to this question gives us the motivation for further research. The second question is *where*—where we need to provide better properties. Here we make a case that we should examine existing operating systems first. The third question is *what*—what new or better properties we need to provide. To answer this question, we use three case studies drawn from our own research to demonstrate what kinds of properties we need to provide. The fourth question is *how*—how we can provide better properties. The answer to this question gives us the realizations of better properties.

In answering these four questions, we put more weight on the last two questions. The reason is that we do not necessarily raise new points in our answers to the first two questions; we mostly include the discussions for the sake of completeness. Nevertheless, we hope that examining these four questions of *why*, *where*, *what*, and *how* gives a comprehensive view of our claim.

4.1 WHY WE NEED TO PROVIDE BETTER PROPERTIES

The short answer to our first *why* question is that new contexts are emerging for operating systems in fog networking, and those new contexts call for better properties that operating systems should provide. Such new contexts include existing mobile

Fog for 5G and IoT, First Edition. Edited by Mung Chiang, Bharath Balasubramanian, and Flavio Bonomi.

devices used in ways that we did not anticipate before (e.g., using Android for medical devices), as well as brand new contexts such as Internet of things (IoT), unmanned aerial vehicles (UAVs), etc. Operating systems used in these contexts need to support new types of applications, which have their own requirements and challenges to deal with. For example, if Android had run on a medical device, it would need to support applications such as a patient health monitoring that has tight timing requirements. Naturally, the need arises for operating systems to provide better properties for the applications they run.

This is by no means new reasoning; in the past few decades, many new contexts have appeared for operating systems, and every time a new context arose, the need for better properties arose as well. Mobile computing is perhaps one of the most recent examples—laptops, smartphones, and tablets arose as a new context for operating systems, and it called for better properties, for example, efficient energy management. But now, we are moving past the traditional mobile computing, and new application contexts in fog networking are emerging such as IoT and UAVs as mentioned earlier. Thus, it is important to examine these new contexts and what better properties we need to provide in operating systems for fog networking.

4.2 WHERE WE NEED TO PROVIDE BETTER PROPERTIES

The next question we want to answer is *where* we need to provide better properties, and we argue that a reasonable starting point is an existing operating system. As with the first question, we are not making a new point here. It has been a general trend—once an operating system is developed, it is leveraged in many different contexts, and adjustments are made along the way. Android is perhaps one of the most recent examples—it adapted an existing operating system (Linux) for smartphones in the beginning, but it evolved as new contexts arose, for example, tablets, smartwatches, in-car infotainment systems, etc. Every time a new context arose, Android added new properties (e.g., better energy management) and adapted to the new context. iOS has had similar adaptations, which adapted OS X first for smartphones and then evolved to work in other contexts.

The likely reason behind this trend is cost—writing a brand new operating system from scratch is a costly undertaking. Reusing and adapting an existing operating system in a new context is a reasonable cost-saving strategy. In addition, one might argue that it is even better to adapt an existing operating system since it has been tried and tested. For example, Linux has withstood years of field usage and proven itself to be a robust working system. By adapting Linux (as Android did), one can gain the proven robustness for the components already implemented in Linux.

Following the same line of reasoning, we anticipate that this trend of adapting an existing operating system will continue in the near future. Thus, we think that new properties that arise in fog networking contexts should also be tested in existing operating systems first.

One thing to note is that providing a new property in an existing operating system does not simply mean that one uses the operating system as an implementation platform. Reasoning about a new property and implementing it within an operating system requires deep understanding of the operating system's internals. It also requires careful evaluation of the operating system with a keen eye on how to provide the new property. As a result, it is often the case that innovation takes place; one needs to adapt existing techniques in novel ways or come up with brand new techniques. We discuss a few examples of this kind in the rest of the chapter.

4.3 WHAT PROPERTIES WE NEED TO PROVIDE AND HOW

The remaining questions are *what* and *how*, that is, what better properties we need to provide and how. Since we recognize that there can be vastly many properties that need to be provided, we highlight three properties that deserve more attention. The first property is transparency; it is the ability for an operating system to tell its users why it is doing what it is doing. The second is predictable performance; it is the ability for an operating system to provide consistent, predictable performance for important applications no matter what the system load is. The third is openness; it is the ability for an operating system to allow innovation from third parties to be easily incorporated and deployed. The rest of the section describes in more detail what these properties are and why we need them, as well as our previous and current efforts to make progress on these fronts.

4.3.1 Transparency

We define transparency as the ability for a system to tell its users why it is doing what it is doing, and the need for this property is uprising in recent years. The primary motivator is the ubiquitous popularity of personal devices such as smartphones and tablets. These personal devices enable users to browse the Web, to send e-mails, to take pictures, etc. anywhere they go with high-bandwidth connectivity to the Internet. Due to this convenience, the penetration of personal devices has shown a startling growth; it has been reported that the global shipments of smartphones have already surpassed the shipments of PCs in 2011 [1]. Facebook, the most popular social networking service, reports that 90% of their daily active users access it via mobile [2], and it is the main target of growth in 2012 [3].

However, this popularity of personal devices not only provided convenience but also brought many *why* questions from users. These *why* questions are not just traditional questions that researchers have asked before such as "why is this device so slow?" or "why does this device cannot connect to the Internet?" but rather new questions more pertinent to personal devices such as "is this application sending a personal phone number to some Internet server?" "is it tracking locations?" or "why is this device not lasting even a day?" Many survey results recognize these questions, and the news media has numerous suggestions and advice to deal with these questions [4–7].

The underlying issue of answering these *why* questions is transparency. If a system is transparent, that is, if a system provides detailed information that reveals its inner working, then we can start answering the *why* questions. For example, if an application or a platform reveals how the application behaves in regard to all the data read from contacts, we should be able to answer "is this application sending a personal phone number to some Internet server?" Traditionally, this issue of transparency has been more relevant to researchers and developers as they need to understand how a system behaves for various purposes such as debugging and performance improvement. Accordingly, a large number of tools have been developed to assist researchers and developers, for example, Kprobes [8] and SystemTap [9] among many others. But now, it is relevant to end users as well, and we need to develop tools and techniques to provide transparency not only for researchers and developers but also for users.

Although such tools and techniques could be developed in any layer of a system, it would be particularly useful if they were provided by the operating system layer, either as a platform tool or as a system-level service. This is because those tools and techniques could benefit the entire set of applications running on top of an operating system if they were implemented within the operating system. In the case of a popular operating system, it would also seamlessly benefit a large number of users at the same time.

However, transparency—as we define it—is a broad issue that has many aspects, for example, performance, energy efficiency, privacy, etc. Thus, it is difficult to tackle all transparency issues at once. Thus, in our own research we are taking some initial steps to develop tools and techniques that pertains to one aspect of transparency, which we call *information flow transparency*.

BlueSeal [10, 11] is our first effort toward providing information flow transparency. This transparency reveals answers to questions on how an application uses the data of its user, for example, "is this application sending a personal phone number to some Internet server?" The reason why we call it information flow transparency is because this type of question can be answered by a technique called *information flow tracking* [12–18], which analyzes how information gets propagated in a system from an information source to an information sink. For example, by tracking how the information from a file that contains personal phone numbers gets propagated within an application, we can check if a phone number gets sent over the network by the application. In this example, the information source is the phone number file, and the information sink is the network. BlueSeal analyzes how various types of information get propagated between many different sources (such as storage, GPS, contacts, camera, etc.) and sinks (such as network, log, storage, inter-process communication channels, etc.) and reveals them to end users.

Revealing an information flow is a first step toward enforcing a flow or preventing a flow from happening. For example, an operating system might want to analyze an application first to make sure that at runtime there is no other information flow occurring. Also, an operating system might want to prevent an application from executing at all if there is a potentially malicious information flow.

However, revealing an information flow is also useful in and of itself, especially to users. If presented at installation time, for example, a user could decide whether the user wants to install an application or not. This is particularly useful when there are many alternatives for an application. For example, there are many file browser

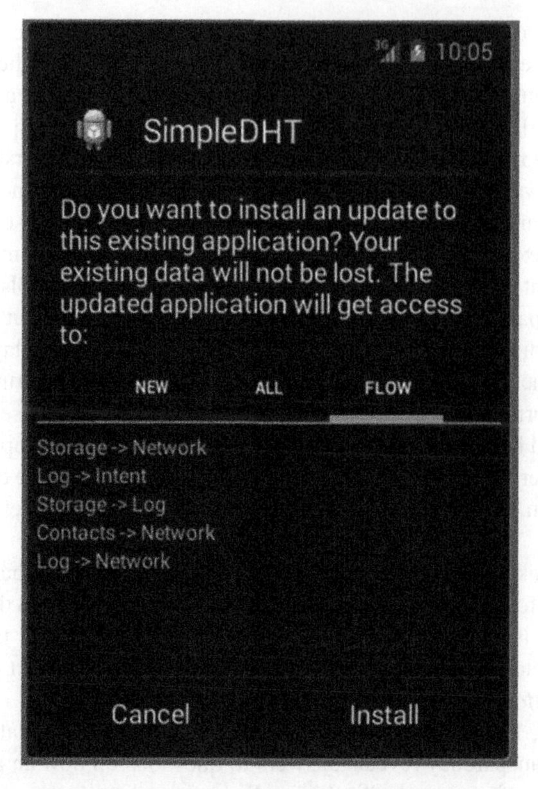

Figure 4.1 BlueSeal information flow permission screenshot.

applications for a smartphone, and a user might want to choose a file browser application with the least number of information flows.

In light of this, the uniqueness of our tool, BlueSeal, is twofold. First, BlueSeal *statically* analyzes an application and displays all potential information flows to end users at installation time. This means that we analyze the binary of an application to detect information flows that exist within the application; this is opposed to *dynamic* information flow tracking, where the tracking happens at runtime. The reason for our static analysis is to give a user a chance to examine the behavior of an application *before installing it* so that the user can make a more informed decision on which applications to install. We do this in the form of permissions; we display at installation time what potential information flows exist in an application for various sources and sinks, and a user needs to approve that the user is still willing to install the application. Two most popular mobile operating systems, iOS and Android, both have such a permission mechanism, and we have chosen Android[1] as the implementation platform. Figure 4.1 shows a screenshot of how BlueSeal displays information flows.

[1]We have chosen Android for its popularity (Android is the most popular mobile OS) and openness (Android is open source, and we can examine the internals).

The second uniqueness of BlueSeal is that it addresses the challenges of adapting existing static information flow analysis techniques to a specific platform. This is due to the fact that there are many practical issues to consider for a specific platform, which general static information flow analysis techniques do not address. This is why we mention in Section 4.2 that providing a new property in an existing operating system does not simply mean that one uses it as an implementation platform. Often times, there are platform-specific issues to address, and it leads to interesting research problems.

In the case of BlueSeal, there are two categories of practical issues we needed to consider for Android when adapting existing static analysis techniques. The first category is the issues caused by the programming model of Android. The fundamental reason is that Android's programming model is highly event driven; an application implements numerous entry points that react to outside events. In other words, an Android application does not have any single entry point where all the logic starts from; an Android application is a collection of handlers that processes events, for example, a button click. This type of programming models is popular for GUI-based frameworks, and Android is no exception. The problem with this programming model for static information flow analysis is that traditional techniques assume that a program has a single call graph starting from `main()`. They do not work if a program is a set of handlers with no single entry point.

Not only that, Android introduced many new constructs such as new thread types, message handlers, and IPC mechanisms. The issue with these new constructs is that their execution is implicit, that is, they are executed not by direct method calls within application code but by the underlying Android platform with some predefined rules. This is the second category of issues we needed to deal with in BlueSeal.

To illustrate this point, we use `AsyncTask`, a popular construct for threading on Android, as an example. `AsyncTask` is a new threading class introduced in Android. It provides a simple way to write a thread that communicates with the UI thread in an asynchronous fashion. An AsyncTask can implement five methods—`onPreExecute()`, `doInBackground()`, `onProgressUpdate()`, `onPostExecute()`, and `onCancelled()`. The interesting aspect about this construct is that there is a predefined order of execution and that there is no explicit method calls that can be found in application code for any of the five methods. Figures 4.2 and 4.3 illustrate this point. Figure 4.2 shows a code snippet written in Java, and Figure 4.3 shows the corresponding order of execution.

As shown in Figure 4.2, a developer can implement a new thread by extending `AsyncTask` and providing the implementation for the five methods previously mentioned. In Figure 4.2, the developer-implemented class is `Task`, and it is started in `onCreate()` by calling `Task.execute()`. Now, as the execution flow shows, the call to `Task.execute()` triggers the execution of `Task.onPreExecute()`, and there is no explicit call to that method in the application code—the execution is triggered by the underlying Android platform. Similarly, other methods (e.g., `doInBackground()`) are called at later times but without any explicit call. `AsyncTask` is just one example; many other constructs in Android have similar

```
public class MainActivity extends Activity {
    protected void onCreate(Bundle savedInstanceState) {
        ...
        new Task().execute("http://www...");
        ...
    }
    ...
    private class Task extends AsyncTask<String, String, Integer> {
        ...
        protected void onPreExecute() {
            ...
        }
        protected  Integer doInBackground(String... strs) {
            ...
            publishProgress("intermediate result");
            ...
            return intObj;
        }
        protected void onProgressUpdate(String...strings) {
            ...
        }

        protected void onPostExecute(Integer intObj) {
            ...
        }
    }
}
```

Figure 4.2 AsyncTask code snippet.

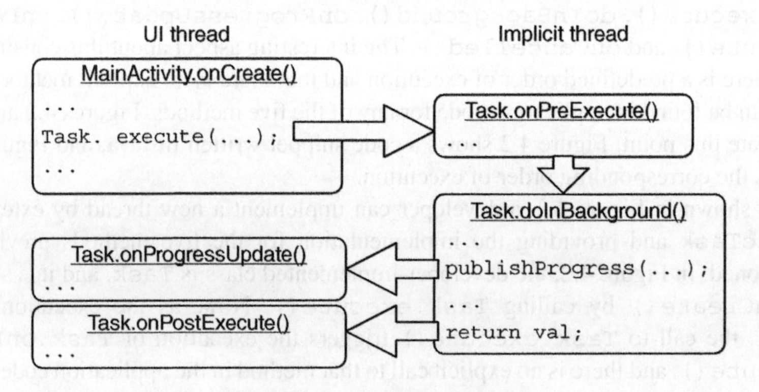

Figure 4.3 AsyncTask flow.

characteristics. The problem of these constructs for static information flow analysis then is how to handle indirect call relationships in various pieces of code.

In order to handle the two categories of issues described so far, BlueSeal implements a mechanism to understand many entry points that exist in an Android application as well as implicit calls introduced by new constructs of Android. In a nutshell, BlueSeal builds a call graph, where all implicit calls are converted to explicit calls, and augments traditional static information flow analysis techniques to handle multiple entry points. Using BlueSeal, we have analyzed 1800 applications (600 top-rated free applications and 1200 malicious applications identified by MalGenomeproject [19]) and found that all but 79 applications take less than 2 minutes to analyze. This shows that if the analysis is done off-line using a server (e.g., an application store server), it is practical to use BlueSeal to provide information flow transparency at installation time.

Our previously published paper [11] reports the full set of results, and here we summarize some of our findings. First, we have observed that malicious applications are heavily interested in unique phone identifiers such as the device ID. Moreover, when normal applications use unique phone identifiers, they just use them internally without sending it to a server or storing it somewhere. In contrast, malicious applications almost always send the identifiers out of the applications themselves either to a server or to a device storage. Second, we have observed that normal applications access the phones' location data more frequently than malicious applications, that is, there is less interest among malicious applications about users' locations. Third, both normal and malicious applications read system content providers often. The most commonly accessed content provider in both normal and malicious applications is the contacts. Thus, reading contacts is probably not a reliable indicator of a malicious activity.

At the time of our publication, the most closely related work to ours was CHEX [20]. It provides a tool for detecting highjack-enabling flows within an Android application. It uses a brute-force permutation approach to tackle the analysis of Android's constructs such as async tasks and handlers, making it susceptible to incorrect disambiguation. Thus, our call graph restructuring technique can refine CHEX's approach since we identify implicit calls in Android's constructs whenever possible. More recently, FlowDroid [21] has proposed more sophisticated techniques to analyze Android applications. It performs context-sensitive analysis that specifically handles Android's constructs.

4.3.2 Predictable Performance

The second property that deserves more attention is what we call *predictable performance*, which we define as the ability for an operating system to provide consistent, predictable performance to important applications. This is coming from the traditional real-time systems domain, where an application consists of multiple tasks with different priorities. An aircraft control system is a well-known case of a real-time system; there are many tasks an aircraft control system needs to carry out, for example, radar control, navigation, flight control, etc., and these tasks have different priorities in terms of their relative importance. In addition, these tasks should perform in a predictable fashion. This means that for each task, there has to be a guarantee about exactly when it starts and exactly when it finishes. For example, if the radar of an aircraft senses a foreign object, the sensed data should be delivered to its data processing unit in a predictable way with a reasonable bound on latency. If there is no such guarantee, then the aircraft will not be able to react to sensed objects in a timely manner. This guarantee should be provided even when there are multiple tasks that compete for shared resources such as CPU, memory, and I/O.

This property of predictable performance has become more important recently, as there are many more types of newer devices that operate in real-time contexts. For example, many new medical IoT devices are appearing, and they perform critical tasks such as patient monitoring. These critical medical tasks typically require predictable performance; otherwise, there will be no guarantee about whether or not the critical tasks can be performed in a timely manner for a patient. Similarly, small UAVs are becoming popular, and they also require predictable performance; otherwise, they will not be able to fly reliably. Thus, the property of predictable performance is becoming more important now.

In our own research, we are investigating how to provide the property of predictable performance in Android. The system we are developing is a variation of Android called RTDroid [22–24]. Our goal for RTDroid is to add real-time support to Android as a whole system so that one can run real-time applications and non-real-time applications on a single system. Having such a system enables scenarios previously not possible. For example, a hospital inpatient can potentially use a tablet to play games, while the table is also being used as a central hub for controlling all medical devices that the patient needs. Or, a smartphone can potentially become an onboard satellite controller as the United Kingdom envisions [25]. Before RTDroid, there were a few discussions on how to add real-time support to Android [26–28], but RTDroid is the first system that explores the question of how to add real-time support to Android as a whole system, to the best of our knowledge.

With RTDroid, we have explored several research questions related to adding real-time support to Android. The questions are (i) whether or not the current Android provides real-time guarantees, (ii) if not, what kinds of techniques we need, (iii) what changes we need to make in Android to implement the techniques identified by the previous question, and (iv) how effective the implemented techniques are. It turns out that Android does not provide adequate real-time guarantees, and while we can reuse existing techniques to add real-time support to Android, namely, priority awareness, implementing these techniques in Android *as a whole system* requires careful examination and strategies. This presents a significant challenge due to the complexities of Android, especially when considering the entire system. In the succeeding text, we elaborate these findings further, starting with the answer for the first question.

To answer the first question, we began by analyzing Android for its real-time capability. To explain our findings, Figure 4.4 shows a simplified view of Android's internal components. As shown, there are three main layers that support Android applications—the Android kernel, the Android runtime (ART), and the application framework. The Android kernel is basically a Linux kernel with some

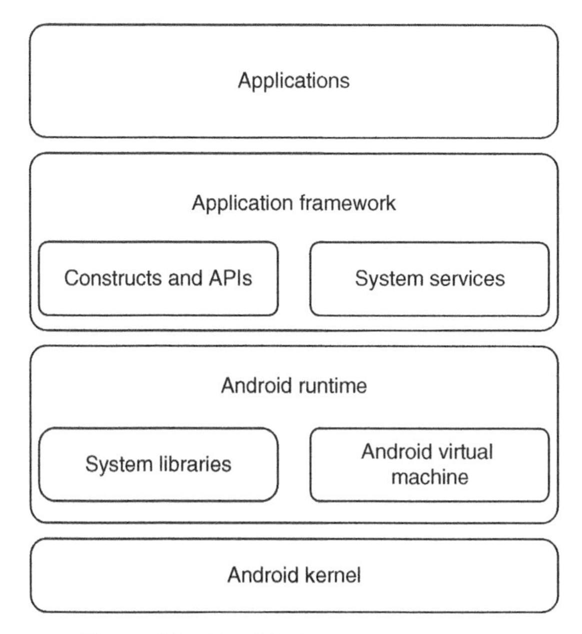

Figure 4.4 Simplified Android architecture.

Android-specific customizations. The ART contains Android's Java virtual machine (previously Dalvik, now ART) as well as system libraries (Java and C libraries, similar to the ones that a typical Linux distribution has). Android applications are mainly written in Java, so this runtime layer is directly responsible for running an application. The last layer—the application framework—has system services and provides APIs to access them as well as convenient constructs.

In order to analyze Android's real-time capabilities as a whole system, we need to examine each layer and see if every layer is capable of supporting real-time applications. We start by examining the bottommost layer, which is the Android kernel. As mentioned earlier, the Android kernel is essentially a Linux kernel with some Android-specific customizations; and it is well known that the Linux kernel does not have real-time capabilities.[2] Instead, there is a patch that one can apply to transform a Linux kernel into a real-time kernel. This patch is called RT-Preempt.[3] There also exist other real-time kernels such as RTEMS [29]. Thus, we can conclude that the current Android kernel does not have any real-time capability, although there are potential alternatives.

The layer above the kernel is the ART that contains Android's Java virtual machine and system libraries. Similar to the Linux kernel case, it is well known that Android's Java virtual machine does not have any real-time capability. In fact, designing a real-time Java virtual machine is a research area of its own, and there has been active

[2] Android-specific customizations do not concern with real-time applications either.
[3] https://rt.wiki.kernel.org/index.php/Main_Page (accessed September 15, 2016).

research in the past few years [30–32]. In order to design a real-time Java virtual machine, specific mechanisms need to be implemented, for example, a real-time garbage collector, but Android's virtual machine does not implement such mechanisms. Thus, we can also conclude that the current runtime of Android does not have any real-time capability.

For our research, these bottom two layers—the Android kernel and the runtime—did not require much examination for us to see that they did not have any real-time capability. It was and still is a known and well-documented fact that a kernel layer as well as a Java runtime layer need additional support to make real-time applications work correctly. However, the application framework layer required more thorough understanding and analysis, since it was not known in the literature if that layer provides any real-time capability. After a closer look at the application framework layer, we concluded that the framework layer also needed much work to add real-time capabilities. We first explain the application framework layer and discuss our findings.

As mentioned earlier, the application framework layer consists of two categories of components. The first category is system services, and the second category is constructs and APIs. It is depicted in Figure 4.4. System services are the operating system services of Android that applications can leverage. Some examples include a sensor service that manages sensor input delivery; an alarm service that provides timers (e.g., one-time timers, periodic timers, etc.); a media service that manages on-device cameras, microphones, and speakers; and a location service that manages GPS and location data delivery. The APIs in the second category allow easy access to these system services. The APIs also provide access to standard programming constructs and facilities such as data structures (e.g., `HashMap`) and OS facilities (e.g., `Socket`). In addition, there are extra programming constructs that Android provides to its application developers. We categorize them as "constructs." Some examples include `Looper`, `Handler`, `AsyncTask` (which we showed earlier in Section 4.3.1), etc. `Looper` and `Handler` provide a convenient way for message passing between different components. `AsyncTask`, as explained earlier, provides a convenient way to create and use a thread.

By examining these constructs, APIs, and system services, we have found out that the application framework layer has limitations for real-time applications, as it is not designed with real-time support in mind. This is true for many constructs and system services such as `Looper`, `Handler`, the sensor service, the alarm service, etc. This is understandable, since Android was designed for smartphones initially that do not require any real-time capability.

To illusrate one such design limitation, Figure 4.5 depicts the overall sensor architecture of Android. It shows two sensors, an accelerometer and a gyroscope, and how their data is passed to applications. It crosses the hardware-kernel boundary as well as the kernel-user boundary. As shown, there are separate data delivery paths for different sensors up to the kernel, and then the data is processed and passed through over one path with the sensor service and the sensor manager. The sensor service and the sensor manager basically pull all sensor data out of the kernel, queue them up, and deliver them in the FIFO order to different applications.

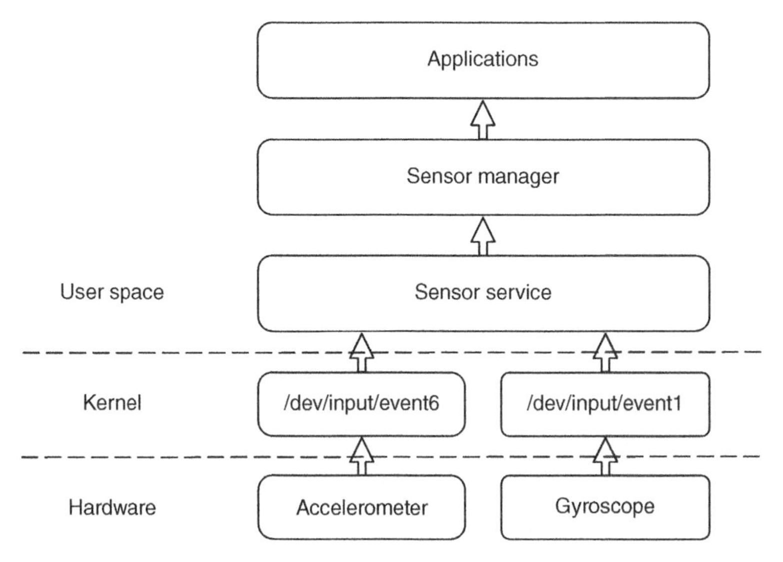

Figure 4.5 Android sensor architecture.

This single path design is a problem for real-time applications if they need to prioritize the delivery of different types of sensor data. For example, suppose a scenario where a real-time application needs accelerometer data for a high-priority critical task, but it also needs gyroscope data for a low-priority task. Since a gyroscope can produce data at a much higher rate than an accelerometer, if there is one path that delivers both types of data in the FIFO order as Android does, the gyroscope data will block and delay the delivery of accelerometer data. For real-time applications, there has to be a mechanism to differentiate important tasks and data from less important tasks and data. The underlying system then needs to guarantee that less important ones do not interfere with more important ones.

Thus, our findings on the real-time capabilities of Android is that in all layers, appropriate support is necessary to make Android real time, and RTDroid is our solution for that. Figure 4.6 shows a quick look at our overall architecture. At the bottommost layer, we replace the Android kernel with a real-time kernel; we currently use Linux with RT-Preempt and RTEMS as our kernel options. We then replace the ART with a real-time Java runtime; we currently use FijiVM [32] as our option and add real-time support for the system C library for Android called Bionic. These components provide essential real-time mechanisms such as priority-aware scheduling, real-time interrupt handling, full preemption, real-time garbage collection, etc. They are mostly off the shelf and require some engineering effort to use on Android.

While these components provide basic real-time building blocks, they do not solve the problems in the application layer design. Thus, we redesign and implement the application framework layer, so that it supports real-time applications. The key principle in our design is to make Android's internal constructs, components, and message delivery mechanisms *priority aware*, that is, if there are more important tasks to run

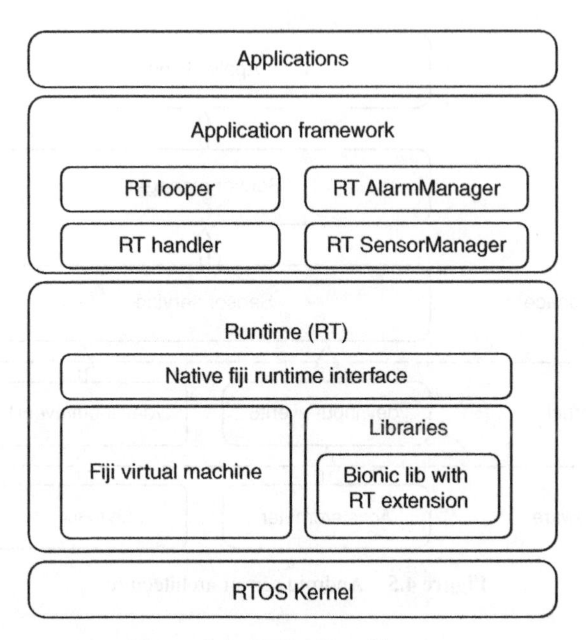

Figure 4.6 RTDroid architecture.

or messages to deliver, the whole system should be able to prioritize them over other tasks or message. Although this is a well-known concept, implementing it in Android requires much careful examination of the whole system and its design, which is a significant challenge. So far, we have redesigned a few constructs and system services to demonstrate the feasibility of our approach. We have redesigned `Looper`, `Handler`, the alarm service, and the sensor service. As an example, Figure 4.7 shows our redesigned sensor architecture. For brevity, we have simplified the figure and do not show the full design.

As mentioned earlier, Android's sensor architecture uses a single path to deliver all sensor events in the FIFO order. This poses a problem for real-time applications, since they cannot prioritize the delivery of important sensor data over less important data. Thus, in our design, we create separate, prioritized paths for different sensors. In Figure 4.7, this is shown with the accelerometer service/manager and the gyroscope service/manager to illustrate how we support the two sensors. Our architecture has the advantage of being able to prioritize the data delivery from one sensor over another. This can be done by assigning different task priorities for different sensor managers (where task priorities are supported by the underlying runtime and the kernel). For example, if we assign a higher priority to the accelerometer manager than the gyroscope manager, the accelerometer manager will have a higher priority for its execution, resulting in the prioritized delivery of accelerometer data over gyroscope data.

Figure 4.8 shows the performance comparison between our sensor architecture and Android's architecture. It shows the result from one set of experiments we

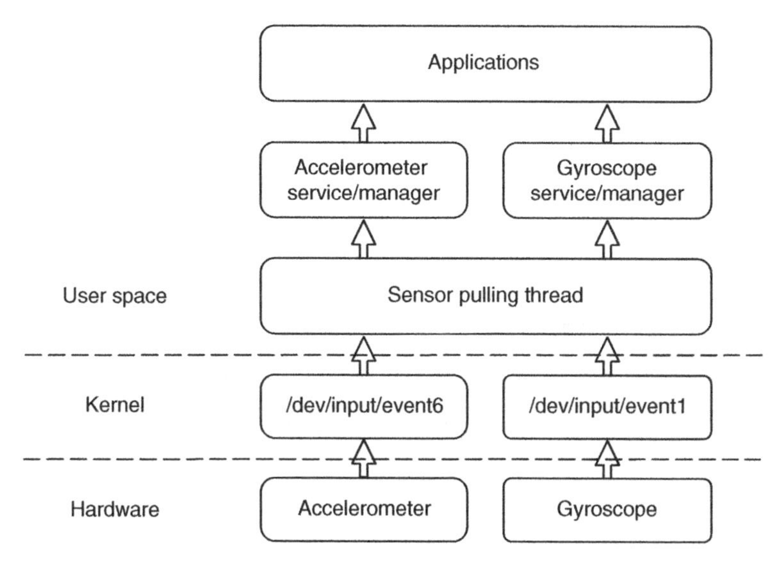

Figure 4.7 RTDroid sensor architecture.

conducted with Google Nexus S smartphone. The workload we ran consists of two parts. The first part is our main workload, and it is a real-time fall detection application. It runs on a phone and implements a fall detection algorithm that alerts a sudden change of height of the phone. It can be used to detect a fall of a soldier, an elderly, etc. The second part is our background workload intended to create much background noise to disrupt our main workload. What we used for the figure is 100 memory-intensive background threads, each of which allocates a 2.5 MB integer array. This creates much memory pressure and stresses the whole system. Since the fall detection application requires accelerometer sensor data, it tests our redesign of the sensor architecture.

Figuire 4.8 illustrates the observed latency of the sensor event delivery for the fall detection application. The x-axis indicates all sensor events, and the y-axis indicates the delivery latency for each sensor event. As shown, RTDroid provides the upper bound of around 30 ms; however, Android does not provide such tight latency bound, and latency varies greatly across different events. This means that the possibility of missing some of the fall events exists on Android. This result demonstrates that our redesign is effective in providing predictable performance for real-time applications.

4.3.3 Openness

The last property that deserves more attention is *openness*, which we define as the ability for an operating system to easily incorporate and deploy innovation from third parties. Traditionally many areas in computer science have benefited from such openness. Perhaps the most recent example is software-defined networking (SDN) (in which Feamster *et al.* [33] provide a good summary for), which essentially provides

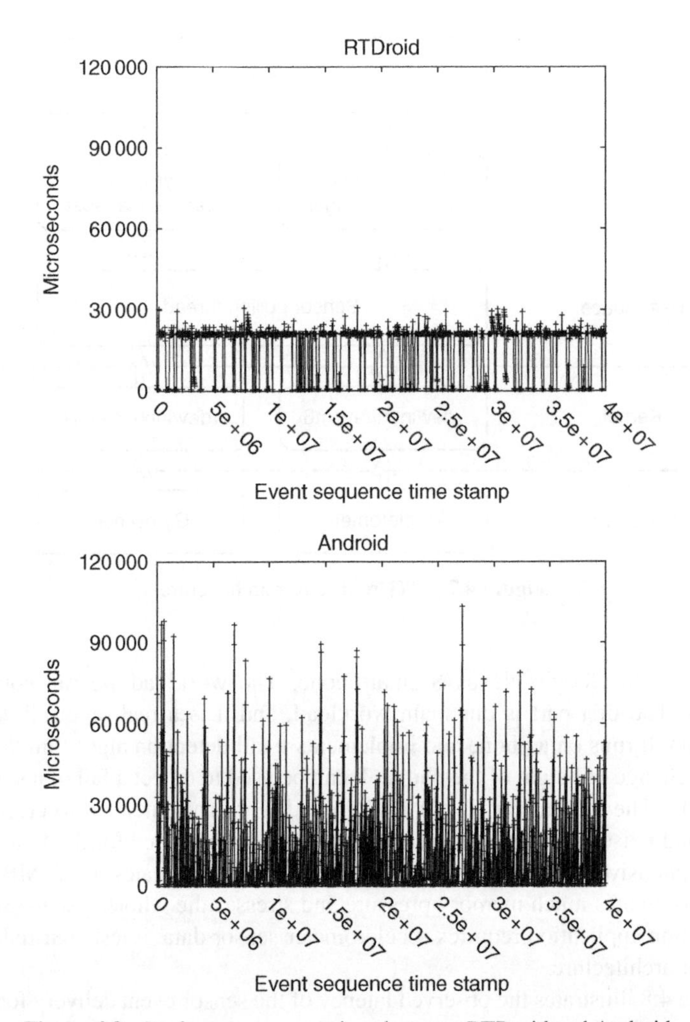

Figure 4.8 Performance comparison between RTDroid and Android.

openness in networking—they open routers and switches so that third-party developers can write software to control them. This openness enabled by SDN has sparked much interest in the networking community, and now there are many innovative solutions being proposed for different problems in networking. There are other popular examples we can find, such as NetFilter[4] and FUSE.[5] NetFilter is a Linux facility for network packet filtering, where a third-party developer can write an extension for packet analysis and transformation. FUSE is also a Linux facility for user-space file system extensions. It allows third-party developers to implement a fully functional file

[4]http://www.netfilter.org/ (accessed September 15, 2016).
[5]http://fuse.sourceforge.net/ (accessed September 15, 2016).

system as a user-space program. These kernel facilities have enabled the development of countless research prototypes.

We believe that such openness is necessary for the operating systems used in fog networking devices such as smartphones and IoT devices. This will enable unobstructed innovation, not only from the vendors of the operating systems but also from third parties. In the current state, even the most open platform—Android—is in fact mostly closed. It is open in the sense that its source code is open. If we want to examine the source, we can simply clone the source repositories and take a look at the source. If we want to contribute to the source, it is also possible; we can review the current list of bugs and submit patches. A patch has a chance to be incorporated in the main distribution. However, it is still closed in the sense that it is not easy to deploy any innovation in the Android operating system, unless one has access to the official Android update channels, which is controlled by Google and other vendors. For example, if a researcher develops an innovative technique for an OS subsystem such as networking or storage, there is no easy way to deploy it and test it with a large number of real users since it requires an operating system update.

Recognizing such a problem, we are taking initial steps in our research. One such effort is a system called BlueMountain [34], which aims to enable open innovation in mobile data management. Though the full development of BlueMountain is in progress, we discuss the system here briefly to give an idea on what kinds of openness might be possible in operating systems for fog networking devices.

The basic idea of BlueMountain is a technique called *storage API virtualization*. It allows a third-party developer to write extensions for storage operations and inject it into existing mobile applications using binary instrumentation. An extension provides a new implementation for storage APIs such as `open()`, `read()`, and `write()` so that, when an application calls those APIs for storage, the extension can intercept the calls and modify the behavior of the calls. In addition, an extension is injected to an application using binary instrumentation. This provides the benefit of deployability; by injecting a storage extension to an application, there is no need to update an operating system—an extension can be distributed as part of an application.

Figure 4.9 illustrates this idea with an extension example. It shows that there is a storage API virtualization layer injected into an existing application, and this API virtualization layer can execute code from an extension; and this extension can

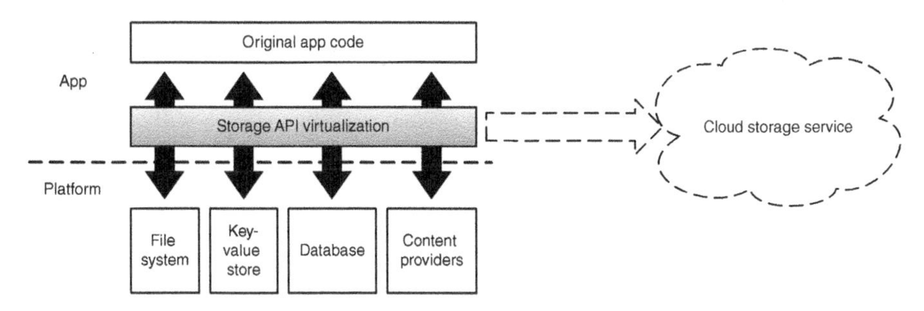

Figure 4.9 Storage API virtualization and its example extension.

modify the behavior of storage API calls, for example, seamlessly intercepting and sending all data written with `write()` API to a cloud storage service for automated backup. There can be many other possibilities for a mobile storage extension; for example, we could automatically enable proximity-based peer-to-peer storage where a user takes a picture with friends and calls `write()` that transparently sends the picture to all friends' devices.

There are several challenges to overcome in order to realize storage API virtualization, some of which are described in our workshop paper on BlueMountain [34]. We encourage interested readers to refer to the paper for further information.

4.4 CONCLUSIONS

This chapter has argued that we need to provide better properties in the operating systems in fog networking. In order to comprehensively support the argument, we have explored four questions—*why* we need to provide better properties, *where* we need to provide better properties, *what* properties we need to provide, and *how* we can provide those properties.

We can summarize the answers to the four questions as follows. First, the reason why we need to provide better properties is that new application contexts are arising in fog networking, and existing operating systems do not provide necessary properties for the new and emerging contexts. Second, when providing those new properties, we need to explore how to provide those properties in existing operating systems, since it is highly likely that existing operating systems will be adapted and leveraged in new contexts. Third, some of the properties we need to provide are transparency, predictable performance, and openness. Transparency allows users, developers, and researchers to understand why a system is doing what it is doing at any point of time. One such example we have discussed is information flow transparency, which reveals how an application uses the data of its user. Predictable performance guarantees that important applications can perform consistently and predictably well, even if there are other applications competing for resources. We has discussed how we could provide such a property in Android. Openness allows innovation from third parties, which has been a successful model for innovation in many areas of computer science. We have briefly discussed our own work in progress that aims to provide such openness for mobile data management.

As new application contexts emerge and new devices appear, operating systems will also need to adapt new technologies and evolve. Thus, we believe that the process of exploring what better properties we need to provide and how we provide them should be a continuous one. For this reason, our own research will continue to examine new contexts and new properties.

ACKNOWLEDGMENT

The research described in this chapter has been supported in part by an NSF CAREER award, CNS-1350883.

REFERENCES

1. The Verge. IDC Forecasts 1.16 Billion Smartphones Shipped Annually by 2016. http://www.theverge.com/2012/3/29/2910399/idc-smartphone-computer-tablet-sales-2011 (accessed October 23, 2016).
2. Napier Lopez. 90% of Facebook's Daily Active Users Access it via Mobile. http://thenextweb.com/facebook/2016/01/27/90-of-facebooks-daily-and-monthly-active-users-access-it-via-mobile/ (accessed October 23, 2016).
3. Bloomberg. Facebooks Zuckerberg Targets Mobile Users For Growth. http://www.bloomberg.com/news/2012-05-11/facebook-s-zuckerberg-addresses-questions-over-mobile-strategy.html (accessed October 23, 2016).
4. PR Newswire. J.D. Power and Associates Reports: Smartphone Battery Life has Become a Significant Drain on Customer Satisfaction and Loyalty. http://goo.gl/ENUOE (accessed October 23, 2016).
5. Rachel Metz. Mobile Summit 2013: Camera Tweaks Should Boost Gadget Battery Life. http://www.technologyreview.com/news/515951/camera-tweaks-should-boost-smartphone-battery-life/ (accessed October 23, 2016).
6. Phila Siu. Privacy Watchdog Warns Smartphone Owners over Personal Data. http://www.scmp.com/news/hong-kong/article/1087787/privacy-watchdog-warns-smartphone-owners-over-personal-data (accessed October 23, 2016).
7. Privacy Rights Clearinghouse. Privacy in the Age of the Smartphone. https://www.privacyrights.org/consumer-guides/privacy-age-smartphone (accessed October 23, 2016).
8. Jim Keniston, Prasanna S. Panchamukhi, and Masami Hiramatsu. Kernel Probes (Kprobes). https://www.kernel.org/doc/Documentation/kprobes.txt (accessed October 23, 2016).
9. SystemTap. http://sourceware.org/systemtap/ (accessed October 23, 2016).
10. Shashank. Holavanalli, Don Manuel, Vishwas Nanjundaswamy, Brian Rosenberg, Feng Shen, Steven Y. Ko, and L. Ziarek. Flow Permissions for Android. In *Proceedings of the IEEE/ACM 28th International Conference on Automated Software Engineering*, ASE '13, pages 652–657, Palo Alto, CA, USA, November 2013.
11. Feng Shen, Namita Vishnubhotla, Chirag Todarka, Mohit Arora, Babu Dhandapani, Eric John Lehner, Steven Y. Ko, and Lukasz Ziarek. Information Flows As a Permission Mechanism. In *Proceedings of the 29th ACM/IEEE International Conference on Automated Software Engineering*, ASE '14, pages 515–526, New York, NY, USA, 2014. ACM.
12. Andrew C. Myers and Barbara Liskov. A Decentralized Model for Information Flow Control. In *Proceedings of the 16th ACM Symposium on Operating Systems Principles*, SOSP '97, pages 129–142, New York, NY, USA, 1997. ACM.
13. Andrew C. Myers and Barbara Liskov. Complete, Safe Information Flow with Decentralized Labels. In *Proceedings of the 1998 IEEE Symposium on Security and Privacy*, pages 186–197, May 1998.
14. Andrew C. Myers. JFlow: Practical Mostly-static Information Flow Control. In *Proceedings of the 26th ACM SIGPLAN-SIGACT Symposium on Principles of Programming Languages*, POPL '99, pages 228–241, New York, NY, USA, 1999. ACM.
15. Petros Efstathopoulos, Maxwell Krohn, Steve VanDeBogart, Cliff Frey, David Ziegler, Eddie Kohler, David Mazieres, Frans Kaashoek, and Robert Morris. Labels and Event

Processes in the Asbestos Operating System. In *Proceedings of the 20th ACM Symposium on Operating Systems Principles*, SOSP '05, pages 17–30, New York, NY, USA, 2005. ACM.

16. Maxwell Krohn, Alexander Yip, Micah Brodsky, Natan Cliffer, M. Frans Kaashoek, Eddie Kohler, and Robert Morris. Information Flow Control for Standard OS Abstractions. In *Proceedings of 21st ACM SIGOPS Symposium on Operating Systems Principles*, SOSP '07, pages 321–334, New York, NY, USA, 2007. ACM.
17. Nickolai Zeldovich, Silas Boyd-Wickizer, Eddie Kohler, and David Mazières. Making Information Flow Explicit in HiStar. In *Proceedings of the Seventh Symposium on Operating Systems Design and Implementation*, OSDI '06, pages 263–278, Berkeley, CA, USA, 2006. USENIX Association.
18. William Enck, Peter Gilbert, Byung-Gon Chun, Landon P. Cox, Jaeyeon Jung, Patrick McDaniel, and Anmol N. Sheth. TaintDroid: An Information-flow Tracking System for Realtime Privacy Monitoring on Smartphones. In *Proceedings of the Ninth USENIX Conference on Operating Systems Design and Implementation*, OSDI '10, pages 393–407, Berkeley, CA, USA, 2010. USENIX Association.
19. Yajin Zhou and Xuxian Jiang. Dissecting Android Malware: Characterization and Evolution. In *Proceedings of the 2012 IEEE Symposium on Security and Privacy*, SP '12, pages 95–109, Washington, DC, USA, 2012. IEEE Computer Society.
20. Long Lu, Zhichun Li, Zhenyu Wu, Wenke Lee, and Guofei Jiang. CHEX: Statically Vetting Android Apps for Component Hijacking Vulnerabilities. In *Proceedings of the 2012 ACM Conference on Computer and Communications Security*, CCS '12, pages 229–240, New York, NY, USA, 2012. ACM.
21. Steven Arzt, Siegfried Rasthofer, Christian Fritz, Eric Bodden, Alexandre Bartel, Jacques Klein, Yves Le Traon, Damien Octeau, and Patrick McDaniel. FlowDroid: Precise Context, Flow, Field, Object-sensitive and Lifecycle-aware Taint Analysis for Android Apps. In *Proceedings of the 35th ACM SIGPLAN Conference on Programming Language Design and Implementation*, PLDI '14, pages 259–269, New York, NY, USA, 2014. ACM.
22. Yin Yan, Sree Harsha Konduri, Amit Kulkarni, Varun Anand, Steven Y. Ko, and Lukasz Ziarek. RTDroid: A Design for Real-time Android. In *Proceedings of the 11th International Workshop on Java Technologies for Real-time and Embedded Systems*, JTRES '13, pages 98–107, New York, NY, USA, 2013. ACM.
23. Yin Yan, Shaun Cosgrove, Varun Anand, Amit Kulkarni, Sree Harsha Konduri, Steven Y. Ko, and Lukasz Ziarek. Real-time Android with RTDroid. In *Proceedings of the 12th Annual International Conference on Mobile Systems, Applications, and Services*, MobiSys '14, pages 273–286, New York, NY, USA, 2014. ACM.
24. Yin Yan, Shaun Cosgrove, Ethan Blanton, Steven Y. Ko, and Lukasz Ziarek. Real-Time Sensing on Android. In *Proceedings of the 12th International Workshop on Java Technologies for Real-time and Embedded Systems*, JTRES '14, pages 67:67–67:75, New York, NY, USA, 2014. ACM.
25. Alexis Santos. Strand-1 Satellite Launches Google Nexus One Smartphone into Orbit. https://www.engadget.com/2013/02/26/google-nexus-one-launched-into-space-cubesat-phonesat-strand-1/ (accessed October 23, 2016).
26. Cláudio Maia, Lúıs Nogueira, and Luis Miguel Pinho. Evaluating Android OS for Embedded Real-time Systems. In *Proceedings of the Sixth International Workshop on Operating Systems Platforms for Embedded Real-time Applications*, Brussels, Belgium, OSPERT '10, pages 63–70, 2010.

27. Igor Kalkov, Dominik Franke, John F. Schommer, and Stefan Kowalewski. A Real-Time Extension to the Android Platform. In *Proceedings of the 10th International Workshop on Java Technologies for Real-time and Embedded Systems*, JTRES '12, pages 105–114, New York, NY, USA, 2012. ACM.
28. Thomas Gerlitz, Igor Kalkov, John F. Schommer, Dominik Franke, and Stefan Kowalewski. Non-blocking Garbage Collection for Real-time Android. In *Proceedings of the 11th International Workshop on Java Technologies for Real-time and Embedded Systems*, JTRES '13, pages 108–117, New York, NY, USA, 2013. ACM.
29. RTEMS. http://www.rtems.org/ (accessed October 23, 2016).
30. Filip Pizlo, Lukasz Ziarek, and Jan Vitek. Toward Java on Bare Metal with the Fiji VM. In *Proceedings of the Java Technologies for Real-time and Embedded Systems (JTRES)*, 2009.
31. Filip Pizlo, Lukasz Ziarek, Petr Maj, Antony L. Hosking, Ethan Blanton, and Jan Vitek. Schism: Fragmentation-tolerant Real-time Garbage Collection. In *Proceedings of the 31st ACM SIGPLAN Conference on Programming Language Design and Implementation*, PLDI '10, pages 146–159, New York, NY, USA, 2010. ACM.
32. Filip Pizlo, Lukasz Ziarek, Ethan Blanton, Petr Maj, and Jan Vitek. High-level Programming of Embedded Hard Real-time Devices. In *Proceedings of the Fifth European Conference on Computer Systems*, EuroSys '10, pages 69–82, New York, NY, USA, 2010. ACM.
33. Nick Feamster, Jennifer Rexford, and Ellen Zegura. The Road to SDN: An Intellectual History of Programmable Networks. SIGCOMM *Comput. Commun. Rev.*, 44(2):87–98, April 2014.
34. Sharath Chandrashekhara, Kyle Marcus, Rakesh G. M. Subramanya, Hrishikesh S. Karve, Karthik Dantu, and Steven Y. Ko. Enabling Automated, Rich, and Versatile Data Management for Android Apps with BlueMountain. In *Proceedings of the Seventh USENIX Conference on Hot Topics in Storage and File Systems*, HotStorage '15, pages 1–5, Berkeley, CA, USA, 2015. USENIX Association.

PART II
Storage and Computation in Fog

5 Distributed Caching for Enhancing Communications Efficiency

A. SALMAN AVESTIMEHR and ANDREAS F. MOLISCH
Department of Electrical Engineering, University of Southern California, Los Angeles, CA, USA

5.1 INTRODUCTION

Wireless data traffic is expected to increase by almost 10 000% over the next 5 years [1]. The implications of these trends for future wireless networks are significant. While continued evolution in spectral efficiency is to be expected, the maturity of air interfaces of current systems (LTE Advanced and IEEE 802.11ac/WAVE 2) means that no major improvements of spectral efficiency can be anticipated from this aspect. Additional measures like the brute force expansion of wireless infrastructure (number of cells) and the licensing of more spectrum, while clearly addressing the problem of network capacity, may be prohibitively expensive, require significant time to implement, or be infeasible due to prior spectrum allocations. Thus, additional innovative solutions are required.

A major driver of the spectrum crunch is wireless video on demand, accounting for the majority (~70%) of the predicted traffic demand increase [1]. This type of data traffic has interesting properties: (i) *the user activity is highly asynchronous*, as users wish to access content when and where they wish (unlike live streaming and digital TV) and (ii) *high content reuse*, in the sense that the users' demands concentrate on a small set of very popular files [2].

In this chapter, we investigate a fog network architecture that is able to exploit the aforementioned features in order to provide unprecedented *bandwidth spatial reuse gains*. This architecture starts from the observation that caching has been used successfully in content distribution networks (CDNs) to solve the scalability problem [3], but the current technology is confined in the core network and does not solve the wireless spectrum crunch problem. On the other hand, storage memory is the fastest growing and cheapest network resource and can be made widely available in

Fog for 5G and IoT, First Edition. Edited by Mung Chiang, Bharath Balasubramanian, and Flavio Bonomi.

small dedicated "helper" nodes [4] as well as in the user devices [5]. This enables caching directly at the wireless edge (in contrast with today's CDNs), such that the asynchronous content reuse can be turned into spectrum spatial reuse.

The architecture that we discuss consists of the following two caching methods. In the first approach, termed *femtocaching*, small dedicated "helper nodes" can cache popular files and serve requests from wireless users by enabling localized wireless communication. Such helper nodes are similar to femto-BSs but critically do not need a high-speed backhaul connection. Additionally, an even higher density of caching can be achieved by using devices themselves as video caches; memory has become cheap enough so that it can be made available also in the user devices. The devices "pool" their caching resources so that different devices cache different files and then exchange them, when the occasion arises, through short-range, highly spectrally efficient, device-to-device (D2D) communications. We will term this approach *user-caching*.

In contrast with today's CDNs, where caching only occurs in the Internet cloud (i.e., the wired part of the network), the aforementioned two architectures for caching at the wireless edge provide two significant advantages. First, femtocaching provides significant opportunities for increasing spectral efficiency by *localizing content delivery*. By properly caching content at the edge of the network, we can increase the probability that the users find their desired content in the caches at the helper stations, which can be delivered efficiently via short-range wireless links. Second, user-caching enables the users to "pool" their caching resources so that different devices cache different files and then exchange them through short-range, highly spectrally efficient, local D2D communications, which results in further gains in spectral efficiency.

We will focus on the *femtocaching* architecture in Section 5.2. We will first discuss the optimal caching mechanism for such architecture, such that the average delay for delivering content to each user in the deliver phase is minimized. We will consider both *uncoded* and *coded* content placement in the caching phase. We will then discuss the delivery phase of this architecture, concentrating on the case that the video files are streamed, that is, replay at the receiver starts before the complete file has been transmitted. The problem thus becomes which user should get a video "chunk," at what quality, and from which helper station.

We will next discuss the *user-caching* architecture in Section 5.3. We will start by considering a "cluster-based" approach for D2D content delivery, in which the network is divided into smaller (disjoint) groups of users called "clusters" and only nodes that are part of the same cluster can communicate with each other. We will discuss the optimal design of system parameters in this architecture so as to maximize total network throughput and maximize the offload of traffic to D2D communication. We will also discuss the scaling laws for this architecture, that is, how the capacity scales up as more and more users are introduced into the network (for a fixed area). We will next consider a detailed model for the underlying physical layer, and develop D2D communication algorithms that take advantage of caching for optimal delivery of content to users. In particular, we will focus on a recently proposed information-theoretic approach for D2D communication, named information-theoretic link scheduling (ITLinQ) [6], and discuss how to optimally utilize caching at the users in order to maximize the *spectral reuse* of ITLinQ. Finally, we will turn our attention to optimal caching at the users to enable multicasting opportunities, in order to convert different demands of the users into a single *coded* multicast transmission, which can be delivered very efficiently over the shared wireless medium.

5.2 FEMTOCACHING

5.2.1 System Model

We first describe the network model that we use for the analysis of femtocaching. As outlined in the introduction, the regular base stations (BS) are helped by a number of helper nodes that have popular video caches stored but have a (wired or wireless) backhaul connection that is so slow that the cache can be refreshed only on a timescale that is much larger than the duration of a video. Specifically, we consider a network with one BS m helper nodes (denoted by the set $\mathcal{H}$) and n users (denoted by the set $\mathcal{U}$) requesting the content. We assume the contents are selected from a library $\mathcal{F}$ of $|\mathcal{F}|$ distinct files, each helper node has a cache of size M files, and each user has a cache of size K files (for the current section, we set $K = 0$ but will consider finite K in the subsequent sections). Without loss of generality, we normalize the size of all files to be 1, noting that larger files can always be broken to subfiles). For later use, we assume that the network area can be partitioned into C clusters. The file requests follow a popularity distribution, and $P_r(f)$ denotes the probability that the file with index f is requested. The predictions for the library ($\mathcal{F}$) and distributions ($P_r(f)$'s) are assumed to be obtained from suitable prediction algorithms (accurate prediction of spatiotemporal demand distribution is a topic of active research). An example of this system model is illustrated in Figure 5.1.

In such a scenario, given ($\mathcal{H}$, $\mathcal{U}$, $\mathcal{F}$) and the $P_r(f)$'s, the goal is to *design the optimal caching mechanism that will maximize the spectral efficiency in the delivery phase, averaged over the spatiotemporal demand distributions.*[1]

In general, every helper $i \in \mathcal{H}$ has a subset $\mathcal{F}(h)$ of files of the library $\mathcal{F}$ in its cache. The femtocaching network is thus represented by the bipartite graph $\mathcal{G} = (\mathcal{H}, \mathcal{U}, \mathcal{E})$, where the set of edges $\mathcal{E}$ denotes which helper is connected to which user and the edge weight $\omega_{i,j}$ is the inverse rate between the helper station and the user; "helper" 0 is the BS that has all files available (but typically a smaller transmission rate). Let furthermore $x_{f,h} = 1$ if file f is cached in helper h and 0 otherwise.

[1]This means that the network topology is assumed to be fixed and known; in reality it is typically time varying with dynamics comparable or faster than the file transmission; therefore reconfiguring the caches at this timescale is definitely not practical. However, additional simulations have also shown that the cache distribution obtained when the mobile stations are in "typical" distances from the helpers also provides good performance for various other realizations of random placement of nodes.

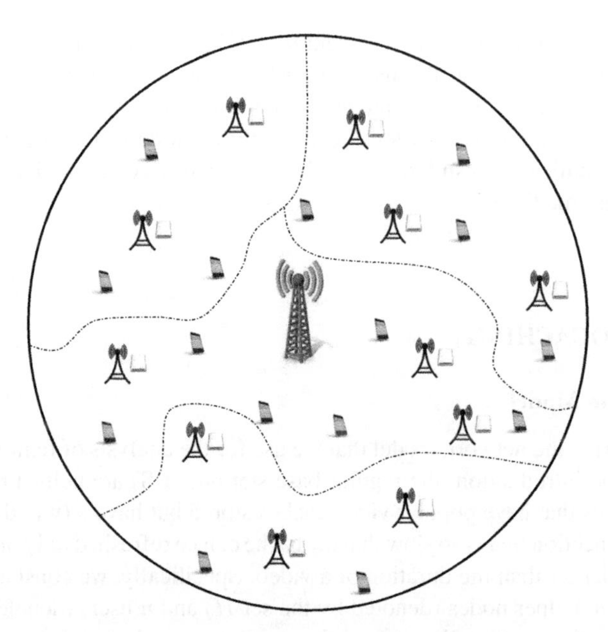

Figure 5.1 System model for femtocaching.

We distinguish between uncoded and coded content placement. In the uncoded case, video files are cached directly, with possible replication.[2] Then the average delay per information bit for user u can be written as

$$\overline{D}_u = \sum_{j=1}^{|\mathcal{H}(u)|-1} \omega_{(j)_u,u} \sum_{f=1}^{|\mathcal{F}|} \left[\prod_{i=1}^{j-1} (1 - x_{f,(i)_u}) \; x_{f,(j)_u} P_r(f) \right] + \omega_{0,u} \sum_{f=1}^{|\mathcal{F}|} \left[\prod_{i=1}^{|\mathcal{H}(u)|-1} (1 - x_{f,(i)_u}) \right] P_r(f). \tag{5.1}$$

where $P_r(t)$ is the request probability distribution and $(j)_u$ indicates the helper index in $\mathcal{H}(u)$ with the j-th smallest delay to user u. The minimization of the sum (over the users) average per bit downloading delay can be expressed as the integer programming problem:

$$\begin{aligned} \text{Maximize} \quad & \sum_{u \in \mathcal{U}} \left(\omega_{0,u} - \overline{D}_u \right) \\ \text{Subject to} \quad & \sum_{f \in \mathcal{F}} x_{f,h} \leq M, \quad \forall h, \\ & \mathbf{X} \in \{0,1\}^{|\mathcal{F}| \times |\mathcal{H}|}. \end{aligned} \tag{5.2}$$

[2]All these nodes are encoded with video codes, such as MPEG or H.264, but no *intra-session coding is used.*

This problem is NP-complete but can be formulated as the maximization of a monotone submodular function over matroid constraints, for which a simple greedy strategy achieves at least one-half of the optimum value [7].

We can simplify the caching problem by using intra-session coding, for example, using the scheme in Ref. [8], such that a file is entirely retrieved when a fraction larger than or equal to 1 of parity bits is downloaded. In particular, let $\boldsymbol{\rho} = [\rho_{f,h}]$, where $\rho_{f,h}$ denotes the fraction of parity bits of file f contained in the cache of helper h. Assuming that user u can download files from its best j helpers, the average delay per information bit necessary for user u to download file f is given by

$$\begin{aligned}\overline{D}_u^{f,j} &= \sum_{i=1}^{j-1} \rho_{f,(i)_u}\omega_{(i)_u,u} + \left(1 - \sum_{i=1}^{j-1} \rho_{f,(i)_u}\right)\omega_{(j)_u,u} \\ &= \omega_{(j)_u,u} - \sum_{i=1}^{j-1} \rho_{f,(i)_u}(\omega_{(j)_u,u} - \omega_{(i)_u,u}).\end{aligned} \tag{5.3}$$

Notice that file f can be downloaded by user u from its best j helpers only if $\sum_{i=1}^{j} \rho_{f,(i)_u} \geq 1$. Since the cellular BS contains all files, we always have $\rho_{f,0} = 1$ for all $f \in \mathcal{F}$.

The delay $\overline{D}_u^{f}$ incurred by user u because of downloading file f is a piecewise-defined affine function of the elements of the placement matrix $\boldsymbol{\rho}$, given by

$$\overline{D}_u^{f} = \begin{cases} \overline{D}_u^{f,1} & \text{if } \rho_{f,(1)_u} \geq 1 \\ \vdots & \vdots \\ \overline{D}_u^{f,j} & \text{if } \begin{array}{l}\sum_{i=1}^{j-1} \rho_{f,(i)_u} < 1, \\ \sum_{i=1}^{j} \rho_{f,(i)_u} \geq 1\end{array} \\ \vdots & \vdots \\ \overline{D}_u^{f,|\mathcal{H}(u)|} & \text{if } \sum_{i=1}^{|\mathcal{H}(u)|-1} \rho_{f,(i)_u} < 1 \end{cases} \tag{5.4}$$

We can show that $\overline{D}_u^{f}$ is a convex function of $\boldsymbol{\rho}$. The average delay of user u is given by $\overline{D}_u = \sum_{f=1}^{|\mathcal{F}|} P_r(f)\overline{D}_u^{f}$. With some further manipulations, the coded placement optimization problem takes on the form:

$$\begin{aligned} &\text{Minimize} \quad \sum_{u=1}^{U}\sum_{f=1}^{|\mathcal{F}|} P_r(f) \max_{j\in\{1,2,\ldots,|\mathcal{H}(u)|\}} \left\{\overline{D}_u^{f,j}\right\} \\ &\text{Subject to} \quad \sum_{f=1}^{|\mathcal{F}|} \rho_{f,h} \leq M, \quad \forall h \\ &\qquad\qquad\quad \boldsymbol{\rho} \in [0,1]^{|\mathcal{F}|\times|\mathcal{H}|}, \end{aligned} \tag{5.5}$$

where the optimization is with respect to ρ. This coded formulation not only is simpler but also gives better performance than the uncoded approach, since the coded optimization is a convex relaxation of the uncoded problem. This is obtained at the price of the complexity of coding/decoding with the MDS codes, and the overhead incurred when the MDS codes are not ideal.

5.2.2 Adaptive Streaming from Helper Stations

We now turn to the delivery phase, concentrating on the case that the video files are *streamed*, that is, the replay at the receiver starts before the complete file has been transmitted. Such streaming is widely used for standard video-on-demand systems, using protocols such as Microsoft Smooth Streaming (Silverlight), Apple HTTP Live Streaming, and 3GPP Dynamic Adaptive Streaming over HTTP (DASH). The problem thus becomes "which user should get a video 'chunk,' at what quality and from which helper station."

Each user $u \in \mathcal{U}$ requests a video file $f_u \in \mathcal{F}$, formed by a sequence of chunks, which are independently decodable stand-alone units [9]. Chunks have a fixed duration T_{chunk} and must be reproduced sequentially at the user end. The streaming process consists of transferring N chunks from the helpers to the requesting users.

Video can be encoded with different quality levels [10], and the quality may vary from chunk to chunk—$D_f(m, t)$ and $B_f(m, t)$ denote the video quality measure and the number of bits for chunk t of file f at quality level m $m \in \{1, \dots, N_f\}$, respectively. $r_{hu}(t)$ denotes the source coding rate (bit per chunk) of chunk t requested by user u to helper h. Hence, the streaming scheduler must satisfy

$$\sum_{h\in\mathcal{H}(u)} r_{hu}(t) = B_{f_u}(m_u(t), t), \quad \forall\, u \in \mathcal{U}. \tag{5.6}$$

Represent the underlying wireless network physical layer through a long-term average rate region $\mathcal{R}(t)$. A *network utility maximization* (NUM) formulation can be used to design an adaptive dynamic streaming scheduler in such a network. Each helper h has a transmission queue pointing at its served users $\mathcal{U}(h)$, which evolves as

$$Q_{hu}(t+1) = \max\{Q_{hu}(t) - WT_{\text{chunk}}R_{hu}(t), 0\} + r_{hu}(t) \tag{5.7}$$

where W is the system bandwidth. The input to queue $Q_{hu}(t)$ is formed by the newly requested $r_{hu}(t)$ video-encoded bits. Then, for large files formed by many chunks, the NUM problem becomes

$$\text{Maximize} \quad \phi(\overline{D}_u : u \in \mathcal{U}) \tag{5.8}$$

$$\text{Subject to} \quad \lim_{t\to\infty} \frac{1}{t} \sum_{\tau=0}^{t-1} \mathbb{E}\left[Q_{hu}(\tau)\right] < \infty \quad \forall\,(h,u) \in \mathcal{E} \tag{5.9}$$

$$\alpha(t) \in A_{\omega(t)}\ \forall\, t, \tag{5.10}$$

where constraint (5.9) corresponds to the queues' *strong stability*; $\alpha(t)$ is the decision policy, including the video coding rate requests $\{r_{hu}(t)\}$, the feasible channel coding rate allocation $\{R_{hu}(t)\} \in \mathcal{R}(t)$, and the video quality selection decisions $\{m_u(t)\}$; and $A_{\omega(t)}$ is the set of feasible policies for network state $\left\{g_{hu}(t), D_{f_u}(\cdot, t), B_{f_u}(\cdot, t) : \forall\ (h, u) \in \mathcal{E}\right\}$. This problem can be solved via a dynamic policy based on the Lyapunov drift plus penalty (DPP) approach [11]. The resulting policy is decentralized and consists of a distributed multiuser version of a DASH-like protocol, where users make adaptive decisions on which helper to request from and at which quality level each chunk should be requested.

Instead of such a "push" strategy, one can also consider a "pull" approach where each user maintains a single *virtual* request queue $Q_u(t)$ and dynamically requests only the chunk at the head of the line [12]. A scheme based on this approach was implemented on a testbed formed by Android smartphones and tablets, using standard Wi-Fi MAC/PHY [13].

5.3 USER-CACHING

5.3.1 Cluster-Based Caching and D2D Communications

We now turn to D2D networks, that is, architectures where the devices themselves act as caches. In other words, we let the number of files cached at each user and K be finite but do not assume any helper stations. If a device cannot obtain a file through D2D communications, it can obtain it from a macrocellular BS through conventional cellular transmission.

For ease of analysis, we consider a network with square-shaped macrocells; the dimensions of each cell are normalized to unity. Intercell interference is neglected, a situation that can be approximated through appropriate cell/frequency planning [14]. Each cell/BS serves n users. The BS is assumed to have full knowledge of the neighborhood graph and channel state information between the users; see Ref. [15] for a discussion of algorithms that can achieve this.

The cell is further subdivided into smaller (disjoint) groups of users called "clusters." Only nodes that are part of the same cluster can communicate with each other. To avoid interference within a cluster, only one D2D link can be active per cluster (we discount here the possibility of FDD or TDD to accommodate more users per cluster). Interference *between* clusters is minimized through a frequency reuse strategy as shown in Figure 5.2b. Communication is possible over a radius r, and interference is created over a distance $(1 + \Delta)r$, and the grey squares represent clusters that are active on a particular frequency. We assume that nodes within a cluster can communicate with each other at a fixed rate, and nodes that are in clusters within the "reuse distance" cannot communicate at all due to interference (highlighted disk),[3] while nodes/clusters outside the reuse distance are not interfered at all. This model is, of

[3]This implies we can pick a frequency reuse factor $K = \left(\left\lceil \sqrt{2}(1 + \Delta)\right\rceil + 1\right)^2$.

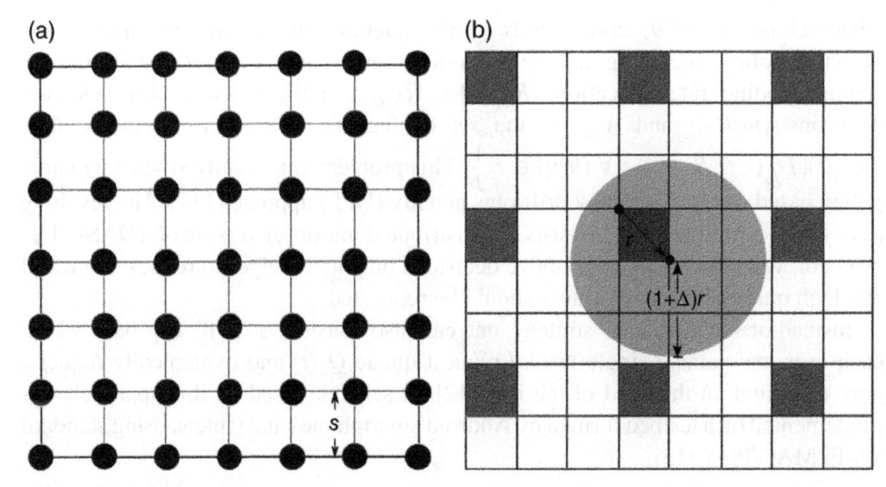

Figure 5.2 (a) Grid network with $n = 49$ nodes (black circles) with minimum separation $s = 1/\sqrt{n}$. (b) An example of single-cell layout and the interference avoidance TDMA scheme. In this figure, each square represents a cluster. The grey squares represent the concurrent transmitting clusters. The highlighted circular area is the disk where the protocol model allows no other concurrent transmission. r is the worst-case transmission range and Δ is the interference parameter. We assume a common r for all the transmitter–receiver pairs. In this particular example, the TDMA parameter is $K = 9$, which means that each cluster can be activated every nine transmission scheduling slot durations.

course, a major simplification whose assumptions do not hold exactly in practice. Yet, it provides a first approximation to the exact solutions. A more detailed discussion of alternative scheduling and power control strategies will be given in the next subsection.

We can optimize two parameters: the caching probabilities, and the cluster size. For the decision of which files to cache, we consider two strategies:

- *Deterministic* caching, where the BS instructs the devices to cache the most popular files in a disjoint manner, that is, no file should be cached twice in devices belonging the same cluster. This approach can only be realized if the device stays in the same locations for many hours; the approach also serves as a useful upper bound.
- *Random* caching, where each device randomly and independently caches a set of files according to a common probability mass function. Ji *et al.* [16] found that the optimal caching distribution P_c^* that maximizes the probability that any user finds its requested file inside its own cluster is given (for a node arrangement on a rectangular grid as described earlier) by

$$P_c^*(f) = \left[1 - \frac{\nu}{z_f}\right]^+, \quad f = 1, \ldots, m, \tag{5.11}$$

where $v = (m^* - 1)/\sum_{f=1}^{m^*} \frac{1}{z_f}$, g_c is the cluster size, $z_f = P_r(f)^{1/(M(g_c-1)-1)}$, $m^* = \Theta\left(\min\left\{\frac{M}{\gamma_r} g_c, m\right\}\right)$, and $[\Lambda]^+ = \max[\Lambda, 0]$.

For the cluster size, we can intuitively judge that there is a trade-off: increasing the cluster size increases the probability that a user will find the file it wants within the cluster (since every device stores at least some files that are different from its neighbors)—in this case we call a cluster "good," and if actual transmission (which is subject to the interference constraints) occurs, the cluster is "active." On the other hand larger clusters reduce the frequency reuse. We can thus anticipate that there will be an optimum cluster size and that it will depend on the video request statistics—the more "concentrated" (i.e., redundant) the requests are, the smaller the optimum cluster size will be. Maximizing the number of active clusters leads to the maximum throughput of the network, though it does not necessarily lead to the maximum offload of traffic from the BS: for that latter goal, we instead wish to maximize the sum of active clusters plus the number of devices that find the file they desire in their own cache.

There are a number of different criteria for optimizing the system parameters. One obvious candidate is the total network throughput. It is maximized by maximizing the number of active clusters. Golrezaei *et al.* [17] showed that for deterministic caching, the expected throughput can be computed as

$$E\{T\} = \frac{1}{r^2} \sum_{k=0}^{n} \left(1 - \prod_{i=1}^{k} (1 - (P_{\mathrm{CVC}}(k) - P_r(f_i)))\right) \times \Pr[K = k]. \tag{5.12}$$

where $P_{\mathrm{CVC}}(k)$ is the probability that the requested file is in the common virtual cache (the union of all caches in the cluster), that is, among the k most popular files. $\Pr[K = k]$, the probability that there are k users in a cluster, is deterministic for the rectangular grid arrangement, and

$$\Pr[K = k] = \binom{n}{k} (r^2)^k (1 - r^2)^{n-k}, \tag{5.13}$$

for random node placement.

We next consider scaling laws, that is, how the capacity scales up as more and more users are introduced into the network (for a fixed area). We consider here a random caching strategy. A system admission control scheme decides whether to serve potential links or ignore them. The served potential links in the same cluster are scheduled with equal probability (or, equivalently, in round-robin), such that all admitted user requests have the same average throughput. Golrezaei *et al.* [18] established lower and upper bounds for the throughput of D2D communications concluding that for highly concentrated demand distribution, $\gamma_r > 1$, the throughput scales linearly with the number of users, or equivalently the per user throughput remains constant as the user density increases. For heavy-tailed demand distributions, the throughput of the system increases only sublinearly, as the clusters have to become larger (in terms of number of nodes in the cluster) to be able to find requested files within the caches of the cluster members.

Ji *et al.* [19] extends the bounds to the case of throughput—outage trade-off. Qualitatively a user is in outage if the user cannot be served via a D2D connection. This can be caused by the following: (i) the file requested by the user cannot be found in the user's own cluster and (ii) the system admission control decides to ignore the request. The outage probability p_o is the average fraction of users in outage.

Consider the case that that $\lim_{n\to\infty} m^\alpha/n = 0$, where $\alpha = (1-\gamma_r)/(2-\gamma_r)$ and $\gamma_r \in (0,1)$ (so that $\alpha < 1/2$). Any scaling of m versus n slower than $n^{1/\alpha}$ is captured by the following result (though the most practically interesting case is where m is sublinear with respect to n) [19].

Theorem 5.1 *Assume* $\lim_{n\to\infty} m^\alpha/n = 0$. *Then, the throughput–outage trade-off achievable by one-hop D2D network with random caching and clustering behaves as*

$$T^*(p) \geq \begin{cases} \frac{C_r}{K}\frac{M}{\rho_1 m} + \delta_1(m), & p = (1-\gamma_r)e^{\gamma_r-\rho_1}, \\ \frac{C_r A}{K}\frac{M}{m(1-p)^{\frac{1}{1-\gamma_r}}} + \delta_2(m), & p = 1-\gamma_r^{\gamma_r}\left(\frac{Mg_c}{m}\right)^{1-\gamma_r}, \\ \frac{C_r B}{K}m^{-\alpha} + \delta_3(m), & 1-\gamma_r^{\gamma_r}M^{1-\gamma_r}\rho_2^{1-\gamma_r}m^{-\alpha} \\ & \leq p \leq 1-a(\gamma_r)m^{-\alpha}, \\ \frac{C_r D}{K}m^{-\alpha} + \delta_4(m), & p \geq 1-a(\gamma_r)m^{-\alpha}, \end{cases} \tag{5.14}$$

where $a(\gamma_r)$, A, B, *and* D *are constants depending on* γ_r *and* M, *which can be found in Ref. [19], and where* ρ_1 *and* ρ_2 *are positive parameters satisfying* $\rho_1 \geq \gamma_r$ *and* $\rho_2 \geq \left(\frac{1-\gamma_r}{\gamma_r^{\gamma_r}M^{1-\gamma_r}}\right)^{\frac{1}{2-\gamma_r}}$. *The cluster size* g_c *is any function of* m *satisfying* $g_c = \omega(m^\alpha)$ *and* $g_c \leq \gamma_r m/M$. *The functions* $\delta_i(m)$, $i = 1,2,3,4$ *are vanishing for* $m \to \infty$ *with the following orders* $\delta_1(m) = o(M/m)$, $\delta_2(m) = o\left(\frac{M}{m(1-p)^{\frac{1}{1-\gamma_r}}}\right)$, $\delta_3(m)$, *and* $\delta_4(m) = o(m^{-\alpha})$. ■

The dominant term in (5.14) can accurately capture the system performance even in the finite-dimensional case shown by simulations in Figure 5.3.

5.3.2 IT LinQ-Based Caching and Communications

We have so far considered a simple, cluster-based physical layer model for the D2D network, in which the area at which users are located is partitioned into clusters, and it is assumed that nodes within a cluster can communicate with each other at a fixed

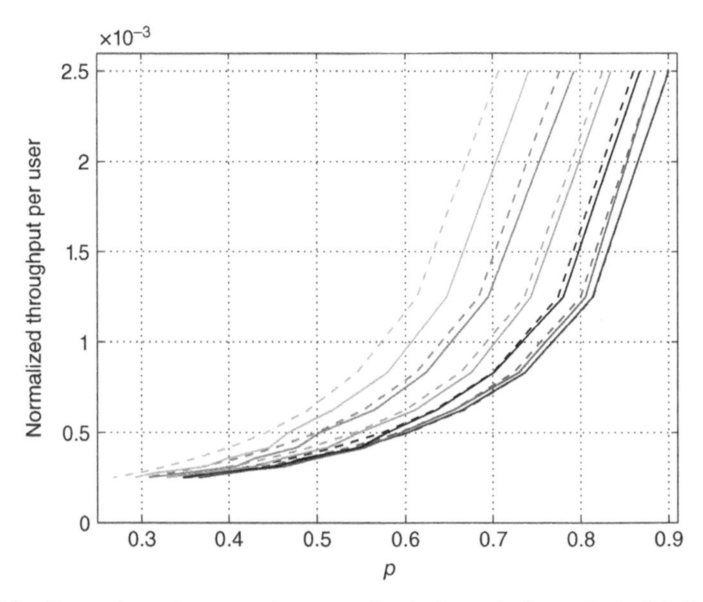

Figure 5.3 Comparison between the normalized theoretical result (solid lines) and normalized simulated result (dashed lines) in terms of the minimum throughput per user versus outage probability. The throughput is normalized by C_r so that it is independent of the link rate. We assume $m = 1000$, $n = 10\,000$, and $M = 1$ and reuse factor $K = 4$. The parameter γ_r for the Zipf distribution varies from 0.1 to 0.6, which are shown from the rightmost curves to the leftmost curves. The theoretical curves show the plots of the dominating term in (5.14) divided by C_r.

rate and nodes that are in clusters within the "reuse distance" cannot communicate at all due to interference, while nodes/clusters outside the reuse distance are not interfered at all. In this section, we consider a detailed model for the underlying physical layer and develop D2D communication algorithms that take advantage of caching for optimal delivery of content to users.

We consider a D2D network consisting of n devices located uniformly and randomly over a circle of radius R. From the set of nodes, $\mathcal{N}$, at any given time n^α are *users* requesting distinct files from other nodes, with $\alpha < 1$. Each of these users may be served by any node that has the desired file. The remaining $n - n^\alpha$ non-user nodes in the network are referred to as *sources*.

We consider a library of n^α distinct files, each of which is cached at multiple source nodes in the network. Specifically, each file is assumed to be cached at n^β source nodes, selected uniformly from the total of $n - n^\alpha$ source nodes and independently for each file. If a source has cached the file f_u desired by user u, then it may potentially act as the source for user u. We emphasize that although files desired by users are assumed to be distinct, sources may (are likely to) cache multiple files desired by multiple users.

In order to receive its desired file, each user must be *associated* with a source that has its file, and then the file must be communicated from the associated source to the user. Multiple users may be associated with the same source. We denote the set of all possible associations of sources to users by Ω. Once a specific association, $\omega \in \Omega$, has been chosen, sources communicate files to their users over a shared wireless medium. Due to the broadcast nature of wireless, interference between concurrent transmissions may occur, and therefore the signal received by each user, u, at each time t is given by

$$y_u[t] = \sum_{s \in \overline{\mathcal{N}}} h_{us} x_s[t] + z_u[t], \tag{5.15}$$

where $\overline{\mathcal{N}}$ denotes the set of source nodes, h_{us} denotes the channel gain between source s and destination u, $x_s[t]$ denotes the transmit signal of source s at time t (subject to a power constraint of $\mathbb{E}[|x_s[t]|^2] \leq P$), and $z_u[t]$ is the additive white Gaussian noise at user u and time t, which is distributed according to a complex Gaussian distribution with zero mean and unit variance.

For the sake of analysis, we assume that the channel attenuation between any two nodes is governed by a path loss model:

$$h_{us} = \sqrt{h_0 d_{us}{}^{-\kappa}}, \tag{5.16}$$

where d_{us} is the distance between two nodes u and s, h_0 is a constant, and κ is the pathloss exponent.

After the association between the sources and the users has been set, the problem will become one of scheduling the file deliveries between sources and users. For this part, we will utilize our recently proposed information-theoretic link scheduling (in short, ITLinQ) scheme [6,20,21] in order to manage the interference and deliver the content to the users. The joint usage of caching and ITLinQ is what we will refer to as *cached ITLinQ* in the rest of this section, and the main question that we will address is how to optimally associate the source and destinations in the network, in order to maximize the *spectral reuse* of cached ITLinQ, as a function of cache redundancy.

We will start by giving a brief overview of ITLinQ. Consider a wireless network consisting of k source nodes $\{\mathrm{S}_i\}_{i=1}^k$ and k destination nodes $\{\mathrm{D}_i\}_{i=1}^k$ where each source S_i intends to communicate to its corresponding destination D_i. Since all the links (i.e., source–destination pairs) share the same wireless spectrum, all the transmissions interfere at all the destinations.

We assume that each source node sends its signal under a power constraint of P, and all the destination nodes, suffering from both interference and additive white Gaussian noise, treat all their incoming interference as noise. In general, treating interference as noise (TIN) is known to be suboptimal for the general interference networks and numerous more sophisticated physical layer coding schemes have been proposed in order to improve it. However, the recent result in Ref. [22] proves that under a general condition in a network consisting of k source–destination pairs, treating interference as noise (in short, TIN) is information-theoretically optimal (to within a constant gap). If SNR_i denotes the signal-to-noise ratio of user i, and for two distinct users i and j, INR_{ij} denotes the interference-to-noise ratio of source j at destination i, then the condition for the optimality of TIN is that for each user i in the network

$$\mathrm{SNR}_i \geq \max_{j \neq i} \mathrm{INR}_{ij} \cdot \max_{k \neq i} \mathrm{INR}_{ki}. \tag{5.17}$$

In words, condition (5.17) can be described as follows:

- *If for each user, the desired channel strength is at least the sum of the strengths of the strongest interference from this user and the strongest interference to this user (all values in dB scale), then TIN can achieve the whole information-theoretic capacity region of the network to within a constant gap.*

We further illustrate this condition in light of the deterministic model proposed in Ref. [23] in Figure 5.4. In this figure, the parallel horizontal lines between S_i and D_i correspond to the quantization of available signal levels in granularity of 1 dB, hence there are a total of SNR_i (dB) of available signal levels for communication between source i and destination i. Also, the dashed lines correspond to the outgoing/incoming interference signal levels from/to user i. Moreover, the highlighted ovals on the left and right correspond to all the most significant signal levels of S_i causing interference at other users and all the least significant signal levels of D_i receiving interference from other users, respectively. Therefore, the aforementioned condition for the optimality of treating interference as noise can be translated to the condition that the highlighted ovals in Figure 5.4 are decoupled and do not overlap. In other words, condition (5.17) implies that as long as these signal levels in an interference network are decoupled, there is no substantial capacity gain from more complicated physical

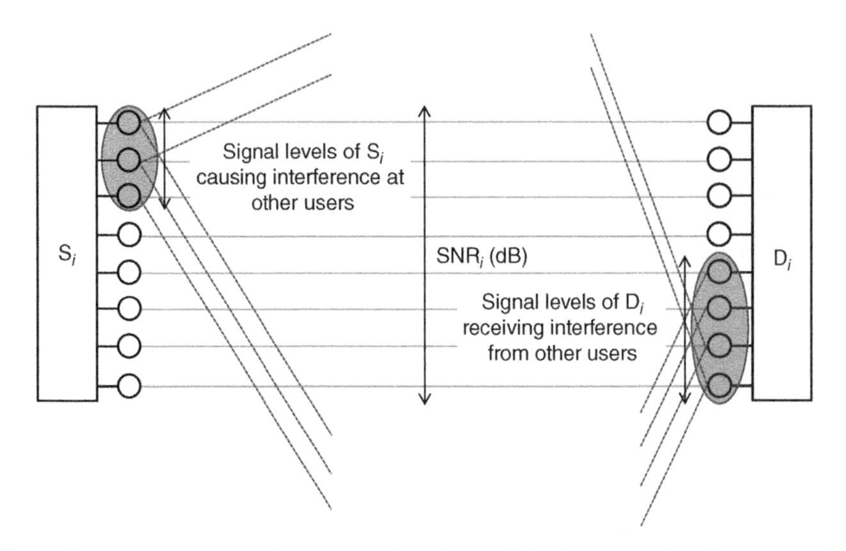

Figure 5.4 A deterministic view of the optimality condition for treating interference as noise.

layer coding schemes for interference management, such as message splitting and successive interference cancellation, and interference alignment.

This is in fact the principle that we use in order to define an *information-theoretic independent set* (ITIS). To be precise:

- *If the aforementioned sufficient condition for the optimality of TIN is satisfied within a subset of users in a wireless network, those users form an ITIS.*

The aforementioned definition of ITIS leads to the spectrum sharing scheme of ITLinQ.

- *The ITLinQ scheme is a spectrum-sharing mechanism, which, at each time, schedules the sources in an ITIS to transmit simultaneously. Moreover, all the destinations will treat their incoming interference as noise.*

We now turn our attention to the main problem of optimal D2D scheduling in cache networks. As mentioned before, for the described problem setting, there is a set Ω of associations of sources to users. Given a specific association $\omega \in \Omega$, we will have an interference network, and we can therefore identify the set of all ITIS in the resulting network, which we denote by S_ω. Then we have the following definition.

Definition 5.1 *The spectral reuse of cached ITLinQ, denoted by ρ, is the asymptotic maximum fraction of the users in the network that can be served simultaneously as an ITIS. More formally,*

$$\rho = \lim_{n\to\infty} \max_{\omega\in\Omega} \max_{S\in S_\omega} \frac{|S|}{n^\alpha}.$$

Having the aforementioned definition, we can now state our main problems as

- Problem 1: optimal association $\omega^* \in \Omega$ of sources to users in order to maximize the spectral efficiency of ITLinQ.
- Problem 2: characterizing the spectral reuse of cached ITLinQ, ρ, in order to determine the fundamental impact of caching on spectral efficiency of ITLinQ.

These two problems are quite challenging, and in fact they are still open problems. However, in Ref. [24], we have been able to make progress by developing a general lower bound on the spectral efficiency of cached ITLinQ, which is stated as follows.

Theorem 5.2 *The spectral reuse of cached ITLinQ as a function of the caching redundancy is almost surely lower bounded as*

$$\rho \geq \frac{2\pi R^2}{\sqrt{3}\gamma^2} \frac{1}{n^{\alpha(1-\frac{\beta}{2})}},$$

where $\gamma = \sqrt[2\kappa]{\frac{P}{N} h_0 R^{\kappa}}$ *is a constant independent of n.*

The association policy for which cached ITLinQ can achieve the lower bound mentioned in Theorem 5.2 is a greedy algorithm in which each user is associated with the closest available source, resulting in a specific type of interference network known as the interference channel. For more details, the reader is referred to the proof of the theorem.

Using the result of Theorem 5.2, we can now compare the maximum fraction of users that can be served by the conventional approach of interference avoidance (i.e., the cluster-based scheduling approach), which is a constant number independent of the value of n, with that of cached ITLinQ. We observe the gain that caching can provide in such a scenario, which is a factor of $1 - (\beta/2)$ in the exponent. Figure 5.5 also shows the spectral reuse gain of cached ITLinQ scheme over the cluster-based interference avoidance scheme (in which we assume that the fraction of users that

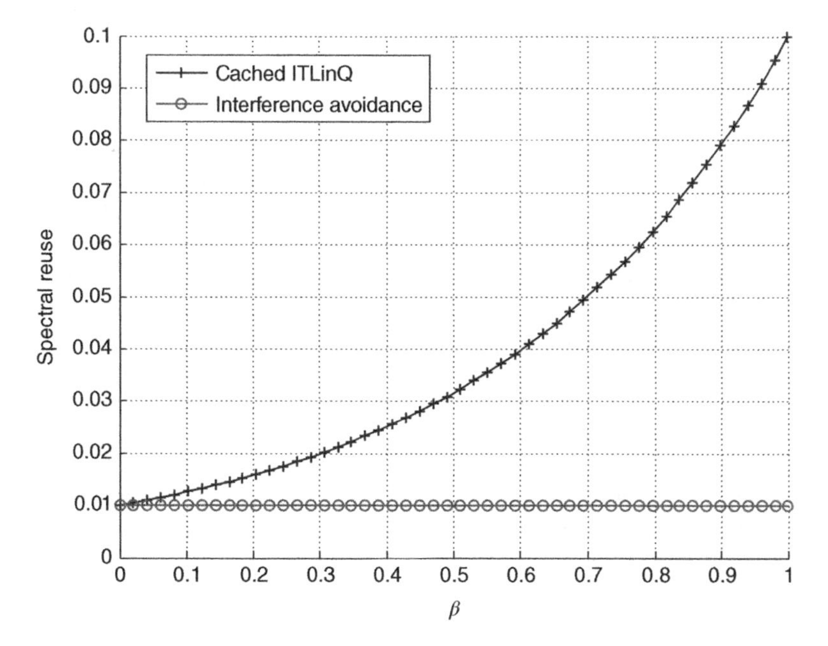

Figure 5.5 Comparison of the achievable spectral reuse of cached ITLinQ and interference avoidance for the case of $n = 10^5$ and $\alpha = 0.4$.

can be served at the same time is at the same level as time sharing: i.e., $1/n^{\alpha}$). In this figure, n is fixed at 10^5 and α is taken to be 0.4.[4] We observe that for higher values of β, each user has a more diverse choice of sources to pick from, which allows cached ITLinQ to enjoy an order of magnitude gain in spectral reuse over the conventional scheme of interference avoidance.

We end this section by considering a practical network setting and providing a numerical analysis of the actual throughput gain that the proposed cached ITLinQ approach is able to provide, over the cluster-based approach in the previous section.

To that end, we consider a square area of side length 2 km inside where n wireless nodes (which can act both as a transmitter and as a receiver) are dropped uniformly at random. We assume there is a library of m distinct files, and each user independently requests a single file out of this library according to a Zipf distribution with parameter $0 < \gamma_r < 1$. In particular, the probability that each user requests file f ($f \in \{1, ..., m\}$) equals $f^{-\gamma_r} / \sum_{i=1}^{m} i^{-\gamma_r}$. Moreover, we assume that each user also caches a single file according to the optimal caching distribution mentioned in Theorem 4 of Ref. [25]. Finally, the channel gins in the network are assumed to follow the ITU-1411 LoS model.

Figure 5.6 illustrates the comparison of the cumulative distribution function of the user rates between the cluster-based and the cached ITLinQ delivery schemes in the small library regime. The number of users, n, is taken to be 1000 and the library size, m, is taken to be 50. As it is clear, the ITLinQ delivery scheme demonstrates a significant improvement in the user throughput distribution over the cluster-based delivery scheme. For example, the probability that the user rate is at most 0.1 bps/Hz is around 55% for cached ITLinQ, while it is over 95% for the cluster-based delivery scheme.

Figure 5.7 compares the distribution of the user rates between the cluster-based and cached ITLinQ delivery schemes for the case of large library size. Here, n is taken to be 100 and m is taken to be 500. Interestingly, ITLinQ shows a considerable throughput gain over the cluster-based delivery scheme in this case as well. A noteworthy point in this plot is the comparison of the fraction of users that cannot be served at all. We observe that while ITLinQ is able to serve all the users, more than 55% of the users are not served in the cluster-based delivery scheme. The main reason for this issue is that in the latter scheme, each user can only look for its desired file within its own cluster (which might be unsuccessful due to the large number of files), whereas in the former scheme, a user can be served by any other user inside the whole area, and the smart scheduling criteria used in ITLinQ is the key to mitigate the interference among the concurrent transmissions.

Finally, we can make use of the rate distribution data in Figures 5.6 and 5.7 to estimate the average user throughput achieved by the two aforementioned delivery mechanisms. In fact, since the throughput values are always non-negative, we can use

[4] The other parameters of the network are chosen such that the constant factor in Theorem 5.2 disappears; that is, $2\pi R^2 / \sqrt{3}\gamma^2 = 1$.

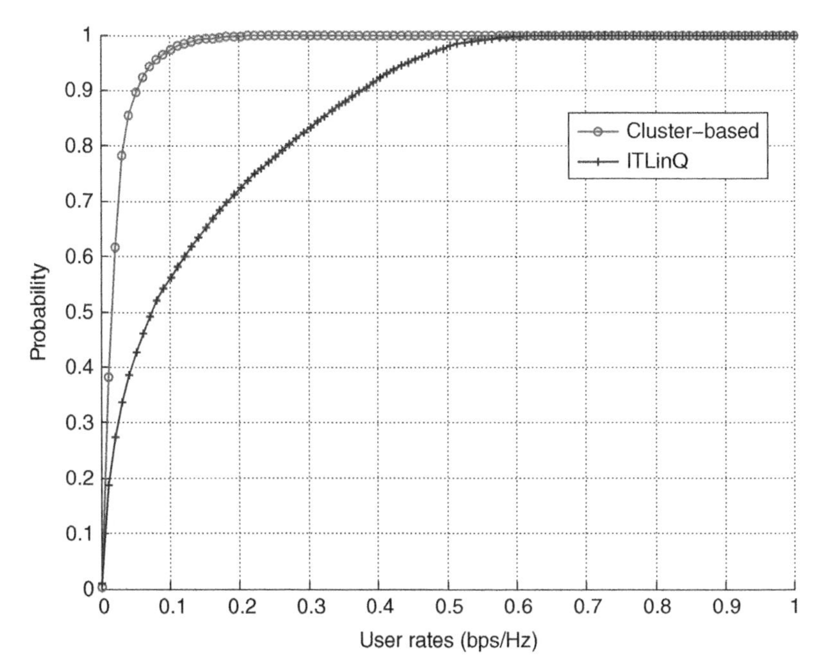

Figure 5.6 Comparison of CDF of the achievable rates of users under the cluster-based and ITLinQ schemes in the small library regime.

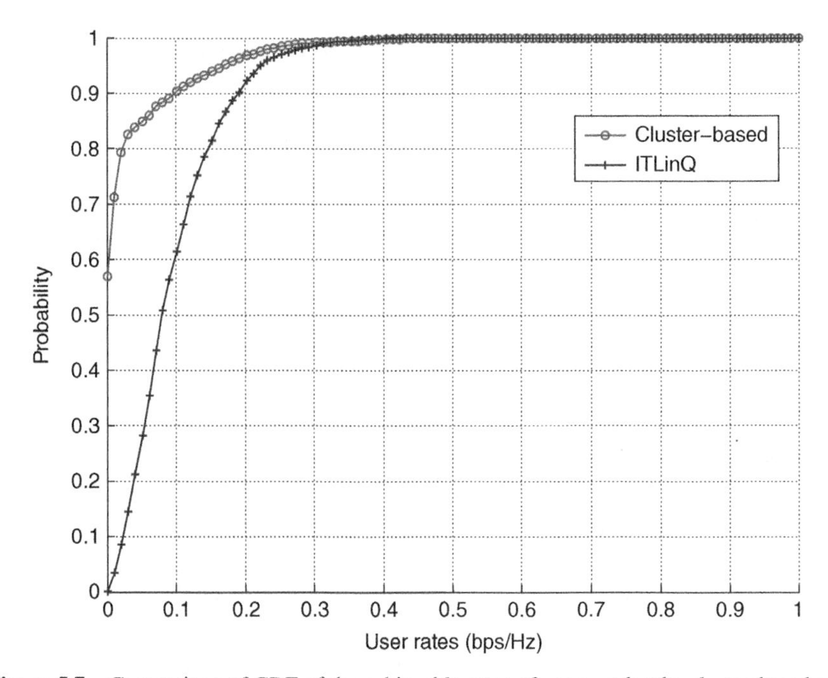

Figure 5.7 Comparison of CDF of the achievable rates of users under the cluster-based and ITLinQ schemes in the large library regime.

TABLE 5.1 The Average User Throughput Comparison Between the Cluster-Based and ITLinQ Delivery Schemes

Library Size	Cluster-Based (bps/Hz)	ITLinQ (bps/Hz)	Gain (%)
Small library	0.0232	0.1367	**488**
Large library	0.0262	0.0958	**265**

the following identity to numerically evaluate the average user rates: for any random variable R that only takes non-negative values, we have

$$\mathbb{E}[R] = \int_0^\infty (1 - F_R(r))\, \mathrm{d}r,$$

where $F_R(r)$ is the cumulative distribution function of R. The results for the average user throughput are reflected in Table 5.1. Interestingly, the average user throughput achieved by cached ITLinQ shows a significant improvement over that of the cluster-based delivery scheme. In particular, the gains for the small and large library sizes are 488 and 265%, respectively.

5.3.3 Coded Multicast

5.3.3.1 An Overiew of Coded Multicast Opportunities Provided by Caching We now turn our attention to optimal caching at the users to enable multicasting opportunities. A conventional approach for caching at users is to store the content that is most likely to be requested so that when a user requests a content, it would be likely that it is already stored in its cache and no delivery would be needed. However, caching at users provides a unique opportunity to turn different demands of the users into a single coded multicast transmission, which can be delivered very efficiently over the shared wireless medium (see, e.g., Refs. [26–28]). We now illustrate this phenomenon through a simple example.

Let there be two users in the network requesting their content from a BS (or helper node). Assume that there are only two files/videos (A and B) that can be requested by users (each file has a 50% chance to be requested) and each user has memory to store only one file. Under the conventional caching approach, each user stores the most popular file, which in this case (equally popular), we choose arbitrarily to be file A. Let us now consider all requests that can happen in the delivery phase:

1. Both users requesting A: Both find A in their cache and no wireless delivery is needed.
2. One user requesting A and the other requesting B: B should be delivered to the user that requests it over the wireless channel.
3. Both users requesting B: B should be broadcasted to both users over the wireless channel and we have a multicast opportunity.

The probability of the aforementioned cases are respectively 25, 50, and 25%. Therefore, the average wireless delivery required for the aforementioned scheme would be $25\% \times 0 + 50\% \times 1 + 25\% \times 1 = 0.75$ fraction of a file.

We now illustrate a more efficient caching scheme that was proposed in Ref. [27]. We first break each file into two parts of the same size, denoted by A_1, A_2, B_1, and B_2. Then, we let user 1 to cache A_1 and B_1, while user 2 caches A_2 and B_2. Note that the memory required for this caching is the same as before. Let us again consider all requests that can happen in the delivery phase:

1. Both users requesting A: In this case user 1 already has A_1 and needs A_2, and user 2 already has A_2 and needs A_1. Note that in this case if the BS broadcasts $A_1 \oplus A_2$ to both users, they can recover the entire file. Therefore, both requests can be satisfied with a "coded multicast transmission" of $A_1 \oplus A_2$. Note that the case in which both users request B can be solved in the same way, by transmitting $B_1 \oplus B_2$ to both users.
2. User 1 requesting A and user 2 requesting B: In this case user 1 already has A_1 and needs A_2, and user 2 already has B_2 and needs B_1. Note that in this case if the BS broadcasts $A_2 \oplus B_1$ to both users, they can recover the entire file. Therefore, both requests can be satisfied with a "coded multicast transmission" of $A_2 \oplus B_1$. Note also that in the same way, when user 1 requests B and user 2 requests A, we can send $A_1 \oplus B_2$ to both users, and both can recover the entire file.

Therefore, with this caching mechanism, all possible requests can be satisfied with a coded multicast transmission of size 0.5 fraction of a file, as opposed to 0.75 fraction of a file average delivery required for the conventional approach.

In a more general case with n users, a library of N files, and cache size of M files at each user, the authors in Ref. [27] have shown that it is possible to design a caching mechanism that provides multicasting opportunities with a multiplicative spectral efficiency gain of $(1 + nM/N)$ over the conventional uncoded delivery scheme. Note that for fixed library size (i.e., N), the overall gain is proportional to nM, which represents the cumulative amount of memory in the network. Therefore, the spectral efficiency gains that caching provides can grow with the global amount of memory in the network.

While this scaling-law gain is almost the same as what can be achieved with D2D communications, we note that in realistic simulations, D2D outperforms coded multicasting considerably; see Figure 5.8.

We note that in the setting of caching for multicasting described earlier, the BS needs to know which files, or parts of files, are cached at each user. This knowledge is required at the BS in order to send a coded bit from which each user could cancel its cached part and infer its requested file. However, since there are many possibilities for what each user can store at a particular time, tracking the exact cache content at the users can be very challenging. Therefore, we considered in a recent work the setting in which the BS only tracks the "amount" of cache at each user and not its exact

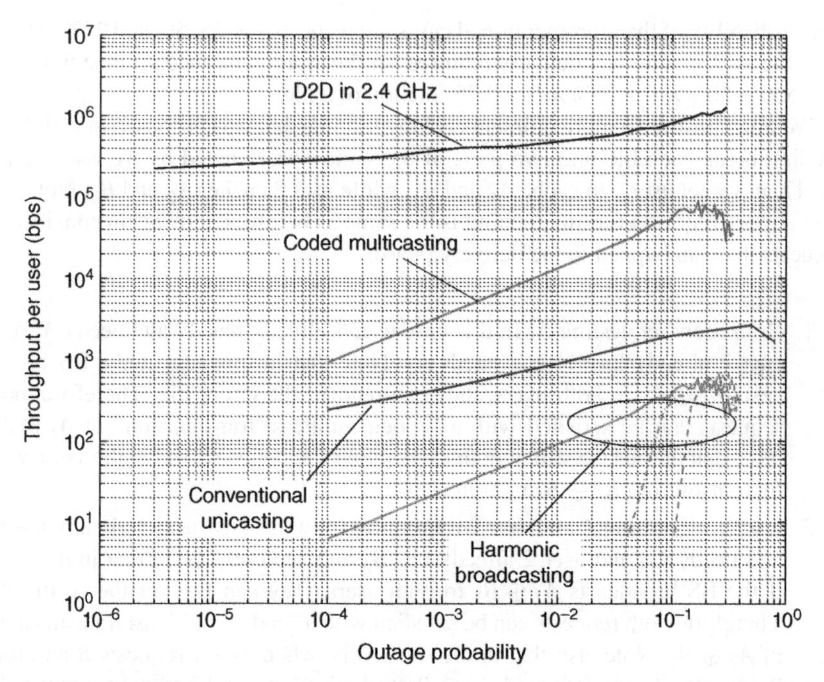

Figure 5.8 Simulation results for the throughput–outage trade-off for different schemes under the realistic indoor/outdoor propagation environment (for details, see Ref. [29]). For harmonic broadcasting with only the m' most popular files, solid line: $m' = 300$; dash–dot line: $m' = 280$; dash line: $m' = 250$. We have $n = 10\,000$, $m = 300$, $M = 20$, and $\gamma_r = 0.4$.

"content." We call this setting "blind index coding." Note that while blind index coding simplifies the implementation of caching protocols, it does not necessarily benefit from the full spectral efficiency gain promised earlier. An important question then is to what extent can caching increase the spectral efficiency, when the BS is only aware of the amount of data in the caches and not its content. Kao *et al.* [30,31] proposed an efficient technique for blind index coding and derived an upper bound on the potential spectral gain when the coding is performed in a single BS. Furthermore, for several important caching cases in the setting of three users, the gain of the proposed coding technique is shown to match the derived upper bound and is thus optimal (Figure 5.9). As a final note, we would like to also point out that "coded multicast" opportunities can also arise in "edge-distributed computing" in order to facilitate data shuffling required for distributed computing. The reader is referred to Refs. [32, 33] for more discussions on this topic.

5.3.3.2 Coded Multicast with D2D At this point, a natural question to ask is whether coded multicasting for D2D transmissions can provide an additional gain or whether the coding gain and the spatial reuse gain can accumulate. This was answered positively in Ref. [16] by designing a subpacketized caching and a

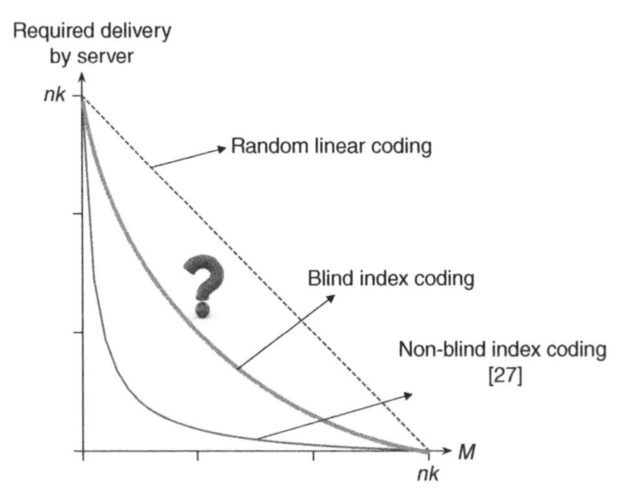

Figure 5.9 Potential spectral gain of blind index coding.

network-coded delivery scheme for the D2D caching networks. We illustrate this via the example shown in Figure 5.10, where we assume no spatial reuse can be used or only one transmission per time–frequency slot is allowed, but the transmission range can cover the whole network. This scheme can be generalized to any n, m, and M. The main insights, as outlined in Ref. [16], are:

1. If each node in the network can reach in a single hop all other nodes in the network, the proposed scheme achieves almost the same throughput of Ref. [27], without the need of a central BS. In other words, for no spatial reuse, every user can successfully decode $m/M\,(1 - (M/m))$.
2. If the transmission range of each node is limited, such that concurrent short-range transmissions can coexist in a spatial reuse scheme, then the throughput has the same scaling law (with possibly different leading term constants) of the reuse-only case [25, 34] or the coded-only case [27]. This result holds even if one optimizes the transmission range and therefore the spatial reuse of the system. Counterintuitively, this means that it is not possible to cumulate the spatial reuse gain and the coded multicasting gain and that these two albeit different type of gains are equivalent as far as the throughput scaling law is concerned. This can be explained by the fact that if spatial reuse is not allowed, a complicated caching scheme can be designed such that one transmission can be useful for as many users as possible. While if we reduce transmission range and perform our scheme in one cluster as shown in Figure 5.2, then the number of users benefitted by one transmission is reduced, but the D2D transmissions can operate simultaneously at a higher rate.
3. The complexity of caching subpacketization and coding can be reduced in coded multicast with D2D, compared to BS-centric coded multicast.

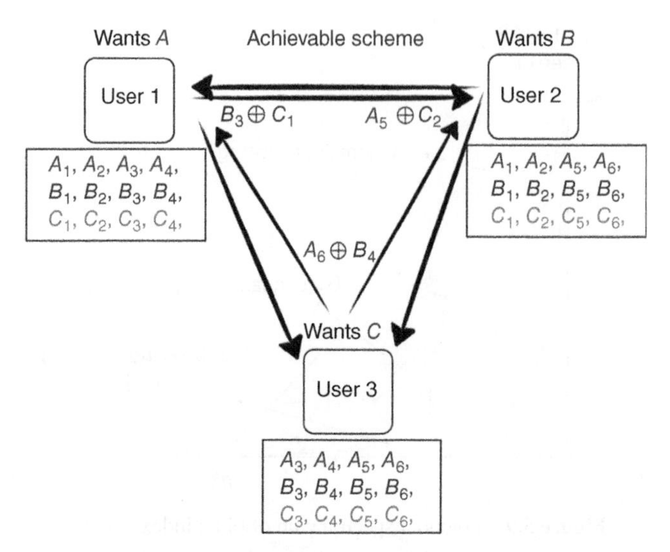

Figure 5.10 Illustration of the example of three users, three files, and $M = 2$, achieving $1/2$ transmissions in term of file. We divide each file into six packets (e.g., A is divided into $A_1, \dots, A_6$). User 1 requests A, user 2 requests B, and user 3 requests C. The cached packets are shown in the rectangles under each user. For the delivery phase, user 1 transmits $B_3 \oplus C_1$, user 2 transmits $A_5 \oplus C_2$, and user 3 transmits $A_6 \oplus B_4$. The normalized number of transmissions is $3 \cdot \frac{1}{6} = \frac{1}{2}$, which is also information theoretically optimal for this network [16].

4. In order to find the best combination of reuse and coded multicasting gains, trading off the rate achieved on each local link (decreasing function of distance) with the number of users that can be reached by a coded multicast message (increasing function of distance) must be sought. This optimization has to be done for the actual throughput and not just the scaling law.

5.4 CONCLUSIONS AND OUTLOOK

In this chapter, we discussed a new architecture for caching content at the edge of the network that consisted of two caching methods. In the first method, termed *femtocaching*, small dedicated "helper nodes" were used to cache popular files and serve requests from wireless users by enabling *localized wireless communication*, hence increasing the spectral efficiency. We discussed the optimal uncoded and coded content placement, as well as the optimal content delivery in such systems. In the second method, termed *user-caching*, the devices pooled their caching resources so that different devices cache different files and then exchange them, when the occasion arises, through short-range, highly spectrally efficient, D2D communications. We discussed the optimal physical layer delivery mechanisms in such systems, in order to maximize the spectral reuse. We also discussed "coded multicast" opportunities that can arise in such systems, in order to further increase the spectral efficiency.

REFERENCES

1. White Paper: Cisco Visual Networking Index: Global Mobile Data Traffic Forecast Update, 2010–2015. http://www.scribd.com/doc/63529506/Cisco-White-Paper-c11-520862 (accessed September 19, 2016), February 2011.
2. M. Zink, K. Suh, Y. Gu, and J. Kurose. Watch global, cache local: Youtube network traffic at a campus network: Measurements and implications. In *Electronic Imaging 2008*, pages 681805–681805. International Society for Optics and Photonics, 2008.
3. E. Nygren, R.K. Sitaraman, and J. Sun. The akamai network: A platform for high-performance internet applications. *ACM SIGOPS Operating Systems Review*, 44(3):2–19, 2010.
4. N. Golrezaei, K. Shanmugam, A.G. Dimakis, A.F. Molisch, and G. Caire. Femtocaching: Wireless video content delivery through distributed caching helpers. In *INFOCOM, 2012 Proceedings IEEE*, pages 1107–1115, 2012.
5. N. Golrezaei, A.G. Dimakis, and A.F. Molisch. Wireless device-to-device communications with distributed caching. In *2012 IEEE International Symposium on Information Theory Proceedings (ISIT)*, pages 2781–2785. IEEE, 2012.
6. N. Naderializadeh and A.S. Avestimehr. ITLinQ: A new approach for spectrum sharing in device-to-device communication systems. *IEEE Journal of Selected Areas in Communications, Special Issue on 5G Wireless Systems*, 32(6):1139–1151, 2014.
7. K. Shanmugam, N. Golrezaei, A.G. Dimakis, A.F. Molisch, and G. Caire. Femtocaching: Wireless content delivery through distributed caching helpers. *IEEE Transactions on Information Theory*, 59(12):8402–8413, 2013.
8. S. Pawar, N. Noorshams, S. El Rouayheb, and K. Ramchandran. Dress codes for the storage cloud: Simple randomized constructions. In *2011 IEEE International Symposium on Information Theory Proceedings (ISIT)*, pages 2338–2342. IEEE, 2011.
9. Y. Sánchez de la Fuente, T. Schierl, C. Hellge, T. Wiegand, D. Hong, D. De Vleeschauwer, W. Van Leekwijck, and Y. Le Louédec. idash: Improved dynamic adaptive streaming over http using scalable video coding. In *Proceedings of the Second Annual ACM Conference on Multimedia Systems*, pages 257–264. ACM, 2011.
10. P. Pancha and M. El Zarki. Mpeg coding for variable bit rate video transmission. *IEEE Communications Magazine*, 32(5):54–66, 1994.
11. M.J. Neely. Stochastic network optimization with application to communication and queueing systems. *Synthesis Lectures on Communication Networks*, 3(1):1–211, 2010.
12. D. Bethanabhotla, G. Caire, and M.J. Neely. Adaptive video streaming in MU-MIMO networks. ISIT 2014, 2014.
13. J. Kim, F. Meng, P. Chen, H.E. Egilmez, D. Bethanabhotla, A.F. Molisch, M.J. Neely, G. Caire, and A. Ortega. Adaptive video streaming for device-to-device mobile platforms. In *Proceedings of the 19th Annual International Conference on Mobile Computing & Networking*, pages 127–130. ACM, 2013.
14. A.F. Molisch. *Wireless Communications, 2nd edition.* IEEE-Press - Wiley, Chichester, 2011.
15. A.F. Molisch, M. Ji, J. Kim, A. Tehrani, and D. Burghal. Device-to-device communications. 2015.
16. M. Ji, G. Caire, and A.F. Molisch. Fundamental limits of distributed caching in D2D wireless networks. In *Information Theory Workshop (ITW), 2013 IEEE*, pages 1–5, 2013.

17. N. Golrezaei, A.F. Molisch, and A.G. Dimakis. Base-station assisted device-to-device communications for high-throughput wireless video networks. In *2012 IEEE International Conference on Communications (ICC)*, pages 7077–7081. IEEE, 2012.
18. N. Golrezaei, A.G. Dimakis, and A.F. Molisch. Scaling behavior for device-to-device communications with distributed caching. *IEEE Transactions on Information Theory*, 60(7):4286–4298, 2014.
19. M. Ji, G. Caire, and A.F. Molisch. The throughput-outage trade-off of wireless one-hop caching networks. *arXiv preprint arXiv:1312.2637*, 2013.
20. N. Naderializadeh and A.S. Avestimehr. Itlinq: A new approach for spectrum sharing. In *IEEE International Symposium on Dynamic Spectrum Access Networks (DySpan)*, pages 327–333, April 2014.
21. N. Naderializadeh and A.S. Avestimehr. Itlinq: A new approach for spectrum sharing in device-to-device communication systems. In *IEEE International Symposium on Information Theory*, pages 1573–1577, July 2014.
22. C. Geng, N. Naderializadeh, A.S. Avestimehr, and S.A. Jafar. On the optimality of treating interference as noise. *IEEE Transactions on Information Theory*, 61(4):1753–1767, 2015.
23. S. Avestimehr, S. Diggavi, and D. Tse. Wireless network information flow: A deterministic approach. *IEEE Transactions on Information Theory*, 57(4):1872–1905, April 2011.
24. N. Naderializadeh, D. Kao, and A.S. Avestimehr. How to utilize caching to improve spectral efficiency in device-to-device wireless networks. In *Allerton Conference on Communication, Control, and Computing*, October 2014.
25. M. Ji, G. Caire, and A.F. Molisch. Optimal throughput-outage trade-off in wireless one-hop caching networks. In *2013 IEEE International Symposium on Information Theory Proceedings (ISIT)*, pages 1461–1465, 2013.
26. M. Ji, A.M. Tulino, J. Llorca, and G. Caire. Caching and coded multicasting: Multiple groupcast index coding. In *2014 IEEE Global Conference on Signal and Information Processing (GlobalSIP)*, pages 881–885, December 2014.
27. M.A. Maddah-Ali and U. Niesen. Fundamental limits of caching. *IEEE Transactions on Information Theory*, 60(5):2856–2867, 2014.
28. K. Poularakis, V. Sourlas, P. Flegkas, and L. Tassiulas. On exploiting network coding in cache-capable small-cell networks. In *2014 IEEE Symposium on Computers and Communication (ISCC)*, volume Workshops, pages 1–5, June 2014.
29. M. Ji, G. Caire, and A.F. Molisch. Wireless device-to-device caching networks: Basic principles and system performance. *IEEE Journal on Selected Areas in Communications*, 34(1):176–189, 2016.
30. D. Kao, M.A. Maddah-Ali, and A.S. Avestimehr. Blind index coding. *submitted to IEEE Transactions on Information Theory*, 2015, arxiv.org/abs/1504.06018.
31. D. Kao, M.A. Maddah-Ali, and A.S. Avestimehr. Blind index coding. In *IEEE International Symposium on Information Theory*, pages 2371–2375, June 2015.
32. S. Li, M.A. Maddah-Ali, and A.S. Avestimehr. Coded MapReduce. In *53rd Annual Allerton Conference on Communication, Control, and Computing*, September 2015.
33. S. Li, M.A. Maddah-Ali, and A.S. Avestimehr. Fundamental tradeoff between computation and communication in distributed computing. *submitted to IEEE International Symposium on Information Theory*, 2016.
34. M. Ji, G. Caire, and A.F. Molisch. The throughput-outage trade-off of wireless one-hop caching networks. *arXiv preprint arXiv:1302.2168*, 2013.

6 Wireless Video Fog: Collaborative Live Streaming with Error Recovery

BO ZHANG,[1] ZHI LIU,[2] and S.-H. GARY CHAN[1]

[1] *Department of Computer Science and Engineering, The Hong Kong University of Science and Technology, Clear Water Bay, Hong Kong*
[2] *Global Information and Telecommunication Institute, Waseda University, Tokyo, Japan*

6.1 INTRODUCTION

With technological advances in multimedia processing and wireless networking, live video content can now be streamed to wireless devices over the Internet. To serve geographically distributed clients,[1] the stream is distributed using a *cloud*, with the wireless clients independently pulling the stream via an access point (AP) or base station. This model suffers from last-hop scalability problem—as the number of clients served by the AP increases, its wireless bandwidth becomes a bottleneck (wireless broadcasting standard is not yet widely used nowadays). To overcome such limitation and scale up the last-hop user capacity, the wireless devices may form a wireless *fog*, where they collaboratively *share* the live stream they pulled from the AP with their neighbors in a multihop manner via a secondary channel (such as Bluetooth, Wi-Fi, etc.). In a wireless fog for live video, devices hence donate their resources (computing power, storage, and communication bandwidth) to scale up the system in a cost-effective manner.

We illustrate in Figure 6.1 a typical live streaming network. The streaming cloud distributes the live stream over the Internet. The cloud is integrated with a wireless fog. In the fog, a *source node* first pulls the stream from a nearby cloud server through an AP or base station. It then distributes its stream to nearby clients. By cooperatively relaying (rebroadcasting) their received packets, nodes can efficiently distribute the live stream within the fog. In this chapter, "broadcasting" means delivering a packet to all the one-hop neighbors of the "broadcaster" by sending just one packet (the so-called broadcast packet).

[1] In this chapter, we use "client", "user", "device", "peer," and "node" interchangeably.

Fog for 5G and IoT, First Edition. Edited by Mung Chiang, Bharath Balasubramanian, and Flavio Bonomi.

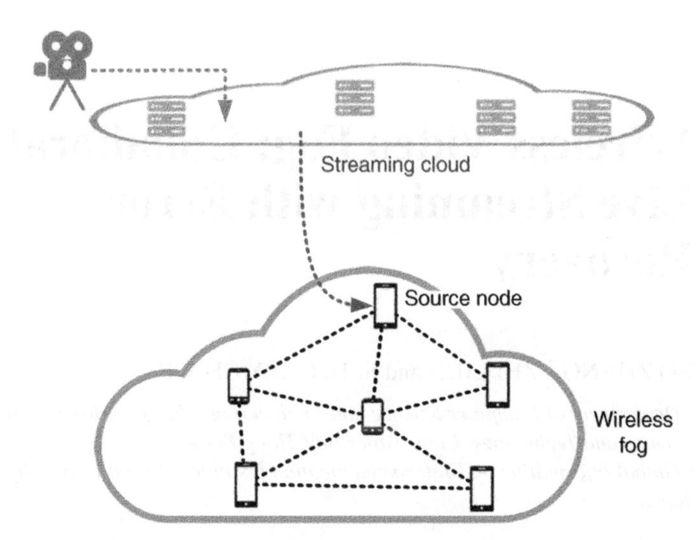

Figure 6.1 Illustration of wireless fog. The *source* node pulls the video stream from the streaming cloud and broadcasts the stream to other clients within the fog in a multihop manner.

In a wireless fog, conventionally the live stream is distributed using the "store-and-forward" approach, where selected broadcasters simply forward their received packets to their neighbors. As packet loss is inevitable in packet broadcasting, this approach suffers from loss propagation, resulting in high loss toward the leaves of the broadcasting tree. In order to achieve good video playback quality, the loss needs to be mitigated through efficient recovery strategies.

Losses in network transmissions are traditionally recovered by feedback-based source retransmission. This is not scalable to a large pool of clients due to the bandwidth limitation of the wireless channel and at the source. Random linear network coding (RLNC) lets the video source perform network coding (NC) on every k source packets and broadcast $n(n > k)$ coded packets to clients. This approach requires the knowledge of network loss condition and is often designed for certain worst-case loss (i.e., $(n - k)/n$). In diverse loss conditions, RLNC therefore often leads to high redundancy for low-loss receivers but "starves" high-loss receivers with insufficient number of packets. In contrast, cooperative loss recovery between neighboring nodes can adapt recovery packets to local loss conditions. One may use source packet sharing, but it limits the recovery efficiency. With the broadcast nature of wireless transmissions, NC can be applied over a selected subset of source packets according to neighbors' loss status to further improve the efficiency and flexibility of cooperative recovery.

In this chapter, we propose a "store–recover–forward" approach for effective video live streaming over wireless multihop networks. A node repeatedly goes through the following steps:

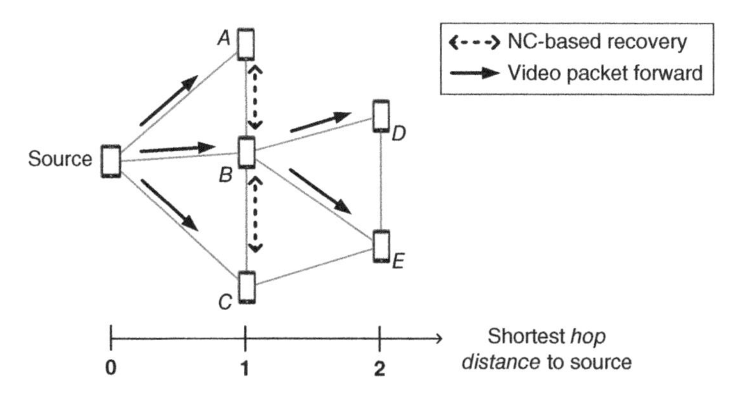

Figure 6.2 A representation of a wireless streaming mesh, with nodes arranged according to their shortest distance to the source node.

1. *Store: Buffer Some Video Packets for Recovery.* A node first buffers received video packets, with the packet availability indicated as a bitmap. As packets may be lost, there may be holes in the bitmap.
2. *Recover: Cooperative Recovery Using NC.* The node shares its bitmap to its neighbors by broadcasting. It also receives the bitmaps from its neighbors. Based on the received bitmaps, the node selects some of its received packets, applies NC to them, and broadcasts the NC packets to its neighbors. Generating NC packets based on packet subset can lead to more flexible recovery.

   After such broadcasting, the node also receives some NC packets from its neighbors. It uses these NC packets to recover some, if not all, of its lost packets in the bitmap. Neighboring nodes hence work together to selectively and effectively recover their losses.
3. *Forward: Video Packets Forwarding.* After the aforementioned NC-based cooperative recovery, the node then exchanges its updated bitmap with neighbors. Based on that, the node decides which packets to forward to its neighbors.

We show in Figure 6.2 an example of wireless live video streaming mesh. Nodes are arranged according to their shortest *hop distances* to the source. Nodes *A*, *B*, and *C* each first receives some video packets from the Source. *B* then shares its bitmap with *A* and *C*. After receiving its neighbors' bitmaps, a node generates NC packets for neighbors with the same hop distance from the source and shares these NC packets. Nodes further away from the source, that is, Nodes *D* and *E*, can buffer the NC packets for their later use. Among Nodes *A*, *B*, and *C*, broadcasters are then selected for each packet. In this example, *B* forwards its video packets to Nodes *D* and *E*. *D* and *E* then perform the same process to cooperatively recover their losses using NC.

In a wireless fog, bandwidth and device energy are scarce resources. Our objective is, therefore, to reduce the total packets broadcasted given a (possibly heterogeneous) target residual loss rate at each node. There are two important decision problems:

- *NC Packet Selection (NCPS).* In order to maximize error recovery, a node generates NC packets based on selected subsets of its packets. The NCPS problem is hence to determine at each node which video packet set should be used for NC coding and how many NC packets to generate, based on neighbor information, so as to achieve the target residual loss rate at each neighbor, with minimum NC recovery traffic.
- *Broadcaster Selection (BS).* This problem determines which packet each node should forward, based on updated neighbor information, in order to achieve the target residual loss rate at each node, with minimum video packet forwarding traffic. This is the so-called broadcaster selection problem.

For high-quality video distribution in a wireless fog, we therefore address the following issues in this chapter:

- *Problem Formulation and Complexity Analysis.* We aim at reducing the total network traffic, which includes both NC recovery traffic and video forwarding traffic, with *NCPS* and *BS* as the two subproblems. We show that both NCPS and BS problems are NP-hard.
- *Video Broadcasting with Cooperative Recovery (VBCR), Distributed Live Video Distribution with NC-Based Cooperative Recovery.* In a wireless fog, network condition is often highly dynamic and unpredictable. It is hence impossible to deploy a central controller to collect network-wide information, compute optimal solution, and distribute the solution in a timely manner. We therefore propose an effective distributed algorithm termed **VBCR** to tackle the problems. Each node in VBCR independently determines the NC packets to code and which video packets to forward, given a certain target residual loss rate at its neighboring nodes. VBCR is fully distributed, with decisions based on only neighbor information and hence is scalable to a large group.
- *Simulation Study.* We have conducted extensive simulation studies on VBCR. Our results show that VBCR is efficient compared to other schemes in terms of total network traffic generated.

The remainder of the chapter is organized as follows. We introduce the wireless cooperative live streaming network in Section 6.3. We formulate the problems in study and analyze their hardness in Section 6.4. We describe our distributed algorithm VBCR in detail in Section 6.5. We present illustrative simulation results in Section 6.6 and conclude in Section 6.7.

6.2 RELATED WORK

We next briefly review related work. Peer-cooperative wireless video streaming has been mentioned in many papers [1–4]. There are two main angles: one is to use peers as relays and the other is to use the local neighbors for cooperative loss recovery.

The work in Ref. [3] is a typical work of the first class, where the device-to-device communication is applied to help improve the received video quality. In both Refs. [1, 2], a source server streams video to users directly, while the packet loss that happened over the primary channel (server-to-user channel) are recovered by local neighbors' help via the secondary channel (user-to-user channel) instead of server retransmission. The work in Ref. [4] proposes cooperative video chunk pulling to reduce redundancy and to reduce chunk loss. However link loss is not considered in the work. A video coding scheme for wireless cooperative video broadcasting is proposed in Ref. [5]. Our work does not depend on the coding scheme optimization; instead we try to optimize application level transmissions. In this chapter we study live video distribution in a wireless cooperative fog video streaming, and the approach is a co-design of the relay node selection and cooperative loss recovery, which distinguishes our work from these schemes.

A large body of research focuses on delay minimization in point-to-point transmission [6–9]. We instead target at bandwidth and energy conservation in broadcast scenario. Schemes in Refs. [10, 11] try to minimize energy cost in an error-free environment by tuning node transmission ranges according to nodes' geographic information. In contrast, we do not require geographic information. We also take link loss and loss recovery into consideration. Le and Liu Tan *et al.* proposed Minimum Steiner Tree with Opportunistic Routing (MSTOR) to construct a minimum Steiner tree based on unicast opportunistic routing [12]. With the utilization of one-hop broadcast and the optimization of BS and NC packets, VBCR enjoys high throughput, low network traffic, and high robustness.

Maximizing data throughput per unit of energy is studied in Ref. [13] by selecting links with better channel conditions. Source packets are retransmitted for loss recovery. A multihop cooperative relay scheme for point-to-point transmission is proposed in Ref. [14], where data retransmission based on incremental redundancy with ARQ is employed in relay nodes. One-hop broadcast nature is however only utilized for the purpose of implicit ACK in Ref. [14]. Others have studied employing NC for cooperative repair in video broadcasting [15, 16]. Most of them assume WWAN as the primary channel so that neighboring nodes naturally enjoy low loss correlation, which is convenient for cooperative recovery. These schemes are not directly applicable to our problem as losses are often propagated and correlated in multihop broadcasting scenarios.

For loss resilience, conventional coding approaches perform all the coding at the source, for instance, LT codes [17] or RLNC, with intermediate nodes performing simple "store-and-forward" operations. Such code is often designed for certain worst-case loss rate. Though effective, such codes are too aggressive due to their "all-or-nothing" nature and therefore do not perform well in networks with diverse regional loss conditions, which is often the case in wireless multihop transmissions. These source coding schemes further require a reliable feedback channel from every node to the source, which is hardly practical in wireless multihop scenarios. In VBCR we allow intermediate nodes to perform NC over a subset of source packets based on local loss conditions, thus achieving high recovery efficiency with low recovery traffic.

MicroCast [18] is a work that tries to achieve goals similar to VBCR but with different approaches. The network considered in MicroCast is a small-scale fully connected graph so that each node can reach every other node in one-hop, which eliminates loss propagation effect. It is also a centralized approach in that a single node is responsible to assign downloading tasks to every other "requester." Through careful task assignment, on-demand retransmission, and the conventional RLNC for sharing, MicroCast significantly reduces network overhead and redundancy in recovering all losses in low-loss network. However such scheme may overwhelm the network with control and retransmission overhead as well as NC redundancy when loss rate is even slightly higher. VBCR on the other hand is designed as a fully distributed algorithm working in a larger-scale multihop network, with only periodical beacon exchange as the control overhead.

6.3 SYSTEM OPERATION AND NETWORK MODEL

We model the network in study as a connected graph $\mathcal{G}(V, E)$, where V is the set of the nodes and E is the set of the links corresponding to the neighbor relationship between node pairs. Link $(i, j) \in E$ if, for example, the signal-to-noise ratio (SNR) measured at node j regarding node i's signal is higher than a predefined threshold. There is a single source node in the network that needs to broadcast its video stream to all other clients.

For a live video stream, each video packet has a playback deadline D (seconds), that is, the maximum tolerable playback delay after the packet is sent by the source. We assume in our network that D is slotted into $\lfloor D/(T_0 + T_1) \rfloor$ slots with a fixed slot duration $(T_0 + T_1)$. The nodes in the network are synchronized in terms of slots.[2] Operations between different slots are therefore independent of each other. Operations of each node in each slot are over the newly received packet set at the beginning of the slot.

We focus on an arbitrary video packet set W and the decisions made within a time slot by the nodes that newly received video packets of this set. We then have two decisions to make: what NC packets should each node code and share and which nodes should forward each of their received/recovered video packets in W.

We show in Figure 6.3 a slot-based live streaming timeline in the system. We introduce the operations in a slot in the following. At the beginning of a slot (i.e., at t_0), a node may receive some packets of a new W from upstream nodes (i.e., nodes that are one hop closer to the source). The node then performs the following operations for its newly received video packets:

- *Initial Information Exchange (IIX).* The node sends a *beacon* to its one-hop neighbors. A beacon message includes the bitmap indicating each locally available video packet of W. If the node has received and buffered some NC packets (from upstream nodes' *Recovery* operation) in previous time slot, information

[2] Time synchronization between clients and the server can be achieved by various protocols such as network time protocol (NTP) or precision time protocol (PTP).

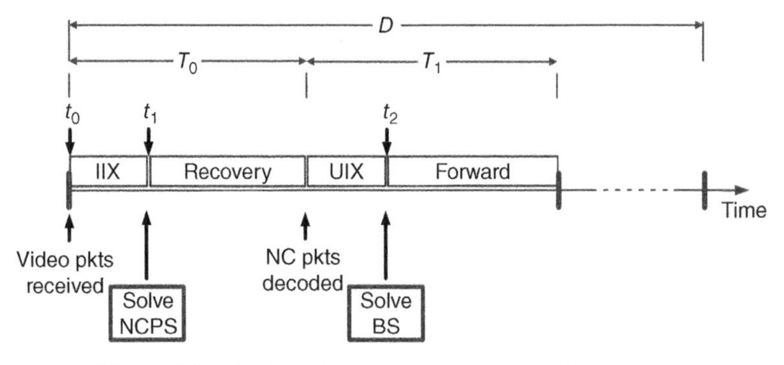

Figure 6.3 Slot-based operations. “pkt” stands for “packet”.

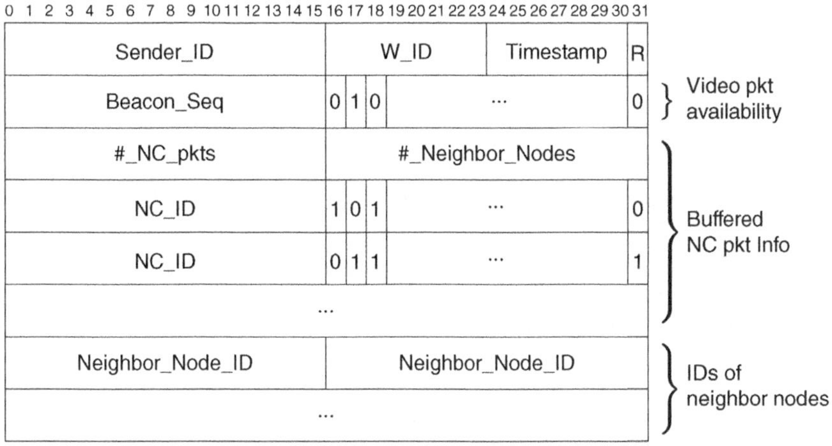

0–15	16	17	18	19–30	31	
Sender_ID	W_ID			Timestamp	R	
Beacon_Seq	0	1	0	···	0	Video pkt availability
#_NC_pkts	#_Neighbor_Nodes					Buffered NC pkt Info
NC_ID	1	0	1	···	0	
NC_ID	0	1	1	···	1	
···						
Neighbor_Node_ID	Neighbor_Node_ID					IDs of neighbor nodes
···						

Figure 6.4 An example beacon packet format for the case that $|W| = 16$. The beacon contains one bitmap for video packet availability and a bitmap for each buffered NC packet. IDs of any downstream nodes are also appended in *UIX* beacon exchange. For different $|W|$ sizes, bitmap lengths can be adjusted accordingly. R is a reserved 1-bit field.

of the NC packets, such as which video packets are carried by each NC packet, is also included in beacon. Figure 6.4 shows an example beacon format. *W_ID* is the unique ID of video packet set W.[3] *NC_ID* is used to identify NC packets the beacon sender buffered and only needs to be unique within the beacon message. IDs of neighbor nodes are learned through last beacon exchange and are appended in the beacon. The node can then compute the effectiveness of its NC sharing.

[3] In this example format, *W_ID* monotonically increases within range $[0, 2^{16-1}]$, and *Src_timestamp* is used for wrap-around handling.

- *Recovery.* The node checks received beacons and determines the NC packets to code and share by solving NCPS problem. To prevent a node from trying to recover packets for downstream nodes (i.e., nodes that are one hop further from the source) with excessive NC traffic, the node only considers beacons with same *W_ID* while treating others as upstream/downstream nodes, who should either be processing a newer packet set or be dealt with by later operations. Note that this does not prevent downstream nodes from receiving NC packets transmitted. The NC packets are then broadcasted. At the end of *Recovery* operation, the node can then try to decode received NC packets to recover its loss.
- *Updated Information Exchange (UIX).* After NC recovery, the node then shares another beacon with updated bitmap, as well as information of any buffered NC packet, to its neighbors. The IDs of neighboring nodes learned from IIX are also appended. This beacon exchange facilitates the next operation.
- *Forward.* The node determines for each of its received/recovered video packets in W, whether it should be a broadcaster by solving BS problem. NC recovery introduces packet heterogeneity at neighboring nodes; BS is hence per packet based instead of for the set W. All beacons with smaller *W_ID* (i.e., downstream nodes) are included for decision making. If the node is selected as a broadcaster of a packet, it then forwards this video packet to its neighbors.

The previous operations are repeated until T expires. Locally available video packets are then used for video playback. In this slot-based system, network losses (due to signal fading, interference, etc.) and topology dynamics (due to node mobility, join/leave, etc.) can be timely reflected by the neighbor beacon exchange during both IIX and UIX periods of each slot.

6.4 PROBLEM FORMULATION AND COMPLEXITY

We summarize major terms used in Table 6.1. There are two operations, **Recovery** and **Forward** in the slot-based system, corresponding to the two problems **NCPS** and **BS**, respectively. We discuss the two optimization problems in the following.

6.4.1 NC Packet Selection Optimization

We try to minimize total NC recovery traffic for video packet set W while achieving the required residual loss rate. Let B_i be the number of NC recovery packets shared by node i to its neighbors. We therefore try to minimize

$$\sum_{i \in V} B_i. \tag{6.1}$$

For collision avoidance, we require that for each node i, only one of its neighbors should broadcast its NC packets, that is,

$$|\{B_j | j \in N_i, B_j > 0\}| \leq 1, \quad \forall i \in V. \tag{6.2}$$

TABLE 6.1 Symbols Used in the Paper

Symbol	Description
E	The set of links in the network
V	The set of nodes in the network
N_i	The set of neighbor nodes of node i
W	The video packet set in consideration
$W_i(t)$	The set of video packets available at node i at time t. $W_i(t) \subseteq W$
$1_i^w(t)$	The binary indicator $1_i^w(t) = 1$ if video packet $w \in W$ is available at node i at time t and $1_i^w(t) = 0$ otherwise
W_i	The subset of video packets that i uses for NC coding. $w_i \subseteq W_i(t_0)$
$\hat{B}_i$	The max number NC packets can be generated at i.
B_i	$B_i \in \mathbb{Z}, 0 \leq B_i \leq \hat{B}$ is the number of NC packets node i shares
S_i	The set of NC packets received by i
l_{ij}	The loss rate of the link (i,j)
I_i^w	Binary indicator, equals to 1 if node i forwards video packet w and 0 if otherwise
T	Playback deadline for video packets in W
Q	The target residual loss rate
T_0, T_1	Durations of NC recovery phase and video packet forwarding phase, respectively
t_0, t_1, t_2	The three time points corresponding to beginning of slot, end of IIX, and end of UIX, respectively

Next we evaluate loss recovery outcome at a node, from its neighbors' NC packet sharing.

First of all, due to limited *Recovery* duration, there is an upper bound $\hat{B}_i$ to the total number of NC packets node i can send. We therefore have

$$B_i \leq \hat{B}_i, \quad \forall i \in V. \tag{6.3}$$

Let w be the index of a single packet in W, that is, $w \in W$. We denote t_0 as the time when IIX begins. Let $1_i^w(t_0)$ be the indicator function of w's availability at node i at time t_0, that is,

$$1_i^w(t_0) \triangleq \begin{cases} 1, & \text{if } i \text{ has } w \text{ at } t_0; \\ 0, & \text{otherwise.} \end{cases} \tag{6.4}$$

So the bitmap of node i at t_0 is the set $\{1_i^w(t_0) | w \in W\}$. The set of video packets that are available at i at t_0, denoted as $W_i(t_0)$, is

$$W_i(t_0) \triangleq \{w | 1_i^w(t_0) = 1\}. \tag{6.5}$$

The set of video packets used for NC coding at node i, denoted W_i, must be the subset of video packets available to i at time t_0. We hence have

$$W_i \subseteq W_i(t_0). \tag{6.6}$$

At the end of *Recovery* operation, node i tries to decode some received NC packets sent by neighbor nodes. Note that due to loss, i may only receive a subset of these NC packets. Denote the link loss rate from i to $j \in N_i$ as l_{ij}. Denote S_j as the NC packet set received by node j. According to Constraint (6.2), S_j is sent by a single neighbor of j. Suppose S_j is sent by node i, that is, $B_i > 0$. Due to loss, we have

$$|S_j| = (1 - l_{ij})B_i, \tag{6.7}$$

Video packet set carried in S_j sent from i, namely W_i, is only useful if S_j can be decoded by j. S_j is decodable only if the total number of video packets it carried *and* are missing at j is no greater than the number of NC packets. Denote $1^{S_j}(T_0)$ as the decodability of S_j at the end of *Recovery* operation, we hence have

$$1^{S_j}(T_0) \triangleq \begin{cases} 1, \text{ if } |W_i - W_j(t_0)| \leq |S_j|; \\ 0, \text{ otherwise.} \end{cases} \tag{6.8}$$

So the set of available video packets at node j after *Recovery*, in the case of i sending B_i NC packets carrying W_i, can be written as

$$W_j^{W_i,B_i}(T_0) = \begin{cases} W_j(t_0) \bigcup W_i, \text{ if } 1^{S_j}(T_0) = 1; \\ W_j(t_0), \text{ otherwise.} \end{cases} \tag{6.9}$$

We can then evaluate the residual loss rate $q_j(T_0)$ at node j at the end of T_0 in this case, as

$$q_j^{B_i}(T_0) = 1 - \frac{|W_j^{W_i,B_i}(T_0)|}{|W|}. \tag{6.10}$$

We want the residual loss rate at every node $i \in V$ after i's NC decoding, denoted as $q_i(T_0)$, to meet the target residual loss rate Q if possible. That is, for all $i \in V$, we want

$$q_i(T_0) \leq Q. \tag{6.11}$$

In case that the loss rate at node i cannot be reduced below Q, we then try to minimize $q_i(T_0)$.

Also we have the integer constraint

$$B_i, |W_i| \in \mathbb{Z}^*, \quad \forall i \in V, \tag{6.12}$$

where $\mathbb{Z}^*$ is the set of all nonnegative integers.

The *NCPS* optimization problem can then be stated as jointly determine W_i and $B_i, \forall i \in V$, such that the Objective (6.1) is minimized, subject to Constraints (6.2–6.12).

6.4.2 Broadcaster Selection Optimization

After *Recovery* operation, nodes exchange their updated beacon with their neighbors in UIX period. Next we need to determine which node should relay each video packet in *Forward* operation (denote its starting time point as t_2). Our objective is to minimize total video packet forwarding traffic for W. Denote $I_i^w = 1$ if node i forwards packet w and $I_i^w = 0$ if otherwise. The total forwarding traffic can be defined as

$$\sum_{i \in V} \sum_{w \in W_i(t_2)} I_i^w. \tag{6.13}$$

where $W_i(t_2) \subseteq W$ is the set of available video packets at i at t_2.

For collision avoidance, we require that

$$\sum_{j \in N_i} I_i \leq 1, \quad \forall i \in V. \tag{6.14}$$

Suppose node i forwards w. Node $j \in N_i$ may not receive it due to loss. The probability that node j receives video packet w from i is

$$P_{ij}^w = 1 - \left(I_i^w \left(1_i^w(t_2) l_{ij} + (1 - 1_i^w(t_2))\right) + (1 - I_i^w)\right) \tag{6.15}$$

Denote the available video packets at node j right after receiving neighbors' forwarding as $W_j'(T_1)$. The expected number of video packets available at node i is

$$E[|W_j'(T_1)|] = |W_j(t_2)| + \sum_{i \in N_j} \sum_{w \in W_i(t_2)} P_{ij}^w (1 - 1_j^w(t_2)). \tag{6.16}$$

Apart from received video packets, node j also tries to use these packets to decode any buffered NC packet. Recall that node j receives and buffers NC packet set S_k sent from upstream neighbor k, with video packet set W_k carried in it, given that $B_k > 0$. According to Equation (6.7), $|S_k| = (1 - l_{kj})B_k$. We now evaluate the probability that j can decode S_k. This probability can be written as

$$P(S_k) = P(|W_k - W_j'(T_1)| \leq |S_k|). \tag{6.17}$$

Then after such decoding, the expected number of available video packets at node j can be written as

$$E[|W_j(T_1)|] = |W_j'(T_1) \cup W_k| P(S_k) + |W_j'(T_1)|(1 - P(S_k)). \tag{6.18}$$

We now evaluate the expected residual loss rate at node j as

$$E[q_j(T_1)] = 1 - \frac{E[|W_j(T_1)|]}{|W|}. \tag{6.19}$$

We again aim at achieving the target residual loss rate Q at every node by the end of T_1 if possible, that is,

$$q_i(T_1) \leq Q, \quad \forall i \in V. \tag{6.20}$$

In case that the target cannot be met, we then try to minimize $q_i(T_1)$.

Also we have the integer constraint

$$I_i^w \in \{0, 1\}, \quad \forall i \in V, w \in W. \tag{6.21}$$

Our BS optimization problem is therefore to minimize Objective (6.13), subject to Constraints (6.14–6.21), via the assignment of I_i^w for every node $i \in V$ and every packet $w \in W$.

6.4.3 Complexity Analysis

We now show that both NCPS problem and BS problem are NP-hard. We consider a simplified scenario for this analysis.

In the simplified case, packet delivery is always successful, and there is only one video packet to distribute. We hence simplify I_i^w to I_i in this case. Our BS problem hence becomes finding min $\sum_{i \in V} I_i$, such that Constraint (6.20) holds. Note that in this "lossless" network, the residual loss rate at any node is either 100% if it is not covered by any broadcaster or 0% if otherwise. We therefore require 0% loss at all nodes, which means every node must be covered by at least one broadcaster. Such constraint is effectively equivalent to

$$\sum_{j \in N_i} I_j \geq 1, \quad \forall i \in V. \tag{6.22}$$

This is effectively the ***maximum leaf spanning tree (MLST)*** problem, which has been shown to be NP-hard [19]. Hence the previously simplified case is NP-hard. So our BS problem in general cases is also NP-hard.

The NCPS problem in the same simplified case effectively becomes BS problem. Following the previous proof, NCPS for the simplified case is also NP-hard. So the NCPS problem in general is also NP-hard.

6.5 VBCR: A DISTRIBUTED HEURISTIC FOR LIVE VIDEO WITH COOPERATIVE RECOVERY

We now present VBCR in detail. The key idea is to let the nodes whose NC/video packet broadcast lead to the highest transmission efficiency to perform the corresponding broadcasting and suppress their neighbors from doing so.

6.5.1 Initial Information Exchange

During IIX operation, each node broadcasts a beacon to its neighbor nodes information about packets in W received at the beginning of this slot. An example beacon packet format has been shown in Figure 6.4. With such information, a node then knows the loss condition of W at each of its neighbors.

6.5.2 Cooperative Recovery

During NC operation, node i makes decisions on which NC packets to code and share, based on beacons received in IIX. There are two decisions to make: which subset of video packets to use for NC coding and how many NC packets to code. In other words, i needs to determine W_i and B_i.

Node i first filters received beacons and only considers neighbors that fulfill all the following three rules:

1. Video packet bitmap in the beacon is not all zeros. The sender of an empty bitmap is still expecting the first packet set. Such neighbors will be treated as downstream nodes and will be dealt with by later operations.
2. *W_ID* in the beacon is equal to that of W. A larger *W_ID* indicates an upstream node, which should now be processing a new packet set. Similarly a smaller *W_ID* indicates a downstream node.
3. Current loss rate is greater than Q. There is no need to consider neighbors with loss rates already lower than Q.

Denote $N_i' \subseteq N_i$ as the subset of i's neighbors with eligible beacons. Node i then tries to reduce the loss rate of each neighbor. Our *NC Recovery* algorithm, as shown in Figure 6.5, operates in the following procedures:

Case 1. $N_i' = \Phi$: Then there is no need to broadcast any NC packet. Node i ends the algorithm.

Case 2. $N_i' \neq \Phi$: Node i then computes W_i and B_i.

To evaluate NC packets, we define a *utility* function $U(W_i, B_i)$ as

$$U(W_i, B_i) \triangleq \sum_{j \in N_i'} \left(\left(1 - \frac{|W_j(t_0)|}{|W|}\right) - \max\left(q_j^{B_i}(T_0), Q\right) \right), \tag{6.23}$$

where $q_j^{B_i}(T_0)$ is a function of W_i and B_i and is defined by Equation (6.10). This utility computes the total loss rate reduction at the neighbors, given W_i and B_i. We only concern residual loss rate down to Q. Node i therefore prefers a high utility. In case of a tie, a smaller B_i is preferred.

Node i aggressively solves it through iterations. In each iteration, i adds video packet w into W_i and evaluates the utility. More specifically, i initializes

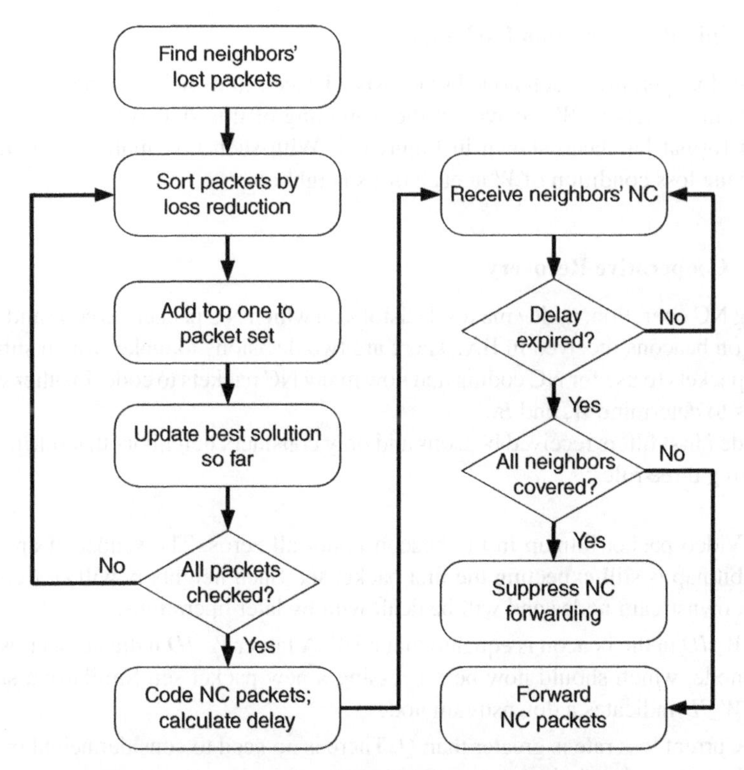

Figure 6.5 Flowchart of NC-based cooperative recovery.

$W_i = \Phi, B_i = 0$ and finds all its packets that are lost by any neighbor in N_i', denoted W_i', by

$$W_i' \triangleq W_i(t_0) - \bigcup_{j \in N_i'} W_j(t_0). \tag{6.24}$$

The following steps are then performed by i:

Step 2.1. *Sorts* W_i'*:* For each video packet $w \in W_i'$, node i tries to add w into W_i and calculates the maximum achievable utility by varying B_i. Note that for any W_i, there is a logical upper bound of B_i, denoted $\overline{B_i}$, as

$$\overline{B_i} = \frac{|W_i|}{1 - \max_{\substack{j \in N_i' \\ W_i - W_j(t_0) \neq \Phi}} l_{ij}}, \tag{6.25}$$

which is the number of NC packet needed for any neighbor wanting some packets in W_i to decode. Therefore for each $w \in W_i'$, different values of $B_i \in [1, \min(\overline{B_i}, \hat{B}_i)]$ are tried for $W_i + w$ to achieve a maximal utility. The packets in W_i' is then sorted by maximum achievable utility if added to W_i.

Step 2.2. *Picks top w and update record:* The top packet w in W_i' is added to W_i and removed from W_i'. If a new high utility is observed, or same as recorded highest but with a smaller B_i, the corresponding W_i and B_i are recorded as the potential solution.

Step 2.3. *Checks termination condition:* The iteration terminates if $W_i' = \Phi$. Otherwise node i begins the next iteration of Steps 2.1–2.3, with one packet less in W_i'.

Step 2.4. *Schedules NC packet broadcasting:* When the aforementioned iteration terminates, node i has the value of the maximal utility, as well as the corresponding W_i and B_i. Node i next schedules its NC packet broadcasting. We prefer nodes with high utility and small B_i to share their NC packets. We define *unit gain* as the loss rate reduction achieved per NC packet, that is, $U(W_i, B_i)/B_i$. So i will postpone its NC broadcasting with a random delay t inversely proportional to the unit gain, for example,

$$t = rand\left(T_{max}\frac{U(W_i, B_i)}{B_i}\right), \tag{6.26}$$

where T_{max} is the system parameter of maximum delay. Function *rand*(x) generates a random number with mean x and a small variance.[4] During such delay, if node i receives NC packets sent by its neighbors who collectively cover N_i', i suppresses its own NC broadcasting.

The pseudo-code of the aforementioned *NC Recovery* algorithm is shown in Algorithm 6.1.

6.5.3 Updated Information Exchange

After NC exchange and decoding at each node, a second beacon is shared by each node. This beacon reflects the updated video packet availability and any buffered NC packet at the sender. A beacon sender can conclude if it has only one upstream node from previous IIX beacon exchange, by counting the number of beacons received in IIX that has a larger *W_ID*, and sets the bit of field *R* in this updated beacon to be 1 if there is only one beacon with larger *W_ID*. The ID of any downstream node identified in IIX is also appended in the beacon.

With such information, node i can then determine whether it should broadcast its video packets in *Forward* operation.

6.5.4 Video Packet Forwarding

Node i then makes forwarding decision regarding each video packet $w \in W_i(t_2)$. i filters received beacons in UIX to consider only the neighbors fulfilling the following two rules:

[4]Other delay functions may also be used, as long as it is inversely proportional to the unit gain.

Algorithm 6.1

```
NC Recovery {
  Find N'_i; W_i ← φ; B_i ← 0;
  if N'_i ≠ φ { \\ Case 2: high-loss neighbor exists
      Find W'_i by Equation (6.24);
      max_util ← 0; W' ← Φ;
      while W'_i ≠ Φ {
          \\ Step 2.1: sort W'_i
          for each w∈W'_i {
              W_temp ← W^f + w;
              compute B̄_temp by Equation (6.25);
              for each B_temp ∈ [1, min(B̄_temp, B̂_i)] {
                  compute U(W_temp, B_temp) by Equation (6.23);
              }
              record max U(W_temp, B_temp);
          }
          sort W'_i by maximal utility achieved by each w∈W'_i;
          \\ Step 2.2: Add the top w
          w ← head(W'_i);
          record B_temp for w;
          W' ← W' + w;
          if (U(W',B_temp) > max_util) OR (U(W',B_temp) == max_util AND B_temp < B_i)
{
              max_util ← U(W^f,B_temp); W_i ← W'; B_i ← B_temp; }
      } \\ Step 2.3: Loop terminates when all packets evaluated
      \\ Step 2.4: schedule NC broadcasting
      Schedule random delay t by Equation (6.26);
      while t not expired {
          N_temp ← N'_i; W_temp ← W_i;
          if received an NC packet from neighbor j {
             N_temp ← N_temp − N'_j; W_temp ← W_temp − W_j;
             if N_temp = Φ AND W_temp = Φ {Suppress NC packet broadcasting;}
          }
      }
      Code NC packets using W_i and B_i; Broadcast NC packets;
  } \\ Case 1: Loss rates of all neighbors are below Q
}
```

1. *W_ID* in the beacon is smaller than that of W. Only downstream neighbors are considered, as those with equal or greater *W_ID* should be expecting newer video packet sets to arrive.
2. Current loss rate is greater than Q. Neighbors with loss rate no greater than Q have already met our target.

We again call the resulting eligible neighbor set N_i'. Among these neighbors, node i further analyzes neighbors according to the following cases and performs corresponding actions as shown in Figure 6.6:

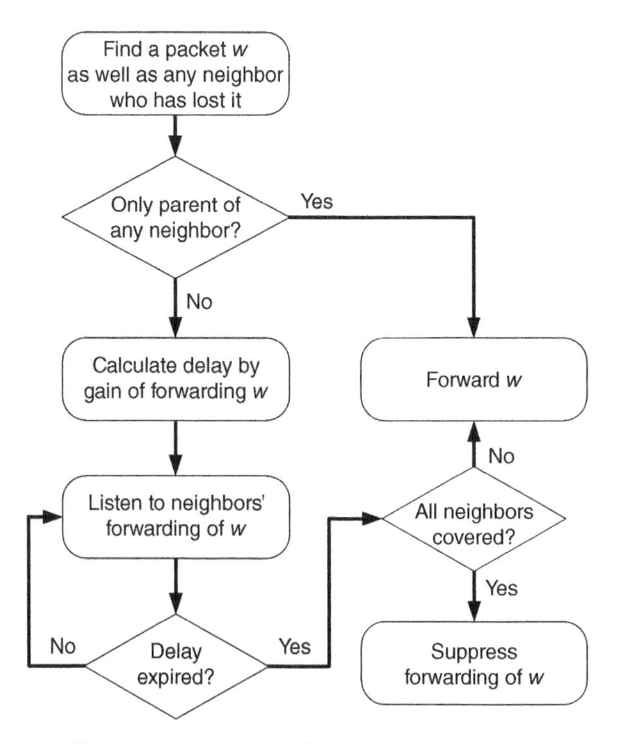

Figure 6.6 Flowchart of video packet forwarding.

Case 1. $N_i' = \Phi$: This means there is no need for node i to broadcast w. So i sets $I_i = 0$ and terminates its *Forward* operation.

Case 2. $N_i' \neq \Phi$: In this case, node i needs to consider broadcasting w to help neighbors in N_i'.

Step 2.1. *Helps neighbors that solely rely on i:* Node i checks if received beacon from j has R field set to 1, meaning i is probably the only upstream node of j. Denote the set of such neighbors as N_i''. If $N_i'' \neq \Phi$, further if $w \in (W_i(t_2) - \cup_{j \in N_i''} W_j(t_2))$, then i should broadcast w irrespective of neighboring nodes' decisions. So i sets $I_i^w = 1$ and a short random delay $t = rand(\epsilon)$ in this case and skips the next step.

Step 2.2. *Helps the rest to approach Q:* If $N_i'' = \Phi$, node i then checks if any neighbor in N_i' has lost w. Node i calculates the *gain* of w, that is, the expected number of innovative packets that can be brought to its neighbors.[5] Such calculation also considers NC packets buffered by neighbors, since sending w to neighbor j may enable the decoding of some of j's NC packets. Denote such gain as $g(I_i^w)$. If $g(I_i^w) > 0$, then i considers a possible broadcast, by setting $I_i^w = 1$ and a random delay t inversely proportional to $g(I_i^w)$, for example,

[5] We call a packet w "innovative" to node j if j has not received w before.

Algorithm 6.2

```
Video Packet Forwarding for w {
  Find N'_i;
  if N'_i ≠ φ { \\  Case 2: has downstream node to help
      Find N''_i;
      if N''_i ≠ Φ AND w ∈ (W_i(t_2) − ∪_{j∈N''_i} W_j(t_2)) {
      \\  Step 2.1: has sole-parent nodes, must forward w
          I_i^w ← 1;  t ← rand(ε);
      } else if w ∈ (W_i(t_2) − ∪_{j∈N'_i} W_j(t_2)) {
          \\  Step 2.2: help others
          Calculate g(I_i^w);
         if g(I_i^w) > 0 {
              Calculate t by Equation  (6.27);
         }
      while t not expired { \\  Step 2.3: schedule forwarding for w
         N_temp ← N'_i;
         if received w from neighbor j {
             N_temp ← N_temp − N'_j;
             if N_temp = Φ {Terminate algorithm;}
         }
      }
      Broadcast w;
  } \\  Case 1
}
```

$$t = rand\left(\frac{T_{max}}{g(I_i^w)}\right). \tag{6.27}$$

Step 2.3. *Schedules broadcasting of w:* After all the neighbors in N'_i have been checked, if $I_i^w = 1$, then node i waits delay t before broadcasting its video packet w. During such delay, if a neighbor j broadcasts packet w, i excludes all the downstream nodes that j covers. Node i suppresses its broadcasting of packet w if all its downstream nodes wanting w have been covered.

The pseudo-code of the previous *Video Packet Forwarding* algorithm is shown in Algorithm 6.2.

6.6 ILLUSTRATIVE SIMULATION RESULTS

In this section, we describe the simulation setup used to evaluate VBCR. Peers are randomly placed in a 500 m × 500 m area with transmission range of 140 m.[6] Each link has an i.i.d. random packet loss with mean loss rate 5%.[7] We consider both node

[6]Although we consider random placement of the nodes, any degree of offline planning and preemptive bandwidth calculation will only improve the performance of VBCR.

[7]Link loss rate is selected to be reasonably practical and for illustrative purposes. VBCR can easily adapt to other reasonable loss rates.

TABLE 6.2 Baseline Parameters of the Simulation

Parameter	Value
Area (m)	500×500
Tx range (m)	140
Network size	35
Link loss	5% i.i.d.
Q	2%
$\hat{B}$	3
$\|W\|$	10

position and link condition to be stable within a slot. Unless stated otherwise, we use the value of parameters summarized in Table 6.2 as our baseline. We've tested with different sets of parameters and obtained qualitatively same results. We consider single source node in our simulation, while VBCR can easily adapt to multisource cases by connecting all source nodes to a virtual super-source with links of infinite bandwidth and zero loss.

We compare VBCR with the following two algorithms:

- *Global Tree*, in which a maximum leaf broadcast tree is built by a centralized algorithm. The source node codes and sends RLNC packets according to the maximum source-to-end loss rate. Each non-leaf node relays each received RLNC packet once.
- *Unstructured Mesh*, a scheme similar to that in Ref. [20] but is modified for single-stream video instead of multiple descriptions coded (MDC) video. Each node makes one-time source packet forwarding decision for each newly received video packet. A node first sums the number of innovative source packets its broadcast could bring to each neighbor with loss rate greater than Q and waits for a delay inversely proportional to this sum. A node's transmission is suppressed if a neighbor's broadcast is overheard during waiting. Comparing to VBCR, the major difference of Unstructured Mesh is its lack of cooperative recovery before forwarding. Sole-upstream-parent case is also not considered, leading to edge nodes likely to be "starved."

We consider the following performance metrics in our study:

- *Network Traffic*. This is to evaluate function

$$\sum_{w \in W} \sum_{i \in V} I_i^w + \sum_{i \in V} B_i.$$

We measure the number of source packet broadcasts in the network, plus the number of NC packet broadcasts. So this metric evaluates the total number of transmissions in the network.

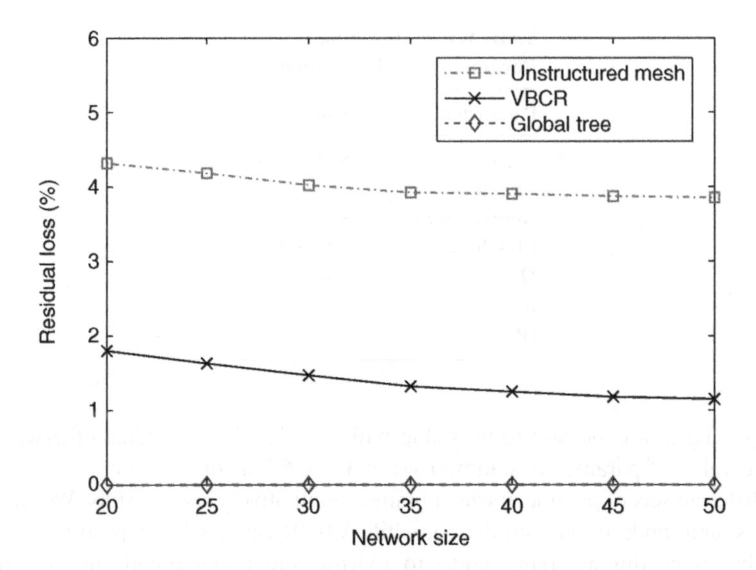

Figure 6.7 Loss rate versus network size.

- *Number of Source Packet Broadcast Parents for Each Node*. We evaluate the number of parents for each node. We show by this metric that the average amount of traffic at each node is low enough for contention avoidance scheduling used in common MAC protocols.

We evaluate residual loss of these algorithms for different network sizes in Figure 6.7. Note that although we set the target loss rate $Q = 2\%$, it is not necessary for the schemes to able to reach this target. RLNC, with the knowledge of link loss rates across the network, produces sufficient redundant packets that every node can successfully decode, leading to full packet delivery. This however comes with the assumption of global knowledge, also at the cost of highly redundant traffic. The residual losses in both Unstructured Mesh and VBCR drop as the network becomes dense, due to the effectiveness of one-hop broadcasting. In Unstructured Mesh, many necessary source packet broadcasts are suppressed, leading to a high residual loss. VBCR achieves lowest residual loss among all the three algorithms because of the effectiveness of the NC cooperative recovery and the consideration of nodes with a single upstream parent.

Figure 6.8 plots the total network traffic generated under different network sizes. As the network size increases within a fixed size area, the network density increases correspondingly. The first thing we notice is the high amount of network traffic of RLNC. The "all-or-nothing" nature of RLNC requires a full delivery at every node to meet the target loss rate. Among the paths from source to each node, a source node needs to produce and send enough packets to cope with the highest loss path, leading to a high volume of redundancy along other paths. Unstructured Mesh achieves the least amount of network traffic, but with high residual loss shown in Figure 6.7.

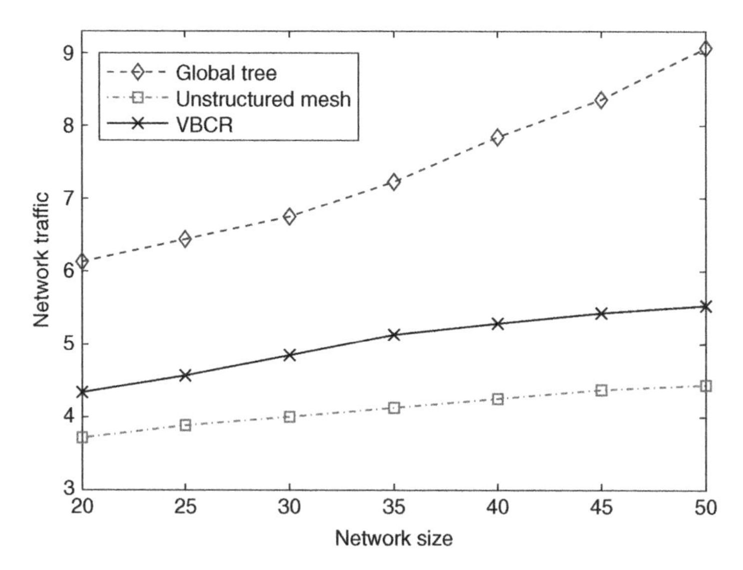

Figure 6.8 Network traffic versus network size.

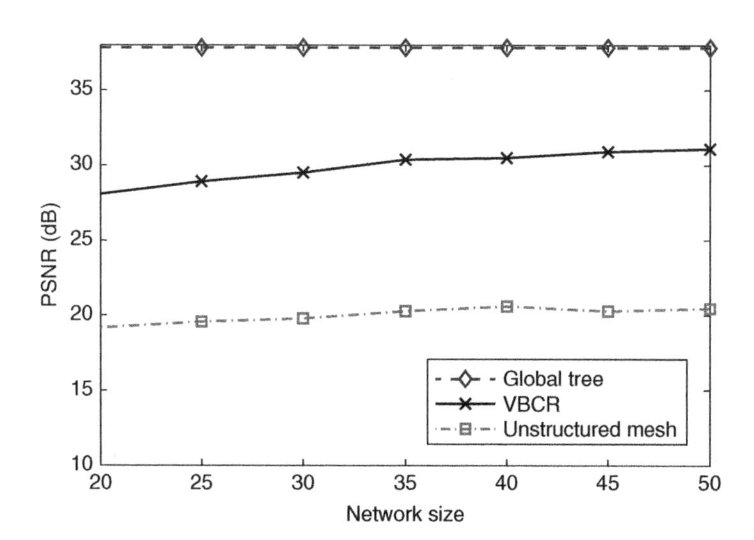

Figure 6.9 Video PSNR versus network size.

VBCR not only uses a fraction more than that of Unstructured Mesh but also leads to much lower residual loss rate.

Figure 6.9 demonstrates the corresponding video quality in terms of PSNR. Unstructured Mesh, with high residual loss shown in Figure 6.7, experiences very low playback quality. VBCR clearly achieves much higher video quality. RLNC is

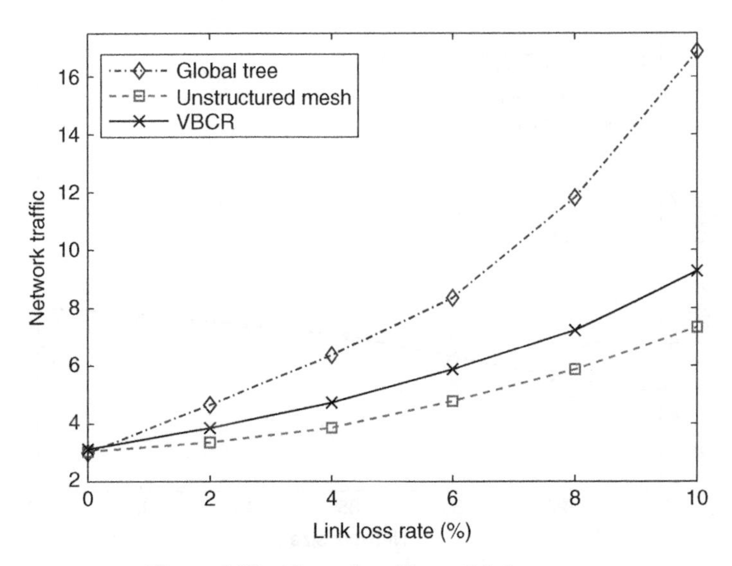

Figure 6.10 Network traffic vs. link loss rate.

able to achieve optimal video quality by 100% delivery ratio, with the expense of high network traffic redundancy shown later.

The total network traffic generated by different algorithms, under different link loss rates, is shown in Figure 6.10. Global Tree generates large traffic that grows very fast as the link loss rate rises. Unstructured Mesh with better sharing leads to the least network traffic, as well as high residual loss and low video quality compared to other two algorithms. The carefully chosen broadcasters as well as the high efficiency of NC recovery in VBCR lead to a network traffic much lower compared to RLNC and only slightly higher than that of Unstructured Mesh while achieving lower-than-target residual loss. This can be further explained by the traffic composition as shown in Figure 6.11. The high efficiency of NC recovery makes a small amount of NC recovery traffic sufficient to recover most losses after video packet forwarding.

Next we evaluate the number of parents for each node. From Figure 6.12, we can see that Global Tree algorithm, due to its optimized one-parent broadcast tree, leads to the least number of relay parents (note that one-hop broadcast is enabled in all cases). However the traffic going through each non-leaf node is very high. Unstructured Mesh results in more relay nodes as each node make individual decisions on whether to relay or not without coordination. VBCR leads to most number of parents for each node, due to the consideration of sole-parent nodes, which are not considered in Unstructured Mesh. Figure 6.13 shows the distribution of the number of parents in VBCR. We can see that the number of upstream nodes is well in control, with most nodes received from 2 to 3 upstream nodes. Today's mobile video streaming rates are normally magnitudes below wireless bandwidth. Thus channel holding time for each packet is quite short, leading to a low interference/collision probability.

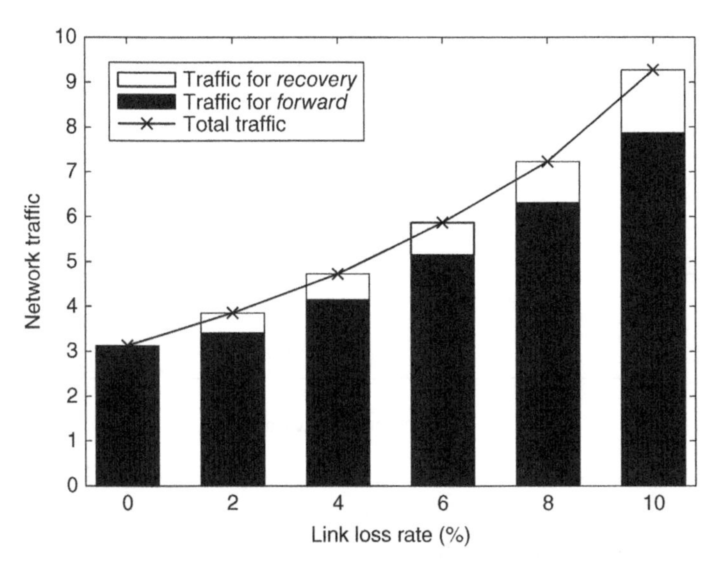

Figure 6.11 VBCR network traffic composition.

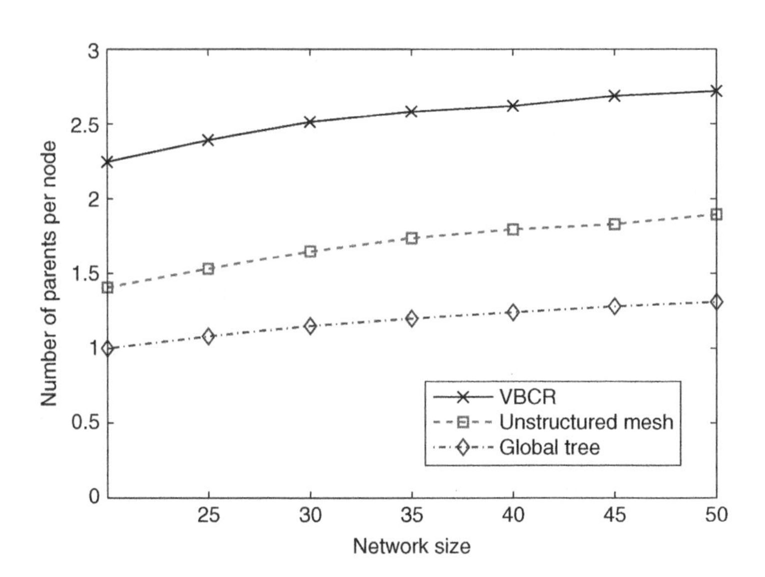

Figure 6.12 Number of parents for each node.

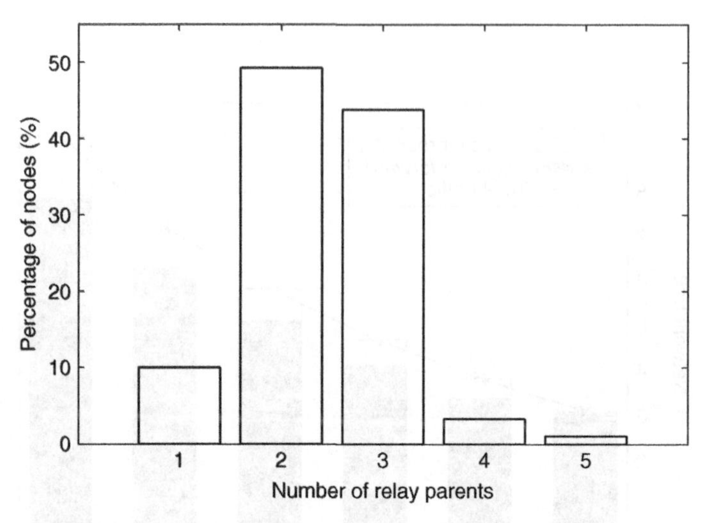

Figure 6.13 Histogram of the number of parents in VBCR.

6.7 CONCLUDING REMARKS

In this chapter we study live video streaming to clients in a wireless fog. In order to save device energy and bandwidth, we need to reduce network traffic during a video streaming session while trying to meet a target residual loss rate to maintain playback quality. The straightforward "store-and-forward" based on video packets approach often fails to reach the target loss rate due to losses and loss propagation. The approach based on source-initiated RLNC, due to its "all-or-nothing" characteristic, is often too aggressive in network with diverse loss conditions. We therefore propose the "store–recover–forward" approach, allowing cooperative recovery that utilizes peer-initiated NC over a subset of received source packets. In our proposed distributed algorithm VBCR, neighboring nodes first performs cooperative NC recovery to recover some of their lost packets, before making source packet forwarding decisions distributedly and independently.

Under such system, nodes′ decisions on (i) which subset of received source packets to code and (ii) which of its video packets to forward directly affect the loss recovery efficiency and hence the total traffic generated. In VBCR, nodes distributedly and independently make the previous decisions based on updated neighbor information. Our simulation results show that VBCR with NC-based cooperative recovery at intermediate nodes significantly outperforms other schemes in that VBCR uses reasonably low network traffic to achieve a low residual loss rate.

REFERENCES

1. Zhi Liu, Gene Cheung, Vladan Velisavljević, Erhan Ekmekcioglu, and Yusheng Ji. Joint source/channel coding for WWAN multiview video multicast with cooperative peer-to-peer repair. In *IEEE International Packet Video Workshop (PV)*, pages 110–117, Hong Kong, December 2010.

2. Zhi Liu, Gene Cheung, and Yusheng Ji. Distributed source coding for WWAN multiview video multicast with cooperative peer-to-peer repair. In *IEEE International Conference on Communications (ICC)*, pages 1–6, Kyoto, Japan, June 2011.
3. Yichao Shen, Wenwen Zhou, Peizhi Wu, Laura Toni, Pamela C. Cosman, and Laurence B. Milstein. Device-to-device assisted video transmission. In *20th IEEE International Packet Video Workshop (PV)*, pages 1–8, San Jose, CA, USA, 2013. URL: http://ieeexplore.ieee.org/stamp/stamp.jsp?tp=&arnumber=6691441&isnumber=6691435 (accessed on October 27, 2016).
4. Kenta Mori, Sho Hatakeyama, and Hiroshi Shigeno. Dcla: Distributed chunk loss avoidance method for cooperative mobile live streaming. In *29th IEEE International Conference on Advanced Information Networking and Applications (AINA)*, pages 837–843, Gwangju, Korea, March 25–27, 2015.
5. Mengyao Sun, Yumei Wang, Yu Hao, and Yu Liu. Distributed cooperative video coding for wireless video broadcast system. In *IEEE International Conference on Multimedia and Expo (ICME)*, pages 1–6, Turin, Italy, June 29–July 3, 2015.
6. S.V.M.G. Bavithiraja and Rathinavel Radhakrishnan. A new reliable broadcasting in mobile ad hoc networks. *Computer Science & Network Security (IJCSNS)*, 9(4):340–348, 2009.
7. Hermann S. Lichte, Hannes Frey, and Holger Karl. Fading-resistant low-latency broadcasts in wireless multihop networks: The probabilistic cooperation diversity approach. In *Proceedings of the Eleventh ACM International Symposium on Mobile Ad Hoc Networking and Computing (MobiHoc)*, pages 101–110, Chicago, IL, USA, September 2010.
8. Xianlong Jiao, Wei Lou, Junchao Ma, Jiannong Cao, Xiaodong Wang, and Xingming Zhou. Minimum latency broadcast scheduling in duty-cycled multihop wireless networks. *IEEE Transactions on Parallel and Distributed Systems*, 23(1):110–117, 2012.
9. Xinyu Zhang and Kang G. Shin. Delay-optimal broadcast for multi-hop wireless networks using self-interference cancellation. *IEEE Transactions on Mobile Computing*, 12(1): 7–20, 2013.
10. Dimitrios Koutsonikolas, Saumitra Das, Y. Charlie Hu, and Ivan Stojmenovic. Hierarchical geographic multicast routing for wireless sensor networks. *Wireless Networks*, 16:449–466, 2010.
11. Yean-Fu Wen and Wanjiun Liao. Minimum power multicast algorithms for wireless networks with a lagrangian relaxation approach. *Wireless Networks*, 17:1401–1421, 2011.
12. Tan Le and Yong Liu. Opportunistic overlay multicast in wireless networks. In *IEEE Global Telecommunications Conference (GLOBECOM)*, pages 1–5, Miami, FL, USA December 2010.
13. Qiang Xue, Anna Pantelidou, and Matti Latva-aho. Energy-efficient scheduling and power control for multicast data. In *IEEE Wireless Communications and Networking Conference (WCNC)*, pages 144–149, Cancun, Mexico March 2011.
14. Ashish James, A.S. Madhukumar, and Fumiyuki Adachi. Spectrally efficient error free relay forwarding in cooperative multihop networks. In *IEEE International Symposium on Personal Indoor and Mobile Radio Communications (PIMRC)*, pages 2255–2259, September 2013.
15. Gene Cheung, Danjue Li, and Chen-Nee Chuah. On the complexity of cooperative peer-to-peer repair for wireless broadcast. *IEEE Communications Letters*, 10(11):742–744, November 2006.

16. Xin Liu, Gene Cheung, and Chen-Nee Chuah. Deterministic structured network coding for WWAN video broadcast with cooperative peer-to-peer repair. In *IEEE International Conference on Image Processing (ICIP)*, pages 4473–4476, Hong Kong, China, September 2010.
17. Michael Luby. Lt codes. In *Proceedings of the Annual IEEE 43rd Symposium on Foundations of Computer Science (FOCS)*, page 271, Vancouver, British Columbia, Canada November 16–19, 2002.
18. Lorenzo Keller, Anh Le, Blerim Cici, Hulya Seferoglu, Christina Fragouli, and Athina Markopoulou. Microcast: Cooperative video streaming on smartphones. In *Proceedings of the 10th International Conference on Mobile Systems, Applications, and Services*, MobiSys '12, pages 57–70, New York, NY, USA, 2012. ACM.
19. Hyojun Lim and Chongkwon Kim. Flooding in wireless ad hoc networks. *Computer Communications*, 24(3–4):353–363, 2001.
20. Man-Fung Leung and Shueng-Han Gary Chan. Broadcast-based peer-to-peer collaborative video streaming among mobiles. *IEEE Transactions on Broadcasting Special Issue on Mobile Multimedia Broadcasting*, 53(1):350–361, 2007.

7 Elastic Mobile Device Clouds: Leveraging Mobile Devices to Provide Cloud Computing Services at the Edge

KARIM HABAK,[1] CONG SHI,[1,2] ELLEN W. ZEGURA,[1]
KHALED A. HARRAS,[3] and MOSTAFA AMMAR[1]

[1] *School of Computer Science, College of Computing, Georgia Institute of Technology, Atlanta, GA, USA*
[2] *Current affiliation: Square, Inc., San Francisco, CA, USA*
[3] *Computer Science Department, School of Computer Science, Carnegie Mellon University, Doha, Qatar*

7.1 INTRODUCTION

Throughout the last decade, mobile devices have become an indispensable part of every aspect of human life. They are increasingly relied on for various services that go beyond simple connectivity and spread across the usage spectrum from gaming and entertainment applications (e.g., video games) [1] to critical health care applications (e.g., health monitoring) [2]. Although these devices are becoming increasingly capable, a large group of these applications still requires resources that exceed the capabilities of a single mobile device. Therefore, Balan *et al.* made a case for mobile devices to cyber forage by finding surrogate (i.e., helper) servers in the environment [3]. Since then, the research community has explored various forms of interaction between mobile devices and fixed, higher capacity infrastructure, including the cloud. The motivation for this exploration has been and remains as articulated by Satyanarayanan [4], namely, that mobile devices are resource constrained in comparison with servers and that users desire high performance applications regardless of the device used to experience the application. By offloading some computation to more powerful servers, mobile devices can offer a user experience beyond what local

Fog for 5G and IoT, First Edition. Edited by Mung Chiang, Bharath Balasubramanian, and Flavio Bonomi.

capabilities can support. Further, offloading may allow mobile devices to save power and extend time between charges.

In addition to questions of performance speedup, energy savings, and cost, the key questions for an offloading system design are where is the higher performance capacity, who provides it and how does it fit into a larger computing ecosystem? In traditional cloud computing, the higher performance capacity is located in data centers reached via the Internet and provided by companies that charge for transient server use. Traditional cloud computing can offer essentially unlimited compute capacity but at the price of latency and bandwidth limitations between the mobile device and the servers in large data centers [5, 6]. In response to these limitations, the cloudlet system moves computation closer to mobile devices, creating a two-tier architecture where a mobile device can offload to a nearby, less capable server, at low latency and high bandwidth rather than (or as a complement to) offloading to the cloud [7]. In the cloudlet vision, these nearby servers would be located in public and commercial spaces where people congregate, such as coffee shops and airport waiting areas [7].

We push further on the vision of edge-based clouds by making two observations. First, while the gap remains between truly mobile devices (handhelds, wearable) and high capacity servers [8], mobile devices have grown increasingly powerful, especially when laptops are included. Second, from an architectural perspective, it is possible to refactor the cloudlet into a *controller*, responsible for receiving tasks, scheduling their computation, and returning results, and a *compute cluster*, responsible for performing the computation. When control and computation are decoupled and mobile devices are recognized as capable to serve as surrogates in the right circumstances, we arrive at the notion of elastic mobile-device clouds. We define an *elastic mobile-device cloud* as a collection of mobile devices configured to provide a computation service. In this chapter we scope the design space for mobile-device clouds and highlight instances of systems that fit in the space.

A key dimension of the design space turns out to be the stability (or *elasticity*) of the mobile devices used to construct the cloud, as reflected in device churn and computation cycles available for cloud computation. In settings of lower stability, the compute cloud is more elastic, and constructing a computation service is extremely challenging; in settings of higher stability, the compute cloud is less elastic, and constructing a computation service is more straightforward. We illustrate this spectrum in Figure 7.1. At one end are extremely stable and deliberately configured clusters such as proposed in the Mont-Blanc project (www.montblanc-project.eu) where, motivated by energy considerations, a large number of mobile CPUs are configured in a single chassis. At the other end are highly mobile and unpredictable devices that are used opportunistically as they are encountered over time [9–11]. Our recent work addresses the middle of the spectrum (coffee shops, public transit, theater/classrooms) where the cluster has some stability based on semantics of the setting [12].

The remainder of this chapter is organized as follows. In the next section we further develop the design space for elastic mobile-device clouds. We sketch a general system architecture that captures functionality common across the design space. We give examples of specific mobile-device cloud systems and review how they achieve system functionality. We then turn in Sections 7.3 and 7.4 to a deeper dive into two

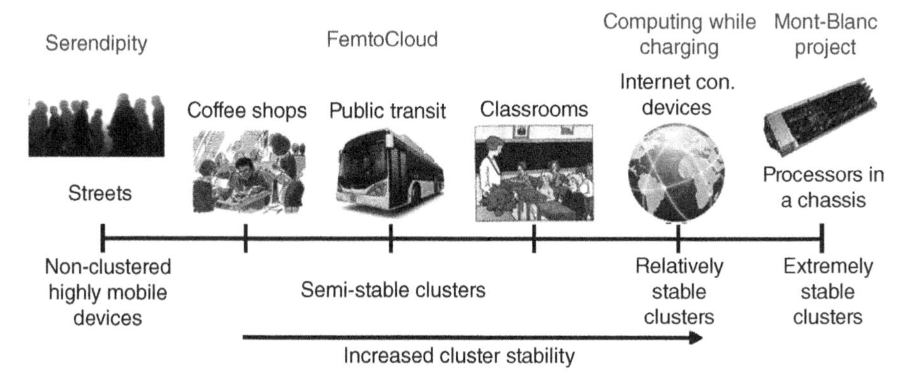

Figure 7.1 Mobile cluster stability spectrum.

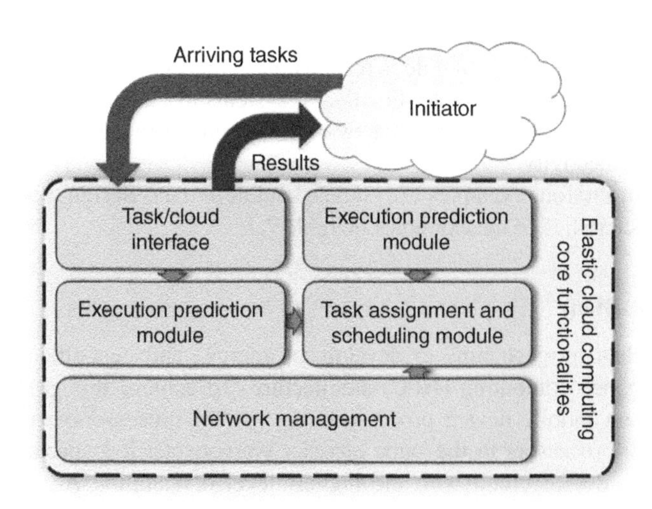

Figure 7.2 Elastic mobile-device clouds architecture.

example systems, FemtoCloud [12] and Serendipity [9], highlighting the evaluation of their performance. We conclude with open challenges in the area of realizing elastic mobile-device clouds.

7.2 DESIGN SPACE WITH EXAMPLES

To scope the design space of elastic mobile-device clouds, we first consider functionality that all such systems must have, regardless of where they reside in the stability space. All systems receive tasks from task initiators and return results to initiators, depicted in Figure 7.2 as the task/cloud interface. To do an effective job assigning

TABLE 7.1 A Summary of Different System Assumptions

System	Stability	Task Origin	Tasks	Network
Mont-Blanc	Extremely stable	Outside	Generic	Ethernet/Myrinet
CWG	Relatively stable	Inside/outside	File processing	Internet
FemtoCloud	Semi-stable	Outside	Generic	Wi-Fi
Serendipity	Unstable	Inside	PNP	Intermittent

tasks to devices for processing, all systems benefit from some knowledge of the task processing requirements. This information may be provided by the task initiator, but more generally, systems will have an execution prediction module responsible for estimating the resources required. All systems keep track of the mobile devices available in the cloud, via a device management function. This function may also conduct ongoing estimation of the current capability of the devices to take on tasks. Given tasks, resource requirements, and a set of mobile devices with capabilities, the task assignment and scheduling module is responsible for assigning tasks to devices and scheduling those tasks to execute. Finally, all systems use a network to move tasks and results around and hence have a network management function to keep track of the current connectivity.

We now turn to four examples of systems that fall into this design space. We summarize characteristics of the examples in Table 7.1.

7.2.1 Mont-Blanc

The Mont-Blanc project aims to develop an energy- and cost-efficient exascale high performance computing (HPC) architecture. To achieve this goal, it utilizes energy-efficient mobile device processors (e.g., ARM processors) and assembles a set of these processors in the same chassis. We consider it a special case of an elastic mobile-device cloud with the highest level of stability. Arguably it isn't particularly elastic or mobile, but it serves as a useful endpoint for our stability spectrum discussion.

The main goals in the Mont-Blanc project are (i) maximizing the performance of every single processor and (ii) efficiently clustering multiple processors. To maximize the performance gains from each processor, the authors address a set of hardware challenges such as the lack of cooling infrastructure as well as software challenges such as using soft-float calling conventions even if hardware floating point operations are supported. The authors also quantify the performance of these processors in both single- and multi-core scenarios. To maximize the performance gains from the whole cluster, the work focuses on providing fast and reliable communication architecture between different cluster nodes and across clusters using Myrinet Express message passing stack on top of Ethernet links.

Unlike the other system examples in this section, Mont-Blanc provides parallel processing abstractions that are used by developers to write parallel-processing friendly code and provide the correct annotations to the compiler. Therefore, the need for task/cloud interface is eliminated, unless implemented as a service on top of this architecture. In addition, fine granularity task scheduling is possible, which avoids the need for estimating task characteristics and requirements. Furthermore, the usage of identical processors and having full control on what they execute enables Mont-Blanc to efficiently manage the existing processors. Full access to the network that connects different cluster nodes, as well as using Myrinet Express over Ethernet links, eases the network management challenges.

7.2.2 Computing while Charging

Computing while charging (CWC) is a distributed computing infrastructure that uses smartphones as the main computing nodes [13]. A CWC system consists of (i) a data center that has tasks to execute and (ii) mobile devices with idle capacities that are charging their batteries and connected to the data center via the Internet, as may be owned by employees of the data center who have gone home for the evening. The CWC architecture is designed to enable a data center to utilize idle capacity in mobile devices to enhance its performance and/or reduce data center energy consumption.

The CWC cluster environment is stable. The mobile devices are connected to power sources, which means they do not move and they have good energy availability. CWC also assumes that these devices are connected to the data center via cellular or Wi-Fi networks that contribute to the stability of the environment. In this architecture the data center controls, these devices and offloads tasks to them. The system estimates the task execution times at various devices and estimates the available bandwidth and latency. The CWC task scheduling mechanism takes these estimates into account to maximize the computational throughput of system. With respect to the generic architecture, CWC realizes the following:

Task/Cloud Interface. CWC relies on the controlling data center to implement the appropriate interface between the computing nodes and the task originators (which can be an entity outside the data center or a process inside the data center).

Execution Prediction. CWC assumes file processing tasks where the execution load introduced by a task is a function of the size of its assigned input file. Therefore, CWC utilizes knowledge of the input file size to determine the computational load introduced by the task.

Device Management. CWC requires knowing the processing capacity of the device and its available storage in order to take them into account while assigning tasks. To avoid profiling all the devices to determine their computational capacity, CWC profiles only the device with the minimum CPU clock speed and use the relative difference in the clock speeds to derive the capacity of faster devices. CWC relies on each device to acquire its available storage information and report it to the controlling data center.

Network Management. To determine the available bandwidth and communication latency between the data center and each mobile devices, CWC uses active probing to estimate available bandwidth and latency. Note that this measurement phase can happen once the device becomes connected to the data center and regardless of the status of other devices because there is generally no network contention from different devices to the data center.

Task Assignment and Scheduling Module. Given certain load, CWC maximizes the cloud's computational throughput by minimizing the time needed by the last mobile device to complete its tasks. Without network bandwidth contention, this problem can be efficiently solved by solving the associated bin packing problem. In this problem, the objective is to pack items in at most P bins with capacity C such that the maximum bin height is minimized, where P is the number of devices, C is the device capacity, and each task has height representing usage of device capacity. Unlike Mont-Blanc, which uses dedicated devices, CWC must deal with the possibility that users need their devices while CWC would otherwise use them. CWC handles this by (i) stopping the CWC task execution on the device once its user uses it and (ii) controlling the CPU utilization in order not to increase the residual battery charging time.

7.2.3 FemtoCloud

FemtoCloud [12] is a system designed and implemented to leverage mobile devices to provide mobile computing services at the edge. It is designed to cluster colocated mobile devices in environments where mobile device presence times can be estimated. These environments include coffee shops, public transit, theater, and classrooms where people carry mobile devices. We situate these environments in the middle of the stability spectrum, not nearly as stable as Mont-Blanc and CWC systems but still with some predictability in device presence time. FemtoCloud's architecture consists of a control device and a set of executing mobile devices. In this architecture, the control device is responsible for managing the Femtocloud, distributing tasks across different devices, gathering their results, and interfacing with task originators. On the other hand, the rest of the mobile devices are responsible for executing tasks and assisting the control device in acquiring the needed environmental information. The FemtoCloud architecture predicts task characteristics, estimates device capacities and presence times, and uses the acquired information to distribute tasks across different executing devices, as follows:

Task/Cloud Interface. In order to increase its usability, FemtoCloud relies on the control device, which is relatively stable compared to the rest of devices in a FemtoCloud to provide a stable and discoverable interface between the femtocloud and its potential users. This control device is deployed in the environment and is responsible for hiding the femtocloud's internal dynamics from their users (task initiators). We highlight that, in some scenarios, this control device could be selected from the femtocloud devices based on their stability, battery level, and available Internet connectivity.

Execution Prediction. FemtoCloud relies on a generic task model where a task is characterized by (i) its input size, (ii) its output size, and (iii) its introduced computational load. Since the task input size is known at the offloading time, FemtoCloud only needs to predict/estimate its output size and its introduced computational load. To acquire this information, FemtoCloud relies on the task originator to send the task with an estimate for its computational load coupled with the its anticipated output size. These estimates can be determined using Mantis system [14], which utilizes application's pre-generated models coupled with runtime execution information (e.g., variable values) to predict the application performance.

Device Management. Managing the devices in Femtocloud and estimating their characteristics is one of the most important functionalities implemented. It is divided into three subtasks implemented by the control device and the computing mobile devices.

Discovery and Registration—To efficiently utilize the available compute resources, it is critical to quickly discover their existence once they arrive. Therefore, FemtoCloud implements discovery and registration mechanisms that can be tuned based on the environment where FemtoCloud is deployed and used. In the discovery protocol, the control device broadcasts beacons to announce its existence and enable neighboring devices to discover it. The beaconing time and frequency is a function of the environment contextual information. When a mobile device receives a beacon, it starts the registration process if it is ready to join the Femtocloud and willing to share its resources. In the registration process, the mobile device sends a registration packet to this control device to join the femtocloud. This registration packet includes some initial estimates of the device characteristics and a user profile shared in a privacy preserving manner. Furthermore, mobile devices send periodic heartbeats to maintain their status with the control device and provide it with real-time update information.

Estimating the Device Capacity—FemtoCloud relies on mobile devices to estimate their shared processing capacity and share these estimates with the control device using the registration and heartbeat messages. Typically, mobile devices use synthetic benchmarks and preassigned tasks to measure their execution speeds. These measurements are coupled with users' willingness to share their resources and power saving status to get a more accurate capacity estimate.

Estimating the Device Presence Time—Accurately predicting the mobile device presence time and intelligently utilizing this information leads to making better resource management decisions. It also can significantly increase the utilization of the existing resources. For instance, device churns can be proactively handled before they actually occur. Devices with higher presence time can be better utilized by assigning more tasks to them. Furthermore, tasks with high computational requirement can be assigned only to devices that can finish them before their departure times.

To predict the device presence times, FemtoCloud relies on a control device that is responsible for gathering environment-specific data to build a generic user profile based on the collective behaviors of the users. This profile is used to estimate the presence times for new users and update these estimates over time. FemtoCloud also combines these estimates with user-specific information (e.g., profile, schedule, social information), acquired in a user privacy-preserving manner, to increase the accuracy of such predictions.

Network Management. Since FemtoCloud utilizes the shared wireless spectrum to send tasks to their executing devices and gather their results, it is critical to estimate the capacity of the spectrum and regulate its access from different devices. FemtoCloud uses the wireless signal strength to derive an estimate for the communication bandwidth between a mobile device and the control device. This estimate is then updated using active probing.

To control spectrum access, FemtoCloud relies on the controller and its scheduling mechanism to determine when to offload a task to a certain mobile device or gather results from others. The FemtoCloud scheduler ensures that only one device is utilizing the spectrum at any point in time to minimize the collisions and maximize the efficiency.

Task Assignment and Scheduling Module. The main goal of the FemtoCloud scheduler and task assignment mechanism is to maximize the computational goodput ("useful computations") inside the femtocloud given task loads, device characteristics, and network parameters. To increase the efficiency, the FemtoCloud scheduler assigns tasks to devices that can complete them and returns their results to the control device before their departure.

FemtoCloud relies on two main heuristics to assign tasks to their executing nodes and gather their results. The first heuristic is to prioritize tasks with higher computation requirement per unit data transfer (inputs and outputs). Tasks are assigned to the device that will first complete them and return their results. The second heuristic is to switch from task-assigning mode to results-gathering mode if (i) the system is unable to assign more tasks to devices while maintaining the feasibility of gathering their results on or before their device's departure time, or (ii) a device with available results is predicted to leave soon. Section 7.3 illustrates performance results for the FemtoCloud system under a variety of simulated and experimental conditions.

7.2.4 Serendipity

Serendipity [9] is a system that enables mobile devices to construct a collaborative community where mobile devices share their computational capacity with one another. It is designed to deal with scenarios where device mobility is relatively high. This mobility leads to frequent connections and disconnections between mobile devices. Serendipity deals with the available contact periods as opportunities to either offload tasks from one device to another or gather the results of completed tasks. The offloading objective in this scenario is enhancing the computational performance of task-initiating mobile devices. In this section, we discuss the main challenges that Serendipity addresses:

Task/Cloud Interface. Due to the dynamic nature of the environment, Serendipity does not offer a cloud service to outsiders. Instead it enables mobile devices to directly communicate with one another to exchange tasks and results. In this scenario, a mobile device can act as a task initiator and/or an executing node.

Execution Prediction. Serendipity assumes a job model where a job consists of a pre-process program, *N* parallel tasks, and a post-process program. Serendipity's task initiator is responsible for (i) executing the pre-process and the post-process programs and (ii) executing or offloading each parallel task. Serendipity requires having a complete profile for each job. This profile includes (i) a complete directed acyclic graph (DAG) that describes the job flow and (ii) the expected input, output, and execution load for each task. To determine this information, Serendipity relies on offline training techniques used by MAUI [6] and CloneCloud [15]. In this case, job profiles are generated using offline code analysis and execution under different circumstances.

Device Management. Managing devices in Serendipity's scenarios is a distributed process where each device discovers its peers and maintains needed information to make good offloading decisions. The device management process is divided into three main subtasks.

Discovery—Device discovery is one of the most important management tasks that Serendipity's devices must handle. In some scenarios, Serendipity uses a control channel to let devices predict their future connectivity and avoid using explicit discovery mechanism. In other scenarios, an explicit discovery mechanism is used. For instance, mobile devices can broadcast periodic messages to enable other mobile devices to discover it when it comes in their vicinity.

Estimating the Device Capacity—Serendipity relies on each mobile device to estimate its processing capacity and share it with its peers. Generally, a mobile device uses synthetic benchmarks to measure its execution speed and capacity.

Estimating the Device Energy Profile—Since most of the mobile devices are battery operated, it is important for Serendipity to take the energy consumption into account while making task assignment decisions. Serendipity relies on PowerBooter [16] to generate a model for device energy consumption. This model is then combined with task parameters to predict their power consumption.

Network Management. Serendipity deals with scenarios where connectivity is intermittent. Therefore, predicting connectivity is one of the most important network management tasks. Serendipity relies on a control channel to predict future connections. If this control channel is not available, predicting the future connectivity is possible via maintaining the historical contacts and mining them [17].

Task Assignment and Scheduling Module. Since tasks are generated at one of the mobile devices, the task assignment mechanism is developed to minimize the job execution time and enhance the job performance on this device. Whenever a mobile device has a task to execute (either received it from an initiator or decided to execute it locally), it starts executing the task and keeps executing until it comes in contact with another device that will be able to complete the task and delivers its results to the task originator faster. In this case, it offloads the task to this device that repeats the whole process till the task is finished and delivered to its originator. Section 7.4 illustrates performance results for the Serendipity system under a variety of simulated conditions.

In the remainder of this chapter, we take a deeper dive into the performance achieved by two of the example systems described thus far, FemtoCloud and Serendipity. Understanding the potential and limitations of performance, especially for systems at the middle and low end of the stability spectrum, is critical for determining how and when mobile-device edge clouds provide value.

7.3 FEMTOCLOUD PERFORMANCE EVALUATION

The FemtoCloud system falls in the middle range of the stability spectrum. In this section we present selected results from FemtoCloud's simulation-based performance evaluation. The reader is referred to [12] for additional results.

7.3.1 Experimental Setup

To have a realistic performance evaluation, we start by identifying the available capacity in real mobile devices and the compute requirements of real applications. Table 7.2 summarizes the results of a measurement study that identifies the average capacity of a background thread running in an Android OS. In this study, a matrix multiplication application is executed with different preset computational load (MFLOPs) on a diverse set of devices. Table 7.3 summarizes another measurement study that determines the resource usage of various applications. These applications are (i) a chess game in high difficulty mode, (ii) a video game called Angry Birds Space, and (iii) an application for object recognition in a video feed (video processing). The results of these studies and a synthetic compute intensive application are used in FemtoCloud experiments to evaluate the system's performance. Table 7.4 provides the parameter values used in the simulations, with the defaults underlined. For the chess application, we experiment with different input sizes. Throughout the evaluation, we use a Poisson arrival process to model the arrival of new users as well as the arrival of new tasks.

TABLE 7.2 FemtoCloud Experimental Device's Characteristics

Devices	Computation Capacity (MFLOPS)
Galaxy S5	3.3
Nexus 7 [2012]	7.1
Nexus 7 [2013]	8.5
Nexus 10 [2013]	10.7

TABLE 7.3 FemtoCloud Experiment Tasks Characteristics and Evaluation Parameters

Task Type	Input (MB)	Computation (MFLOPs)	Output (MB)	Arrival Rate (task/s)
Chess	2	10	0.2	1
Video game	0.2	30	2	2
Video processing	3.125	60	1	1
Compute intensive	8	100	0.5	0.5

TABLE 7.4 FemtoCloud Experiment Parameters

Parameter	Values
Chess input size (MB)	[0.5, 2, 16]
Average user arrival rate (user/min)	[2, 8]
Average user presence time (min)	[0.25, 2, 5]
Average device's available bandwidth (Mbps)	20
Average presence error ratio (%)	[−50, 0.0, 50]

The underlined values are the defaults.

We use the following performance metrics:

- *Compute Resource Utilization.* This is the average utilization of the compute resources in our cluster. To calculate this utilization, we only consider useful computations, which belong to tasks completed by the femtocloud.
- *Network Utilization.* This is the average busy time of the network for sending tasks or receiving results.
- *Computational Throughput.* This is the average amount of useful computations finished by the femtocloud per second (MFLOPS).

We conduct two different sets of experiments. The first set of experiments (Section 7.3.2) aims for understanding the effect of different environmental parameters on the performance of FemtoCloud. The second set of experiments (Section 7.3.3) sheds light on the performance of our developed prototype.

7.3.2 FemtoCloud Simulation Results

In this section, we study the impact of changing different environmental parameters on the performance of FemtoCloud. We start by studying the impact of user arrival rate and presence time followed by the true effect of stability in the system. We also study the impact of changing task characteristics and robustness to estimation errors. In a subset of these experiments, we compare FemtoCloud to a presence time oblivious scheduler (PreOb). Such scheduler uses the same task assignment heuristic used by FemtoCloud but without taking the presence time of a device into account. Due to its unawareness of the presence time, it requests the results from the device once they become available.

Impact of Changing User Arrival Rate and Presence Time. Figure 7.3 shows the effect of changing the average user presence time and the average user arrival rate on the performance of FemtoCloud. Figure 7.3a shows that the increase in the presence time or the user arrival rate significantly enhances the performance and increases the femtocloud's computational throughput. This increased computational throughput saturates for large values of the arrival rate or the presence time. To explain the reason behind this saturation, we refer to Figure 7.3b, which clearly shows that the network utilization increases as more tasks get assigned to our devices until it becomes highly utilized and unable to support more task assignments.

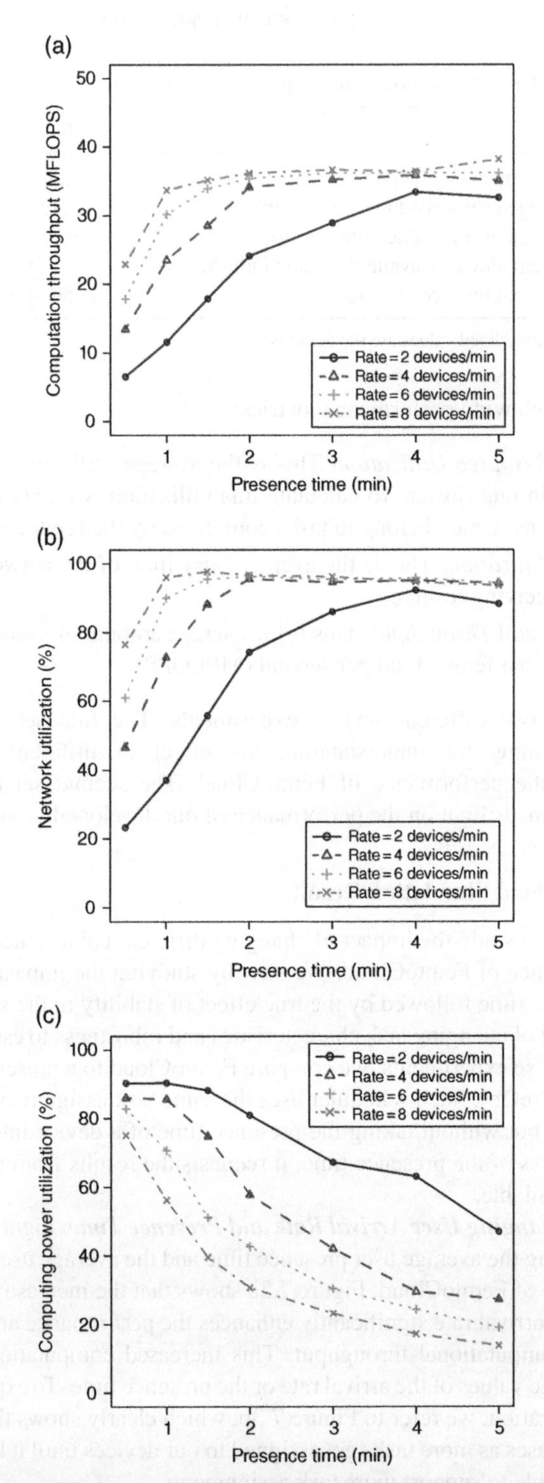

Figure 7.3 Impact of device arrival rate and presence time. (a) Computational throughput, (b) network utilization, and (c) computational resource utilization.

Figure 7.3c shows that the devices' utilization decreases with the increase of the presence time or the arrival rate. This decrease is due to having a lot of available devices in the system, which enables distributing the load on them and minimizing their utilization and overhead.

Stability Impact. Guided by the observations we draw from Figure 7.3c, it is critical to understand the true impact of the increased user presence on the system independent of the changes to the available compute resources. Therefore we construct an experiment in which we fix all the parameters in the system except the presence time of the devices. In this experiment we have three devices (a Nexus 10 and 2 Nexus 7 devices), and we change the average user presence time from 15 seconds to 1 hour. To isolate the effect of presence time, once a device leaves our cluster, an identical copy arrives and joins the cluster. Figure 7.4 shows the results of these experiments and compares FemtoCloud to the PreOb. Figure 7.4a and 7.4c show that with the increase of the presence time, both algorithms utilize the stability to gain more performance. FemtoCloud's awareness of the presence time enabled it to achieve higher performance than PreOb for low presence time values. Figure 7.4b shows that the FemtoCloud's increased performance comes with lower network utilization. The main reason is that without the knowledge of the presence time, PreOb assigns tasks to devices that may not be able to execute them, and, thus, it may have to reassign them again to another device. This behavior keeps the network unnecessarily busy. Figure 7.4b also shows that with the increase of presence time, FemtoCloud becomes able to execute tasks that require high compute resource and low network usage. Therefore, the more stable the devices in the femtocloud are, the less it consumes from the network resources.

Task Characteristics Impact. To study the impact of changing the task characteristics, we conduct an experiment that has only a single task type (chess). While maintaining the average computational requirements and average output size as constants, we vary the input size from 0.5 to 16 MB. Figure 7.5 shows the impact of increasing the task input size on the performance of FemtoCloud. It is clear that with the increase of the input size, the task characteristics moves from being CPU bounded, which enables increasing the compute resource utilization to be fully network bounded. Therefore the compute resource utilization decreases and the network utilization increases.

Robustness to Estimation Errors. To measure the impact of errors in estimating the presence time of the user on the system, we conduct an experiment in which we introduce a Gaussian error and change the mean from −50% of the presence time to +50% of the presence time. Figure 7.6 shows that when the error mean is 0, FemtoCloud is able to achieve the highest utilization of the available devices. When the error is negative, we have a conservative estimate about the presence time that limits FemtoCloud usage of a device, which leads to decreasing the compute and network utilization. When the error is positive, the computational utilization degrades because FemtoCloud fails to gather all the executed task results. Note that our early gathering heuristic is responsible for minimizing this effect. Figure 7.6b shows that the network utilization keeps increasing because FemtoCloud keeps assigning more tasks while moving from a conservative estimate to a less conservative one. Furthermore,

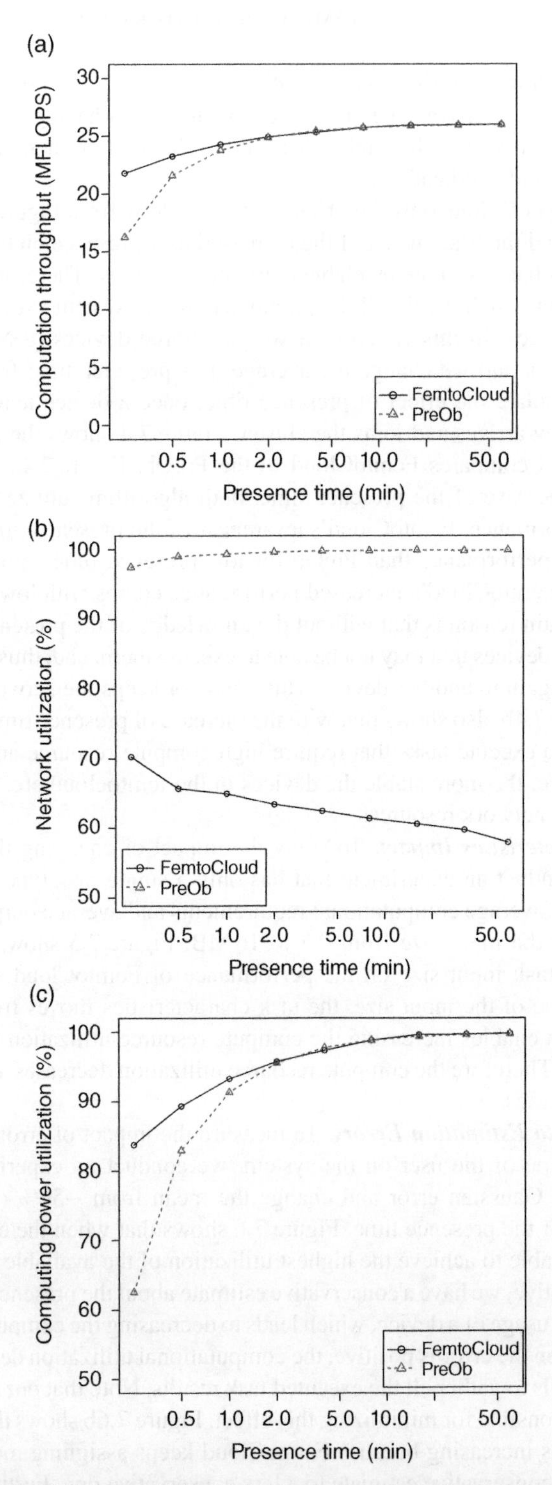

Figure 7.4 Impact of cluster stability. (a) Computational throughput, (b) network utilization, and (c) computational resource utilization.

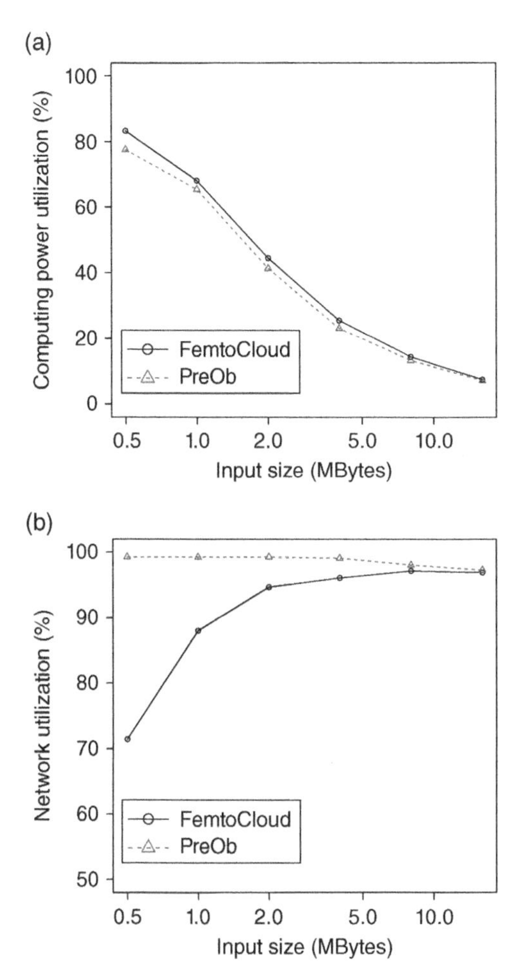

Figure 7.5 Impact of task characteristics. (a) Computational resource utilization and (b) network utilization.

when the error becomes positive, the network utilization will further increase due to reassigning tasks after a device leaves without sending their results.

7.3.3 FemtoCloud Prototype Evaluation

In this section, we discuss the results we gathered while using our prototype. In our experiment, we use three devices, a Galaxy S5 running Android 4.4.4 in addition to a Nexus 10 [2013] and Nexus 7 [2013] tablets running Android 5.0.2. In this experiment, we compare the performance of FemtoCloud to an oracle that assumes accurate knowledge of all connectivity and execution time for every task on every device. Since this oracle is impossible, we gather measurements from all the devices and use after the fact analysis get the results.

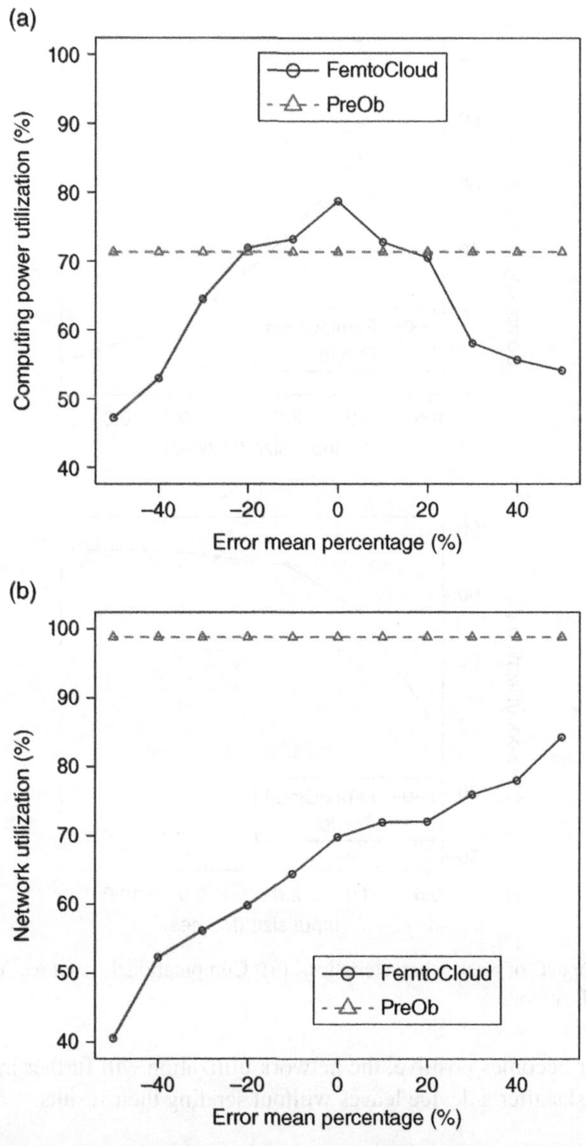

Figure 7.6 Robustness to estimation errors. (a) Computational resource utilization and (b) network utilization.

In our experiment, we compare the achieved compute throughput by the oracle and FemtoCloud under two scenarios: (i) full presence scenario and (ii) emulated arrival/departure scenario. In the first scenario, we assume that the three devices existed during the whole period of experiment (1 hour). The main goal of this scenario is comparing the maximum achievable performance of FemtoCloud to the one

TABLE 7.5 Prototype Performance Measurements

Scenario	Oracle (MFLOPS)	FemtoCloud (MFLOPS)
Full presence	16.54	14.23
Emulated arrival/departure	10.31	8.86

achieved by the oracle. In the second scenario, we emulate average presence time of two minutes for each device. We emulated the arrival of new devices by returning the device to the cluster after average of 1 minute from its last departure. Table 7.5 summarizes these experiment results and shows that FemtoCloud achieved more than 85% of what the oracle achieved in both scenarios.

7.4 SERENDIPITY PERFORMANCE EVALUATION

To complement the performance results from FemtoClouds, we provide a selected set of performance results for Serendipity. The reader is referred to [9] for additional results. In evaluating Serendipity, we assess the performance of three different algorithms for task assignment and scheduling. WaterFilling (WF) represents an ideal case where all future contacts are known and a low bandwidth control channel allows a central algorithm to schedule computation across all devices. Using a greedy task assignment, WF attempts to minimize the completion time of all tasks.

Computing-on-Dissemination with predictable contacts (pCoD) and Computing-on-Dissemination with unpredictable contacts (upCoD) both assume that there is no control channel and hence no centralized opportunity to create a schedule. Instead, when two devices encounter one another, they exchange task lists and make a decision on whether to pass any tasks to the other device. Predictable contacts assumes it is possible to know when the result could be returned in the future; unpredictable contacts makes no such assumption.

7.4.1 Experimental Setup

To evaluate Serendipity in various network settings, we have built a testbed on Emulab [18] to easily configure the experiment settings including the number of nodes, the node properties, etc. In our testbed, a Serendipity node running on an Emulab node has an emulation module to emulate the intermittent connectivity among nodes. Before an experiment starts, all nodes load the contact traces into their emulation modules. During the experiments, the emulation module will control the communication between its node and all other nodes according to the contact traces.

To emulate various contact scenarios, we use both real-world contact traces and synthetic traces. First, we use two real-world contact traces, a 9-node trace collected in the Haggle project [19] and the RollerNet trace [20]. In the RollerNet trace, we select a subset of 11 friends (identified in the metadata of the trace) among the 62 nodes so that the number of nodes is comparable to the Haggle trace. The Haggle trace represents the user contacts in a laboratory during a typical day, while RollerNet represents the contacts among a group of friends during the outdoor activity. These two traces demonstrate quite different contact properties. RollerNet has shorter contact intervals, while Haggle has longer contact durations. Second, we synthesize a set of contact traces to analyze the impact of different mobility factors on Serendipity. To synthesize our traces, we rely on three mobility models, namely, the Levy walk model [21], the random waypoint (RWP) model [22], and the time-variant community mobility model (TVCM) [23].

In our testbed, we are running a speech-to-text application, which we implement based on the Sphinx library [24] that translates audio to text. We implement it as a single PNP-block job where the pre-process program divides a large audio file into multiple 2 Mb pieces, each of which is the task input. The post-process program collects and combines the results.

To demonstrate how Serendipity can help the mobile computation initiator to speed up computing and conserve energy, we primarily compare the performance of executing applications on Serendipity with that of executing them locally on the initiator's mobile device. Previous remote-computing platforms (e.g., MAUI [6], CloneCloud [15]) don't work with intermittent connectivity and, thus, cannot be directly compared with Serendipity.

In all the following experiments, every machine has a 600 MHz Pentium III processor and 256 MB memory, which is less powerful than mainstream PCs but closer to that of smart mobile devices. Every experiment is repeated 10 times with different seeds. The results reported correspond to the average values.

7.4.2 Serendipity's Performance Benefits

We begin the experiments with the speech-to-text application using three workloads in each of the three task allocation algorithms on both RollerNet and Haggle traces. The sizes of the audio files are 20, 200, and 600 Mb. The baseline wireless bandwidth is set to 24 Mbps. We also assume that all nodes have enough energy and want to reduce the job completion time.

Figure 7.7 demonstrates how Serendipity improves the performance compared to executing locally. We make the following observations. First, with the increase of the workload, Serendipity achieves greater benefits in improving application performance. When the audio file is 600 Mb, Serendipity can achieve as large as 6.6 and 5.8 time speedup. Considering the number of nodes (11 for RollerNet and 9 for Haggle), the system utilization is more than 60%. Moreover, the ratio of the confidence intervals to the average values also decreases with the workload, indicating that all nodes can obtain similar performance benefits. Second, in all the experiments WF consistently performs better than pCoD, which is better than upCoD. In the Haggle trace of Figure 7.7c, WF achieves 5.8 time speedup while upCoD achieves 4.2 time speedup. The results indicate that with more information Serendipity can perform better. Third, although Serendipity achieves similar average job completion times on both Haggle and RollerNet, the confidence intervals on Haggle are larger than those on RollerNet. This is because the Haggle trace has long contact interval and duration, resulting in the diversity of node density over the time.

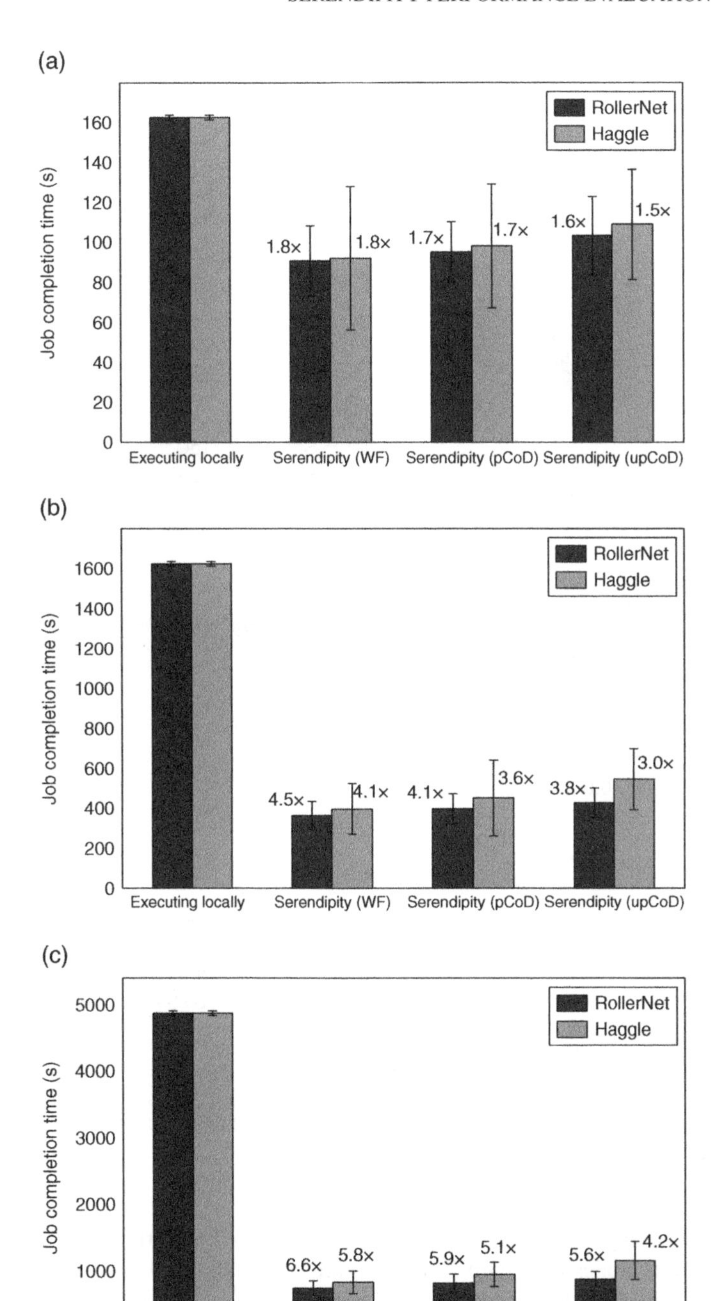

Figure 7.7 A comparison of Serendipity's performance benefits. The average job completion times with their 95% confidence intervals are plotted. We use two data traces, Haggle and RollerNet, to emulate the node contacts and three input sizes for each. (a) 10 tasks, (b) 100 tasks, and (c) 300 tasks.

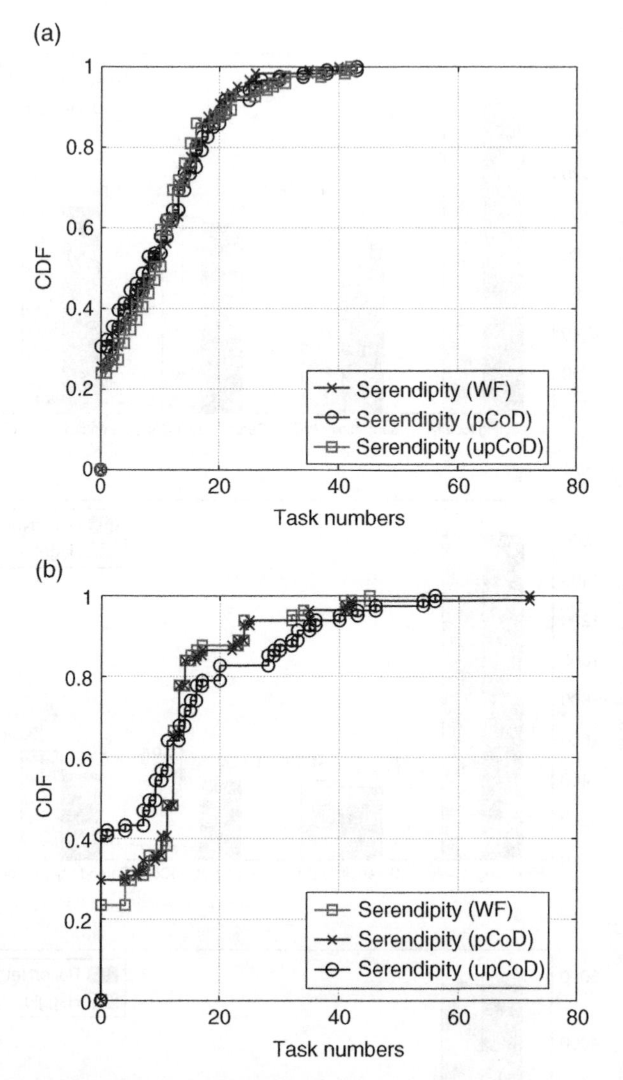

Figure 7.8 The load distribution of Serendipity nodes when there are 100 tasks total, each of which takes 2 Mb input data. (a) RollerNet and (b) Haggle.

To further analyze the performance diversity, we plot the workload distribution on the Serendipity nodes of Figure 7.7b in Figure 7.8. In the RollerNet trace, all three task allocation algorithms have similar load distribution, that is, about 25% nodes are allocated 0 tasks while about 10% of the nodes are allocated more than 20 tasks. In the Haggle trace, WF and pCoD have similar load distribution, while upCoD's distribution is quite different from them. The long contact intervals of the Haggle trace makes the blind task dissemination of upCoD less efficient. In such an environment, the contact knowledge will be very useful to improve the Serendipity performance.

7.4.3 Impact of Network Environment

Next, we analyze the impact of the network environment on the performance of the three task allocation algorithms by changing the network settings from the base case: *Wireless Bandwidth.* We first consider the effect of wireless bandwidth on the performance of Serendipity. The wireless bandwidth is set to be 1, 5.5, 11, 24, and 54 Mbps, which are typical values for wireless links. The audio file is 200 Mb, split into 100 tasks. We plot the job completion times of Serendipity with three task allocation algorithms in Figure 7.9.

We observe the following phenomena. First, in RollerNet, all three task allocation algorithms achieve similar performance. Because these nodes have frequent contacts with each other, using the locality heuristic (upCoD) is good enough to make use of the nearby computation resource for remote computing. Second, when the bandwidth reduces from 11 to 1 Mbps, the job completion time experiences a large increase. This is because RollerNet has many short contacts that cannot be used to disseminate tasks when the bandwidth is too small. Third, in the Haggle trace, the job completion time of upCoD increases from 545.0 to 647.6 seconds when the bandwidth reduces from 24 to 11 Mbps. Meanwhile WF achieves consistently good performance in all the experiments. This is because in the Haggle laboratory, environment users are relatively stable and have longer contact durations. Thus, the primary factor affecting the Serendipity performance is the contact interval. On the other hand, since the contact distribution is more biased, only using locality is hard to find the global optimal task allocation.

Node Mobility. The aforementioned experiments demonstrate that contact traces impact the performance of Serendipity. To further analyze such impact, we use mobility models to generate the contact traces for 10 nodes. Specifically, we use the Levy Walk model [21], RWP [22], and TVCM [23]. These models represent a wide range of mobility patterns. RWP is the simplest model and assumes unrestricted node movement. Levy Walk describes the human walk pattern verified by collected mobility traces. TVCM depicts human behavior in the presence of communities. The basic settings assume a 1 km by 1 km square activity area in which each node has a 100 m diameter circular communication range.

In this set of experiments, we focus on the two most important aspects of node mobility, that is, the mobility model and the node speed. The wireless bandwidth is set to 11 Mbps.

The results of this comparison are shown in Figure 7.10. Figure 7.10a shows that Serendipity has larger job completion time with all the mobility models than it had on Haggle and RollerNet traces. This is because their node densities are much sparser than Haggle and RollerNet traces. Thus it's harder for the job initiator to use other nodes' computation resources. We also observe that Serendipity achieves the best performance when the RWP model is used. This is because RWP is the most diffusive [21] and, thus, results in more contact opportunities among nodes.

Node speed affects the contact frequencies and durations, which are critical to Serendipity. We vary the node speed from 1 m/s, that is, human walking speed, to 20 m/s, that is, vehicle speed. As shown in Figure 7.10b, when the speed increases

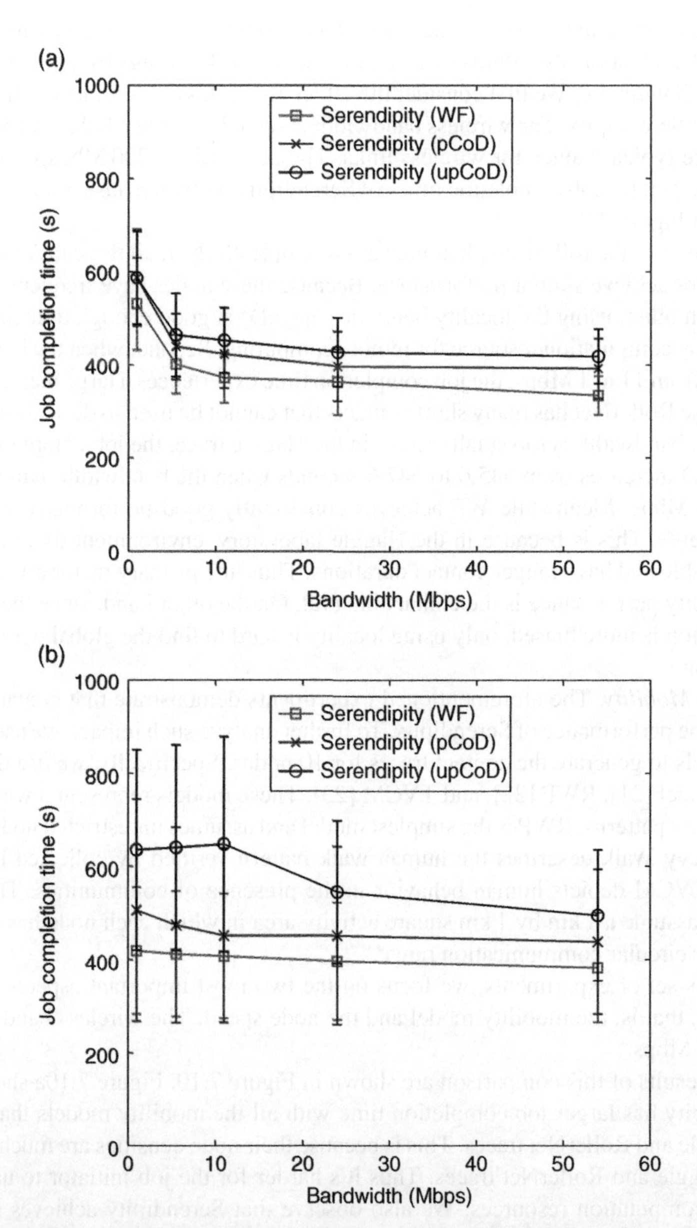

Figure 7.9 The impact of wireless bandwidth on the performance of Serendipity. The average job completion times are plotted when the bandwidth is 1, 5.5, 11, 24, and 54 Mb/s, respectively. (a) RollerNet and (b) Haggle.

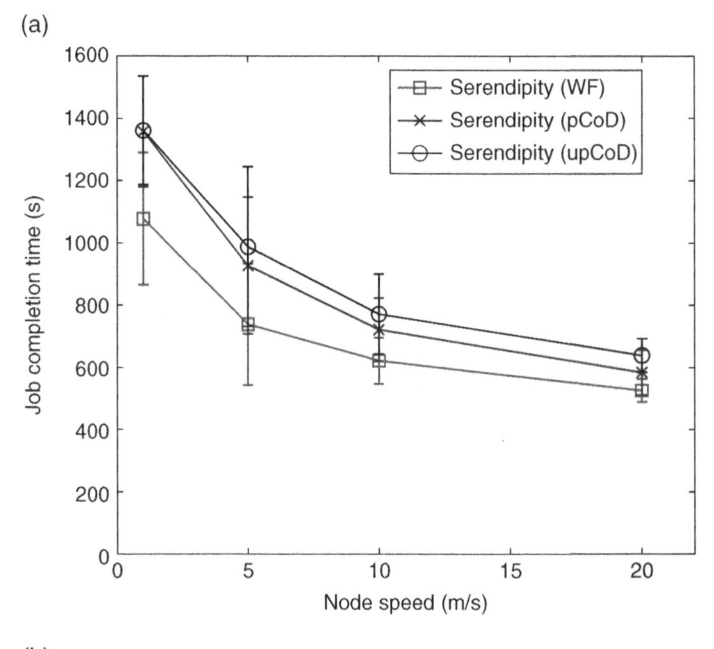

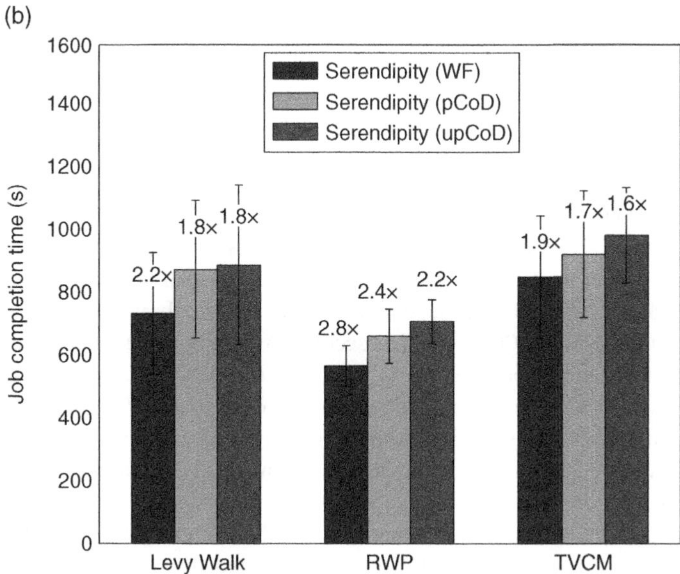

Figure 7.10 The impact of node mobility on Serendipity. We generate the contact traces for 10 nodes in a 1 km × 1 km area. In (a) we set the node speed to be 5 m/s, while in (b) we use Levy Walk as the mobility model.

from 1 to 10 m/s, the job completion times drastically decline, for example, from 1077.1 to 621.6 seconds for WF. This is because the increase of node speed significantly increases the contact opportunities and accelerates the task dissemination. When the speed further increases to 20 m/s, the job completion time is slightly reduced to 526.4 seconds for WF.

Number of Nodes. We finally examine how the quantity of available computation resources impacts Serendipity. To separate the effect of node density and resource quantity, we conduct two sets of experiments. In the first set, the active area is fixed, while in the second one, the active area changes proportionally with the number of nodes using the initial setting of 20 nodes in 1 km × 1 km square area. Figure 7.11 shows the results where nodes follows RWP mobility model with wireless bandwidth at 2 Mbps.

As shown in Figure 7.11a, with the increase in the number of nodes in a fixed area, the job completion times of the three task allocation algorithms are reduced by more than 50%, from 550.0, 647.0, and 748.7 seconds to 273.0, 311.7, and 325.0 seconds for WF, pCoD, and upCoD, respectively. Meanwhile, in Figure 7.11b, the job completion times are almost constant despite the increase in node quantity.

7.4.4 The Impact of the Job Properties

Next we evaluate how the job properties affect the performance of Serendipity:

Multiple Jobs. A more practical scenario involves nodes submitting multiple jobs simultaneously into Serendipity. These jobs will affect the performance of each other when their execution duration overlaps. In this set of experiments, nodes will randomly submit 100 task jobs into Serendipity. The arrival time of these jobs follows a Poisson distribution. We change the arrival rate λ from 0.0013 (its system utilization is less than 20%) to 0.0056 (its system utilization is larger than 90%) jobs per second. Figure 7.12 shows the results on the RollerNet and Haggle traces.

As expected, the job completion time increases with the job arrival rate. In both sets of experiments, the job completion time gradually increases with the job arrival rate until 0.005 jobs per second and, then, drastically increases when the job arrival rate increase to 0.0056 jobs per second. According to queueing theory, with the system utilization approaching 1, the queueing delay is approaching infinity. However, even when the system utilization is larger than 90% (i.e., $\lambda = 0.0056$), the job completion times of Serendipity with various task allocation algorithms are still less than 54% of executing locally, showing the advantage of distributed computation.

DAG Jobs. The previous experiments show that Serendipity performs well for single PNP-block jobs, since DAG jobs are executed iteratively for all dependent PNP-blocks while parallel for all independent PNP-blocks. The previous experiment results also apply to DAG jobs. In this set of experiments, we will evaluate how PNP-block scheduling algorithm further improves the performance of Serendipity.

We use the job structure shown in Figure 7.13, where the processing of one image impacts the processing of another. We use the PNP-blocks of speech-to-text

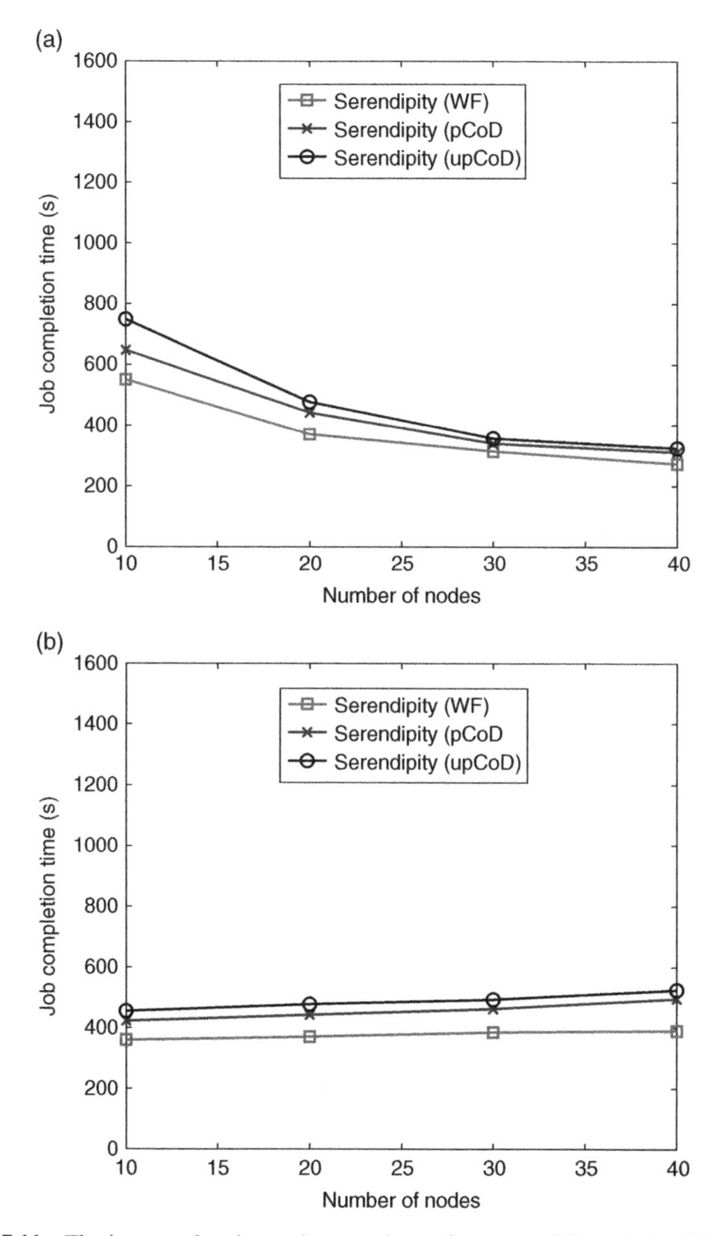

Figure 7.11 The impact of node numbers on the performance of Serendipity. We analyze the impact of both node number and node density by fixing the activity area and setting it proportional to the node numbers, respectively. (a) Fixed active area and (b) fixed node density.

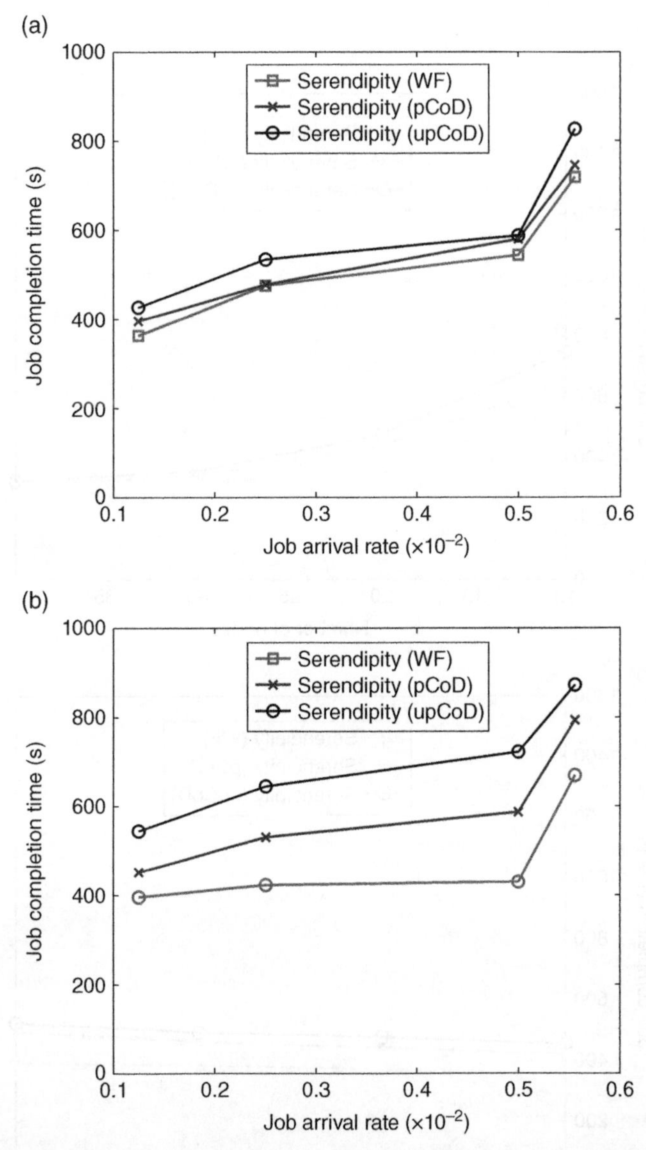

Figure 7.12 Serendipity's performance with multiple jobs executed simultaneously. The job arrival time follows a Poisson distribution with varying arrival rates. (a) RollerNet and (b) Haggle.

application as the basic building blocks. PNP-block A has 0 tasks; B has 200 tasks; C has 50 tasks; D has 100 tasks; E has 100 tasks; F has 0 tasks. The performance difference between our algorithm and assigning equal priority to the PNP-blocks is shown in Figure 7.14.

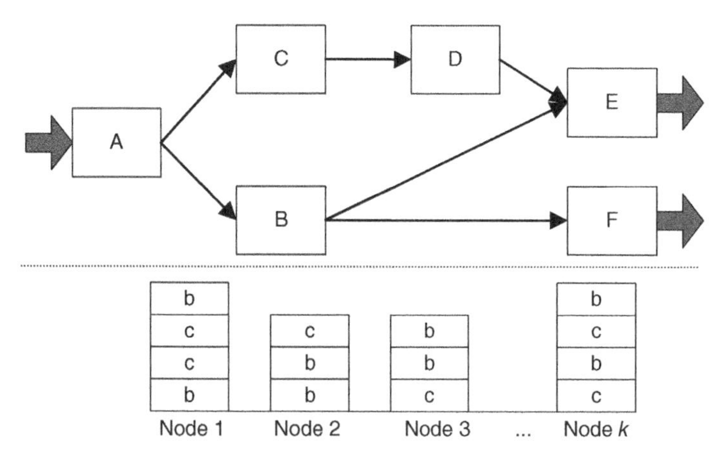

Figure 7.13 A job example where both PNP-block B and C are disseminated to Serendipity nodes after A completes. Their task positions in the nodes' task lists are shown blow the DAG.

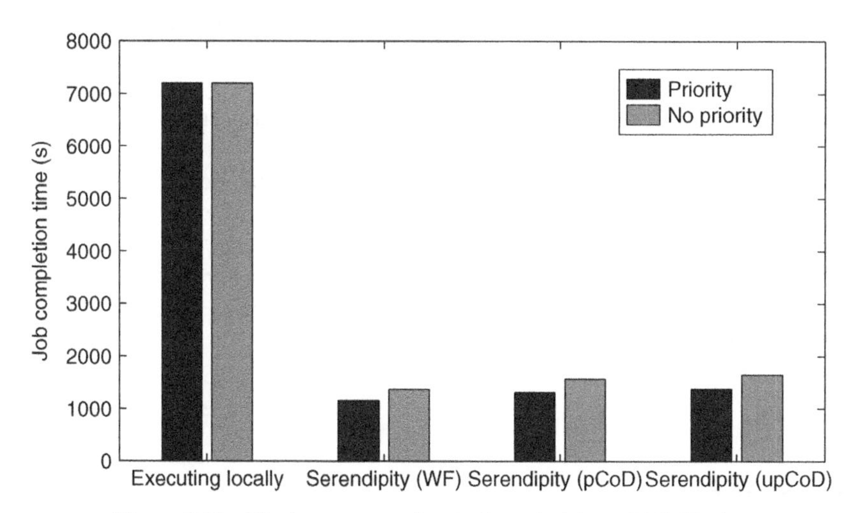

Figure 7.14 The importance of assigning priorities to PNP-blocks.

Our priority assignment algorithm achieves the job completion time of 1155.8, 1315.8, and 1383.2 seconds for WF, pCoD, and upCoD, consistently outperforming that of 1369.2, 1573.4, and 1654.4 seconds when all PNP-blocks have the same priority. These experiments demonstrate the usefulness of priority assigning. Further evaluation of our algorithm on diverse type of jobs will be part of our future work.

7.5 CHALLENGES

There are a number of challenges that need to be addressed to ultimately utilize mobile devices to provide elastic clouds at the edge. These challenges include:

User Incentives. To have a large-scale deployment of elastic cloud computing systems, incentives are needed to encourage users to share their idle devices' computational capacity, especially in situations where these devices are not plugged in. Incentive models in such cases need to be revisited depending on the context and applications for which such systems are used. Possibilities include user-to-user ad hoc offloading that may require variations of reputation or credit-based systems, to corporate-to-user offloading applications requiring a negotiated pricing and payment system.

Security and Privacy. Depending on the scale of deployment and the nature of the applications running or being offloaded, along with potential incentive system solutions that would involve some form of payment, security and privacy are major concerns. Running unknown code, sharing data with other entities, and routing content and code through unknown entities certainly opens the door for various malicious challenges that need to be addressed accordingly.

Context Awareness and Fault Tolerance. Better knowledge about the context in which an elastic cloud computing systems is deployed, about application usage profile or about the stability of other devices within proximity, would all lead to better decisions that would significantly enhance performance. Thus, dynamic adaptation to changes in the system context and failures is one of the key aspects for any large-scale deployed system.

Portability. The ability to run computations on various mobile platforms will be hindered by the same portability challenges faced earlier by the grid computing community. Finding mobile-specific solutions through the creation of virtual machine-like platforms or developing web-browser-based sandboxed-like solutions, along with the classical need to write code modular enough to be easily offloaded, are challenges that will need to be addressed.

REFERENCES

1. Ha, I., Yoon, Y., and Choi, M. (2007) Determinants of adoption of mobile games under mobile broadband wireless access environment. *Information & Management*, **44** (3), 276–286.

2. Chib, A., van Velthoven, M.H., and Car, J. (2015) mHealth adoption in low-resource environments: A review of the use of mobile healthcare in developing countries. *Journal of Health Communication*, **20** (1), 4–34.

3. Balan, R., Flinn, J., Satyanarayanan, M., Sinnamohideen, S., and Yang, H.I. (2002) The case for cyber foraging, in *Proceedings of the 10th Workshop on ACM SIGOPS European Workshop*, ACM, pp. 87–92.

4. Satyanarayanan, M. (2015) A brief history of cloud offload: A personal journey from odyssey through cyber foraging to cloudlets. *SIGMOBILE Mobile Computing and Communications Review*, **18** (4), 19–23, doi:10.1145/2721914.2721921. URL http://doi.acm.org.prx.library.gatech.edu/10.1145/2721914.2721921.

5. Shi, C., Habak, K., Pandurangan, P., Ammar, M., Zegura, E., and Naik, M. (2014) Cosmos: Computation offloading as a service for mobile devices, in *ACM MobiHoc*.

6. Cuervo, E., Balasubramanian, A., Cho, D.K., Wolman, A., Saroiu, S., Chandra, R., and Bahl, P. (2010) MAUI: Making smartphones last longer with code offload, in *Proceedings of the Eighth International Conference on Mobile Systems, Applications, and Services*, ACM, pp. 49–62.

7. Satyanarayanan, M., Bahl, P., Caceres, R., and Davies, N. (2009) The case for VM-based cloudlets in mobile computing. *IEEE Pervasive Computing* **8** (4), 14–23.

8. Flinn, J. (2012) Cyber foraging: Bridging mobile and cloud computing, in *Synthesis Lectures on Mobile and Pervasive Computing*, Morgan and Claypool Publishers, San Rafael, CA, pp. 1–103.

9. Shi, C., Lakafosis, V., Ammar, M.H., and Zegura, E.W. (2012) Serendipity: Enabling remote computing among intermittently connected mobile devices, in *Proceedings of the 13th ACM international Symposium on Mobile Ad Hoc Networking and Computing*, ACM, pp. 145–154.

10. Teo, C.L.V. (2012) Hyrax: Crowdsourcing Mobile Devices to Develop Proximity-based Mobile Clouds. Master's thesis, Carnegie Mellon University, Pittsburgh, PA.

11. Mtibaa, A., Harras, K., and Fahim, A. (2013) Towards computational offloading in mobile device clouds, in *IEEE Fifth International Conference on Cloud Computing Technology and Science (CloudCom)*, IEEE, pp. 331–338.

12. Habak, K., Ammar, M., Harras, K.A., and Zegura, E. (2015) Femtoclouds: Leveraging mobile devices to provide cloud service at the edge, in *IEEE International Conference on Cloud Computing (IEEE CLOUD)*.

13. Arslan, M.Y., Singh, I., Singh, S., Madhyastha, H.V., Sundaresan, K., and Krishnamurthy, S.V. (2012) Computing while charging: Building a distributed computing infrastructure using smartphones, in *Proceedings of the Eighth International Conference on Emerging Networking Experiments and Technologies*, ACM, pp. 193–204.

14. Kwon, Y., Lee, S., Yi, H., Kwon, D., Yang, S., Chun, B.G., Huang, L., Maniatis, P., Naik, M., and Paek, Y. (2013) Mantis: Automatic performance prediction for smartphone applications, in *Proceedings of the 2013 USENIX conference on Annual Technical Conference*, USENIX Association, pp. 297–308.

15. Chun, B.G., Ihm, S., Maniatis, P., Naik, M., and Patti, A. (2011) Clonecloud: Elastic execution between mobile device and cloud, in *Proceedings of the Sixth Conference on Computer Systems*, ACM, pp. 301–314.

16. Zhang, L., Tiwana, B., Qian, Z., Wang, Z., Dick, R.P., Mao, Z.M., and Yang, L. (2010) Accurate online power estimation and automatic battery behavior based power model generation for smartphones, in *Proceedings of the Eighth IEEE/ACM/IFIP International Conference on Hardware/Software Codesign and System Synthesis*, ACM, pp. 105–114.

17. Burgess, J., Gallagher, B., Jensen, D., and Levine, B.N. (2006) MaxProp: Routing for vehicle-based disruption-tolerant networks, in *Proceedings of the IEEE International Conference on Computer Communications (INFOCOM)*, pp. 1–11.

18. White, B., Lepreau, J., Stoller, L., Ricci, R., Guruprasad, S., Newbold, M., Hibler, M., Barb, C., and Joglekar, A. (2002) An integrated experimental environment for distributed systems and networks, in *Proceedings of the Fifth Symposium on Operating Systems Design and implementation*, USENIX Association, Berkeley, CA, USA, pp. 255–270.
19. Scott, J., Crowcroft, J., Hui, P., and Diot, C. (2006) Haggle: A networking architecture designed around mobile users, in *WONS 2006, Third Annual Conference on Wireless On-demand Network Systems and Services*, pp. 78–86.
20. Tournoux, P.U., Leguay, J., Benbadis, F., Conan, V., de Amorim, M.D., and Whitbeck, J. (2009) The accordion phenomenon: Analysis, characterization, and impact on DTN routing, in *IEEE International Conference on Computer Communications (INFOCOM)*, pp. 1116–1124.
21. Rhee, I., Shin, M., Hong, S., Lee, K., and Chong, S. (2008) On the levy-walk nature of human mobility, in *IEEE INFOCOM 2008. 27th Conference on Computer Communications.*
22. Saha, A.K. and Johnson, D.B. (2004) Modeling mobility for vehicular ad-hoc networks, in *Proceedings of the First ACM International Workshop on Vehicular Ad Hoc Networks*, ACM, pp. 91–92.
23. Hsu, W.-J., Spyropoulos, T., Psounis, K., and Helmy, A. (2007) Modeling time-variant user mobility in wireless mobile networks, in *IEEE INFOCOM 2007, 26th IEEE International Conference on Computer Communications*, IEEE, pp. 758–766.
24. Lee, K.F., Hon, H.W., and Reddy, R. (1990) An overview of the SPHINX speech recognition system. *IEEE Transaction on Acoustics, Speech and Signal Processing* **38** (1), 35–45.

PART III
Applications of Fog

8 The Role of Fog Computing in the Future of the Automobile

FLAVIO BONOMI,[1] STEFAN POLEDNA,[2] and
WILFRIED STEINER[2]

[1] *Nebbiolo Technologies, Inc., Milpitas, CA, USA*
[2] *TTTech Computertechnik AG, Wien, Austria*

8.1 INTRODUCTION

The modern automobile is a computing-rich electronic system on wheels, with more than 100 computers per vehicle, and it will become much more powerful in the not too distant future. This trend is motivated by a number of converging requirements and developments, including the need for connectivity of automobiles to sources of travel information and entertainment, the need for vehicle-to-vehicle and vehicle-to-infrastructure exchanges for accident prevention, the move toward more dynamic and modern vehicle maintenance, the need to rationalize the electronic vehicle control architecture by reducing control system weight and cost of software development, the evolution toward electric vehicles, and, most importantly, the need to assist or even replace drivers.

Indeed, our life today strongly depends on the automobile, a finely engineered element in today's critical infrastructures. While the primary function of the car is obvious, engineers and scientists have aimed since the automobile's origins to improve various important factors (we will refer to them as the "four factors" in this chapter): economic efficiency, environmental sustainability, safety, and passenger comfort. Over the last decades, the advancements in computer technology have enabled breakthrough improvements to all four factors. Most remarkable at this moment, the automobile industry is about to radically transform the driving experience—autonomous cars will step by step transfer the driver's responsibilities to the car. Indeed, the first fully functional prototypes of autonomous systems are available by established automotive OEMs as well as by their rising competitors originating from the information industry. These prototypes have successfully

Fog for 5G and IoT, First Edition. Edited by Mung Chiang, Bharath Balasubramanian, and Flavio Bonomi.

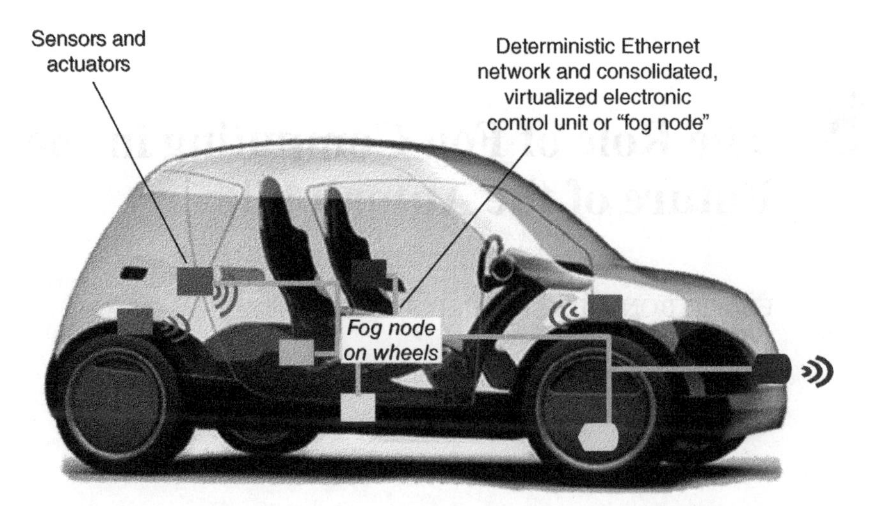

Figure 8.1 The future automobile as fog computing on wheels.

validated the autonomous system functionality. However, there are still various technical challenges to master on the way toward making autonomous systems available to the general public, because the evolution of automotive technologies is bringing together the most advanced academic and industrial knowledge in the fields of communications, computing, systems control, artificial intelligence, security and safety, and energy storage and management.

The evolution of thought at the origin of fog computing was in many ways motivated by the critical requirements foreseen in the automotive and intelligent transportation sectors, which have contributed to crystallize fog computing at the convergence of the most advanced embedded system virtualization and of modern real-time, deterministic, and safety-aware computing and networking.

We can now safely claim that future automobiles will look like powerful, compact, scalable data centers on wheels, or fog computing nodes on wheels, as illustrated in Figure 8.1, traveling within highways and cities equipped with powerful fog computing capabilities at their intersections and along their pathways.

In this chapter we focus on challenges in a key technological area of the car—the electrical/electronic (E/E) architecture—and discuss the role of fog computing in addressing these challenges. Note that, in this chapter, we will use the broad definition of fog computing introduced in Ref. [1], which is closely related, even if not identical, to other fog concepts discussed in this book.

In Section 8.2, we will describe the current automobile electronic architecture, while in Section 8.3 we will discuss the challenges of such electronic architecture as it faces near future requirements. In the same section, we will introduce key technologies that will be fundamental in the car electronics evolution. Section 8.4 motivates the vision for a scalable, more centralized computing architecture for the car of the future, while Section 8.5 motivates the roles of deterministic networking and virtualization in the future car control architecture. Section 8.6 hosts our conclusions.

8.2 CURRENT AUTOMOBILE ELECTRONIC ARCHITECTURES

Besides infotainment systems, the automotive industry almost exclusively uses specialized automotive technology. Thus, the automotive industry uses dedicated chips, operating systems (OS) and software, and specialized network technologies.

Automotive chipsets, for example, greatly differ from typical consumer electronics, such as the chips in cell phones or other embedded devices. Automotive chips are able to operate in a temperature range of about −40 to +125°C (sometimes +170°C and higher), while consumer electronics target a temperature range of about 0 to +85°C. Frequently automotive chips implement more safety functions that monitor the behavior of the chip and will disable the chip as a whole or in parts in the event that the monitor detects a failure.

Automotive chips also implement interfaces to automotive-specific networks, like CAN, LIN, FlexRay, and MOST. These automotive networks have been developed to enable robust communication in harsh environments at reasonably low cost. Furthermore, some of these protocols provide guaranteed real-time transmission latencies, which are highly relevant for control applications. For years Ethernet was not able to compete with the automotive networks, but recent advancements in the physical layer as well as protocol extensions for real-time communication make Ethernet a viable technology for future in-vehicle use. Indeed, at the time of this writing, there are the first automobiles in series production that use Ethernet.

Automotive OS and software development in general are quite distinct from typical consumer products. Automotive open system architecture (AUTOSAR) [2] has evolved as a common software architecture and OS for the automotive market. While in the past it has been difficult to migrate software modules from one system to another, AUTOSAR now enables a straightforward transferability as well as scalability to different vehicles and platform variants. AUTOSAR executes typical OS tasks, like scheduling and inter-task communication. However, as opposed to consumer OS, AUTOSAR pre-allocates exact resources to the computations and communication in advance (at design time) to minimize dynamic decisions during runtime. Sometimes, quite typical for embedded systems, applications need to execute on "bare metal" meaning that there is no OS in use at all.

Since the automobile is a safety-critical vehicle that must guarantee passenger safety, rigorous development processes are applied in the development of the automotive software and hardware (including networking equipment). The ISO 26262 standard ("Road vehicles—functional safety"), in particular, has been developed for this purpose. For example, the ISO 26262 specifies a risk-based approach to determine the integrity levels of automotive applications. These integrity levels are called automotive safety integrity levels (ASIL) and range from ASIL A, representing the least stringent level, to ASIL D, representing the most stringent level. Based on the calculated ASIL level, ISO 26262 defines functional safety requirements for each application as well as validation and verification procedures.

An example of a current E/E architecture is sketched in Figure 8.2. A clear distinction can be seen between the control systems of the car (on the left) and the

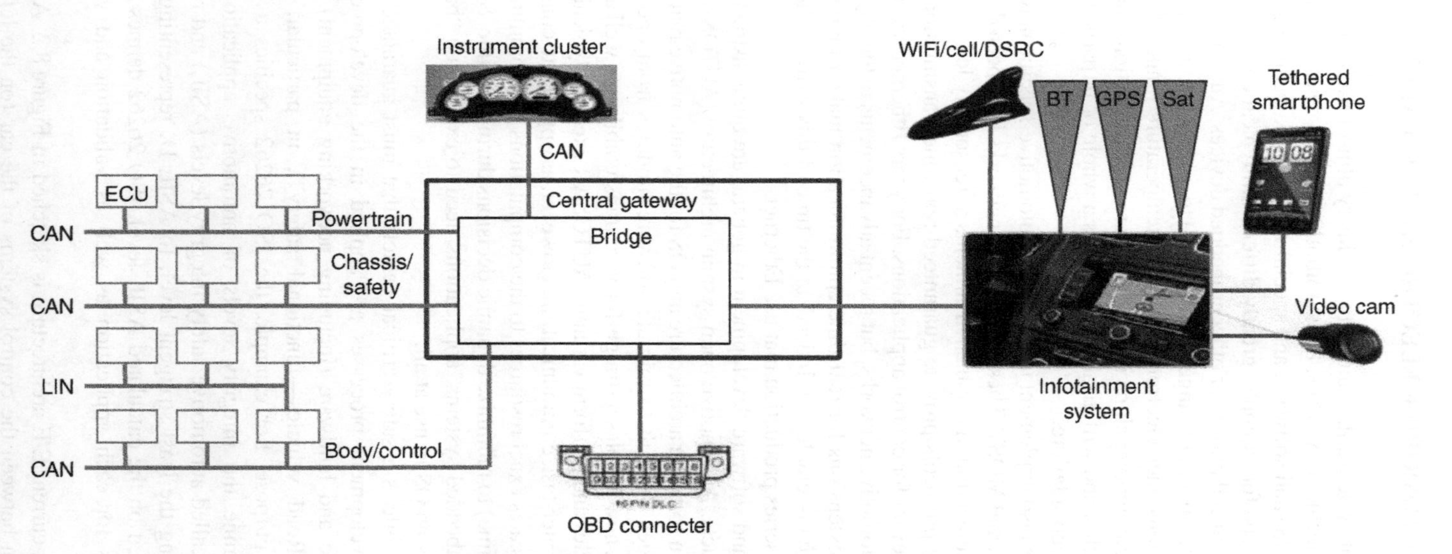

Figure 8.2 Traditional E/E automobile architecture.

infotainment systems (on the right). The control systems are grouped into domains as, for example, powertrain, chassis/safety, and body control.

A central gateway can be used for information transfer between the domains and the infotainment systems. A strict separation of the control systems from the infotainment systems is of upmost importance. For example, as demonstrated by a recent real-world experiment [3], scientists were able to utilize a vulnerability inside the head unit (being part of the infotainment system) to infiltrate the car's control systems. Furthermore, Figure 8.2 shows an onboard diagnostic (OBD) connection to the central gateway. OBD is mandated by law and provides a unified diagnostic access to the car's E/E systems.

8.3 FUTURE CHALLENGES OF AUTOMOTIVE E/E ARCHITECTURES AND SOLUTION STRATEGIES

While today's cars manage to balance well the four factors (economic efficiency, environmental sustainability, safety, and passenger comfort), automotive electronics face various challenges in the future. First, there is an issue of scalability: with the enormous pace of innovation in the automotive E/E area, both functionality and the number of electronic control units (ECUs) are growing rapidly. (Figure 8.3 shows an example of ECU.) Recently, a German OEM announced that one of their top-of-the-line vehicles includes 130 ECUs in total. This growing number of ECUs adds weight and power, consumes precious space, and ultimately adds considerable cost. Therefore, for efficiency reasons, novel architectural approaches are necessary to slow down and ideally reverse this trend. Fog computing can enable higher integration of functions per ECU, through virtualization, resource sharing, and multiplexing, and, therefore, helps to mitigate the scalability challenge.

Secondly, upcoming applications will require higher levels of connectivity inside the car between the various ECUs as well as outside the car to infrastructure and/or to other cars (car-2-x communication). For example, inside the car, the communication requirements may be driven by real-time video applications that need to communicate

Figure 8.3 Example of an automotive ECU, the TTA drive platform from TTTech (inner structure and with automotive-grade housing).

uncompressed video streams to ECUs hosting image recognition processes. Car-2-x communication is not only a valuable instrument for maintenance actions such as software updates but also a fast path to receive emergency signals from outside the car used, for example, for accident prevention [4]. A fog node can act as a local coordinating entity for the car-internal communication and function allocation as well as a proxy for the communication to and from the outside of the car.

Thirdly, security and privacy topics become more and more critical. This is partly a consequence not only of the increasing connectivity to the car but also by the car operating in smarter environments. Again, a fog node element can be equipped with security and privacy mechanisms.

Finally, as we strive to achieve the autonomous drive goal, it is clear that more computing and storage are needed to fuse and analyze in real time the growing amount of information collected by advanced sensors. A flexible and scalable fog node is the natural architectural element satisfying the evolving requirements leading to fully autonomous vehicles.

These four challenges, (i) higher density of functions per ECU, (ii) improved communication inside the car as well as outside, (iii) security and privacy, and (iv) advanced processing needs for autonomous vehicles, are quite complex.

The response to these four challenges, shaping the future of the automobile, is articulated around two main themes:

1. *Fog Computing*, as an ideal bridge between modern information technologies (IT) and operational technologies (OT)
2. *Time-Triggered Technologies*, based on precise time distribution, time-sensitive networking and computing resource allocation making up a collection of design patterns, have been applied in critical computer-based systems such as space vehicles, airplanes, and new energy production systems. The combined use of these technologies forms the time-triggered architecture (TTA). Note that fog computing, in its full expression, needs to embrace and manifest a TTA.

In addressing the four challenges listed previously, we can heavily leverage modern IT technologies. There are existing solutions in IT on which the automotive industry could build matching solutions. IT encompasses, for example, advanced networking technologies, like Ethernet, TCP/IP, and higher layered networks, software-defined networking (SDN), big data technologies, and virtualization technologies, as well as proven security solutions.

In the meantime, the embedded OT domain has in many ways remained separate and even isolated, developing its own networking protocols, which are often specific for each vertical (CAN, MODBUS, Profibus). OT is slowly adopting Ethernet but with nonstandard modifications (Ethernet IP, Profinet) and has not adopted standard wireless technologies (frequently using proprietary radio protocols). The OT world has enjoyed some of the progress in computing but often uses older generation products due to their more restrictive environmental and lifecycle requirements. Existing OT systems limit their connectivity to each other and to the broader Internet infrastructure by using older generation operating systems, by using no open-source software (with some good justification), and by assigning a computer (an ECU or a PLC, or an industrial computer) to solve a limited task, justified by real time and reliability requirements. Virtualization is only now making its baby steps in the OT arena, and lifecycle management of software is often based on older practices (e.g., local and complex, often by manual software updates).

Security is a strong concern for the OT domain and has in many ways deterred the agile adoption of technologies and practices, which have produced fast progress in the IT domain. In particular, the security issue, as driven by safety concerns and requirements, has maintained a strong separation between the IT domain and the OT domain in the industrial and transportation verticals, as examples.

The cultures in IT and OT are radically different, and communication across the two domains is challenging. Also, the electronic upgrade cycles are naturally very different between the two domains, with very fast refresh cycles in IT and much slower cycles in OT. On the other hand, the potential behind a more organic integration of the worlds of IT and OT is immense, particularly for the future of the automobile, and there is an accelerating trend throughout various industries toward this IT/OT convergence and integration.

Some of the "general" implications of the IT and OT convergence powerfully affecting the future of most vertical industries in the Internet of Things (IoT), including automotive and transportation, are the following:

- Introduction of standard networking technologies, such as Ethernet, Wi-Fi, and Bluetooth
- Adoption of modern security approaches typical of IT
- Exposure in OT to advanced models of computation and resource virtualization (GPUs, virtualization, etc.)
- Exposure in OT to advanced models of application development and deployment
- Exposure in OT to advanced methods of software system automation, management, distribution, and update
- OT closed systems can now leverage a large pool of computing resources and information to improve their operations
- Modern analytics approaches may be adopted, even close to the edge in OT systems
- Exposure in OT to the most advanced smartphone advances
- Sharing of economy of scale in OT typical of IT environments
- Pressure to accelerate rate of refresh in OT electronic systems
- Real-time, reliability, security, safety and system acceptance requirements typical of critical OT systems are now imposed to technologies coming from the IT world. In particular, these new requirements have brought about important advances in real-time computing and deterministic networking.

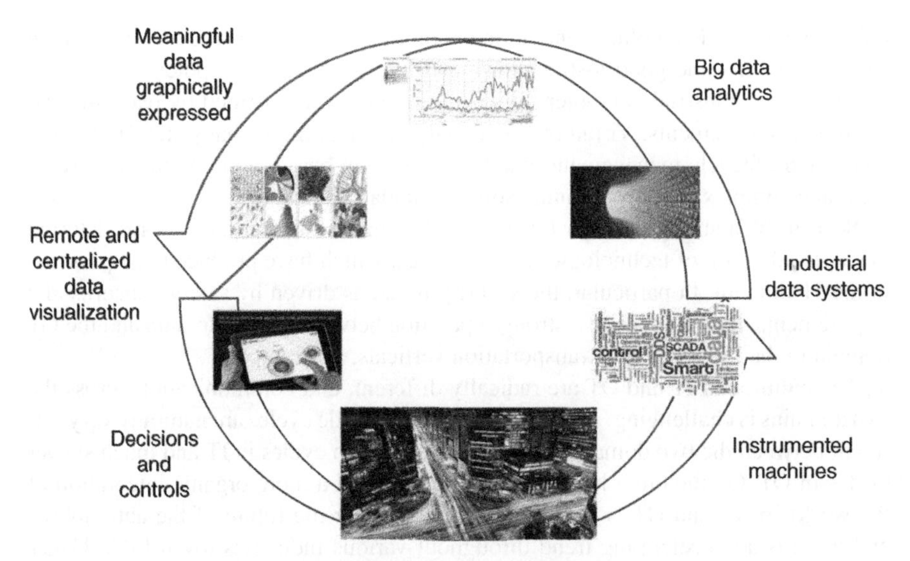

Figure 8.4 The IoT virtuous information cycle.

The overall implication of this convergence is well described by Figure 8.4, which depicts the IoT virtuous information cycle, from extraction to analysis to visualization to control, as inspired by General Electric [5]. This cycle definitely not only applies to the automotive vertical and manifests in a number of fast cycles within a vehicle, but also on broader, slower optimization cycles involving powerful analytics in the cloud.

Fog computing is a key enabler of IT and OT convergence, based on its position, technologies, and features. In fact, fog computing is built on some of the same technologies characterizing cloud computing applied to more compact, embedded computing systems, such as virtualization of all resources, resource management automation, and application lifecycle management, and will lend itself to the application of software-defined networking approaches.

On the other hand, fog computing, with its "edge of the network" positioning and more constrained resources, extends cloud computing in a nontrivial way by introducing:

- Hard real-time and more deterministic behavior in its networking, computing, and storage
- Focus on the direct support of a much wider set of networking technologies, including wireless and sensor networking and legacy-wired networking typical of OT deployments
- Relevance of mobility
- Focus on the interoperability with nonhomogeneous data sources, creating a data mediation functionality enabling applications to have more agile and flexible access to a wide variety of data sources

- Support of compact, streaming, and real-time capable data analytics
- Extended system, networking, and physical security and safety
- Renewed interest in hardware support of functionality, motivated by energy, space, and real-time requirements

Fog computing naturally mediates between the IT and the OT domains, since it inherits elements from both domains. Fog computing provides sufficient resources to achieve this mediation at the various levels of the stack, from low level networking to security, data, and the application levels.

In a later section we will describe more specifically how the same convergence of IT and how OT is manifesting in the automotive space and will drive the future automotive architecture.

Although IT for OT use has a high potential to become a cornerstone of future automotive E/E architectures, there are also new aspects in the four challenges that are not adequately addressed by IT solutions so far. The most distinctive characteristic of an automobile E/E architecture in contrast to a typical IT application is safety. The car, after all, is a powerful machine that can kill if not handled with care. Therefore, automotive software and hardware solutions typically need to be developed according to rigorous development processes as discussed in the previous section. Furthermore, the implementation according to well-defined design patterns prevents design errors of automotive E/E architectures. The TTA [6] is such a collection of design patterns and has been applied in critical computer-based systems such as space vehicles, airplanes, and new energy production systems. In its core, the TTA's design patterns focus to guarantee real-time behavior of subsystems and the system as a whole and inference-free composition of subsystems, two aspects not addressed by typical IT products. On the contrary, "never touch a running system" is a frequent motto in IT and, at least partly, comes from the typically incomplete knowledge of an inherent IT system's structure and composition.

The TTA enforces the system structure and composition by use of synchronized time. In its purest form, all subsystems are synchronized to each other and execute a pre-configured schedule that defines the points in time when subsystems use shared resources, such as computation or communication resources, and interact with each other. An example application of the TTA is an automotive network as depicted in Figure 8.5. Here, three sensors and four actuators connect to four switches. For simplicity we assume that switches 1 and 2 also incorporate processing elements like CPUs or GPUs and, in fact, could be defined as fog nodes. In real automotive networks switches, end points, sensors, and actuators will be partially integrated in single ECUs. Thus, the network in Figure 8.6 should be seen as simplified example only.

In such a network data delivery can be guaranteed while preserving real-time communication by implementing the TTA as depicted in Figure 8.6. Here, the devices, that is, switches and end points that incorporate the sensors and actuators, synchronize their local clocks to establish a system-wide synchronized time. The communication in the network then follows a communication schedule that is defined at system design time and is locally pre-configured in the devices.

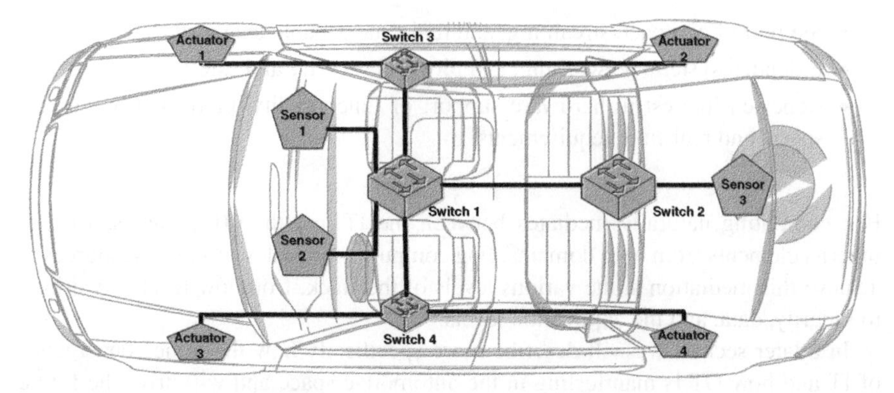

Figure 8.5 Example of automotive network.

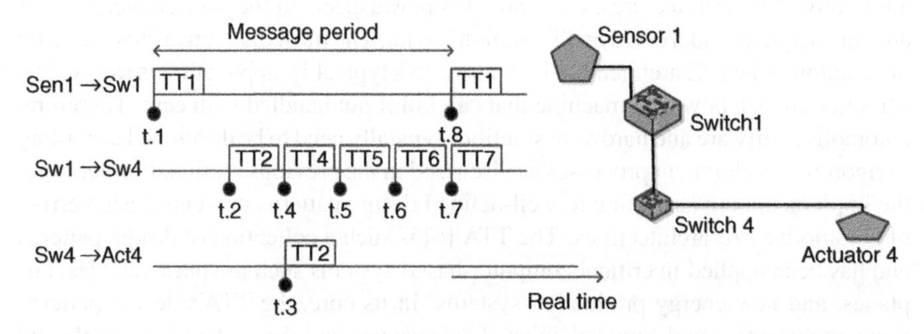

Figure 8.6 Example of communication schedule in time-triggered communication.

The example shows that the transmission of message TT1 from sensor 1 to the switch 1 is scheduled at the point in time t.1. The switch may then compute a command based on this input and generate a new message, TT2, for actuator 4. Switch 1 sends this TT2 at t.2 to switch 4, which in turn forwards TT2 to the receiving actuator 4 at t.3. As depicted, communication links can be tightly scheduled, like the link from switch 1 to switch 4, with messages TT4–TT7 being forwarded at t.4–t.7. Finally, the complete communication repeats at t.8 and continues in a cyclic fashion. Since all devices use the synchronized time to execute the communication schedule, messages do not have to compete for transmission on the links and queueing delays are therefore avoided.

8.4 FUTURE AUTOMOBILES AS FOG NODES ON WHEELS

As we suggested in the previous section, the future automobile architecture will be consistent with and, in some ways, will inspire the general trends characterizing the evolution of IoT and the digitization of many fundamental industrial verticals, such as industrial automation.

IT technologies will reach deep into the traditionally OT automotive world in new and significant ways. More sensors, connected via wires or, in higher numbers, wirelessly, will collect and report more sophisticated information. Video, laser, and radar technologies will be keys in the support of assisted drive. More microphones will help in preventive maintenance, voice activated control, and sound management. Driver health sensors will monitor key vital parameters.

Communications with other vehicles and the infrastructure will see the full adoption of Wi-Fi DSRC [4] with its use for both collision avoidance and general meshed vehicular communications. Multiple cellular connections, including new long-range and low-power connections, will be pervasive. The vehicle cabin will become an entertainment and information center, as well as a mobile office, served by Wi-Fi, Bluetooth, NFC, low-power sensor networks, with high bandwidth available for video, voice, and data over IP. A rich computing and storage capability will be required to support high quality experiences in music, video, and gaming. Networking will move more in the direction of IPv6.

In-vehicle storage requirements will continue to grow. More data, even "Big Data," will be collected on both the vehicle health and on the passengers' health and experience. Some of this data will need to be processed, compressed, or analyzed in real time on the vehicle, and some will need to be uploaded toward data centers and clouds. Large amounts of navigation, entertainment, control data, and software will be downloaded into the vehicle.

Naturally, driven by the evolution of smartphones, the automobile will need to become a platform for the delivery of applications and become more open to the judicious use of open-source software.

All the trends above point in the natural direction toward fog computing.

Fog computing, in its full manifestation, with its scalable computing and storage architecture, rich wired and wireless connectivity support, virtualization for both non-real-time and real-time services, sophisticated data management and analytics support, secure computing and networking, and modern management and application deployment features, will enable:

- The convergence of key functions, today hosted in different, poorly communicating subsystems and ECUs, into virtual functions hosted in virtual machines (VM) or virtual containers (VC), running on server class, but low-power CPUs and GPUs, and potentially supported by accelerators (e.g., FPGAs), all supporting hardware-level virtualization functionality
- Secure software management, with non-service impacting upgrades
- Secure hosting of open-source software and modern application management, with multiple OS
- Centralization of networking and security functions, through SDN approaches. Centralization of the security function (e.g., encryption, IDS, IPS, Firewall) is particularly important, since it would be unmanageable, costly, and even unfeasible to allocate them to them to distributed ECUs
- Natural interplay between cloud and fog on the vehicle activities, particularly in the navigation, entertainment, and data management and analytics areas.

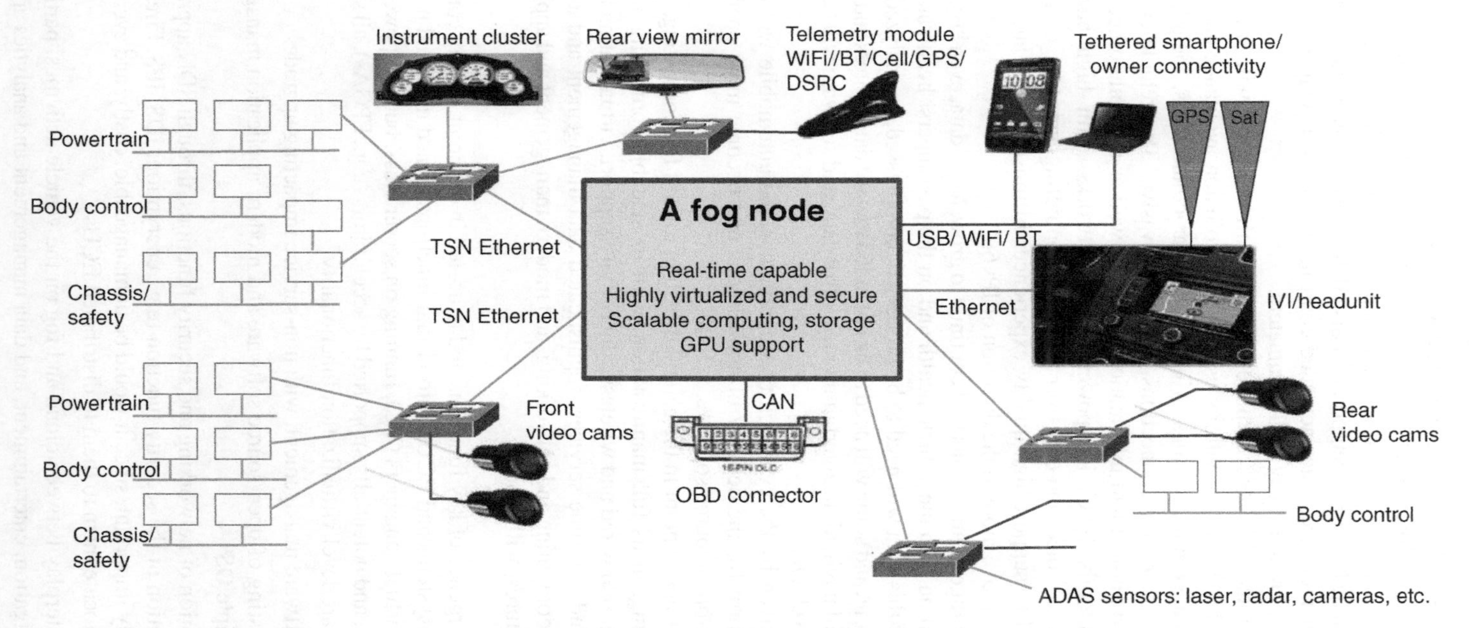

Figure 8.7 Future automotive E/E architecture.

Figure 8.7 illustrates the E/E architecture of future automobiles. Note that the full enjoyment of the rich features of the future highly connected and IT-rich automobiles can only be achieved once such automobiles can truly be autonomous. The evolution toward this goal is the topic of the next section.

8.5 DETERMINISTIC FOG NODES ON WHEELS THROUGH REAL-TIME COMPUTING AND TIME-TRIGGERED TECHNOLOGIES

As discussed in the previous sections, opportunities arise from more tightly coupling the IT and automotive-specific computer technologies. Inventions in the IT domain then can be leveraged by the automotive industry much faster and more efficiently, and they can help to solve the challenges of future automotive E/E architectures. The fog node provides capabilities of utmost importance to address the four challenges outlined before (scalability in the number of ECUs, connectivity within the car and to the outside, security, rich analytics for autonomous drive). Two of the most important capabilities of the fog node are virtualization of resources and managing/participating in deterministic communication.

In the following two subsections, we discuss virtualization on the ECU-level and deterministic communication on the in-vehicle network in more detail. Toward the end of this section, we address an emerging use case of fog computing in the car—vehicle-wide virtualization (VWV)—that combines virtualization on the ECU-level and deterministic communication. This use case provides a relevant step toward the ultimate vehicle E/E architecture.

8.5.1 Deterministic Fog Node Addressing the Scalability Challenge through Virtualization

Virtualization is a well-known concept in the general-purpose computing industry for more than a decade. It introduces an abstraction layer, such that applications (and indeed complete OSs) execute on VMs or VC, instead of their direct execution on the underlying hardware (as depicted in Figure 8.8). VWV is an attempt toward general-purpose and automotive-specific technology coupling by introducing virtualization holistically to the automobile. VWV simplifies the integration of innovations onto a single physical platform while guaranteeing freedom of unintended interference with other systems. Therefore, it allows to significantly cut back the turnaround time from invention to prototyping and even to series-production readiness.

There are plenty of benefits of virtualization in general-purpose computing and many translate easily to benefits also in embedded systems [7]. For example, (i) virtualization supports multiple OSs. Thus, rich OS like Linux can coexist with real-time OS on the same hardware platform. Furthermore, (ii) services that used to run on dedicated hardware can be consolidated into a single hardware platform. Indeed, this transition from federated to integrated architectures has been leveraged by other industries like the avionics industry for many years. In the automotive domain, AUTOSAR is one approach in this direction. Also, (iii) virtualization is a means

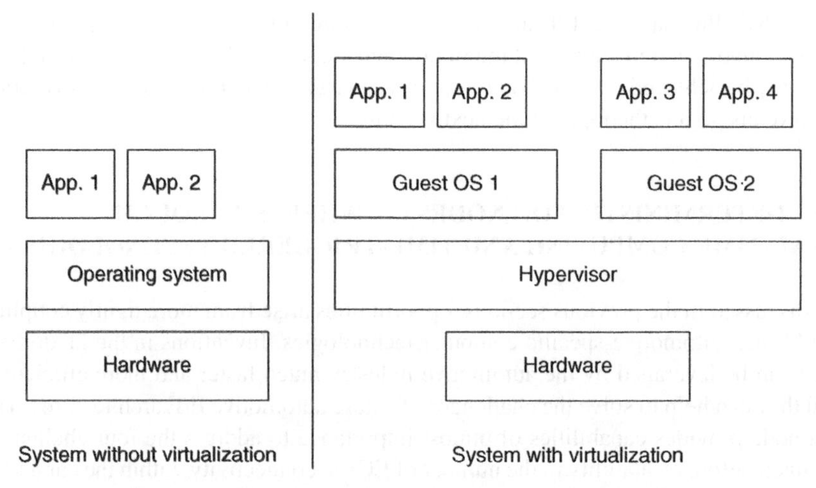

Figure 8.8 Classical system architecture versus virtualization.

to maintain security properties—a security breach in one VM remains isolated and does not affect applications executing in other VMs.

Virtualization techniques allow a single ECU hardware to run multiple so-called VMs in parallel. For example, it is possible to run AUTOSAR (and even multiple instances thereof) in parallel with rich OS like embedded Linux. The hypervisor, as the core element of virtualization, guarantees bounded interference or even interference freedom from one VM to the respective others. Provided the right system architecture is in place, it can be argued that safety-related applications can be even colocated with other noncritical or lesser critical applications on the same ECU and furthermore connected via a deterministic Ethernet to even more ECUs. While virtualization of computing nodes is a well-adopted technique (at least in the general-purpose computer industry), the virtualization of the network is primarily associated with virtual LANs (VLANs). Only recently, the development of network virtualization with the aim of minimizing interference has gained momentum. Such a highly integrated ECU equipped with superior communication technologies, like deterministic Ethernet, is, indeed, a fog computer.

8.5.2 Deterministic Fog Node Addressing the Connectivity and Security Challenges

The fog node for automotive use needs to interact and manage deterministic networks to guarantee upper bounds on transmission latencies. Furthermore, there is a need for the network itself to be used for applications with varying time- and safety-critical requirements. Thus, the underlying network will be virtualized such that the same physical network can support a multitude of applications while minimizing the interference of different messages on each other as well as precisely characterizing such interference. Recently, automotive Ethernet has evolved as the

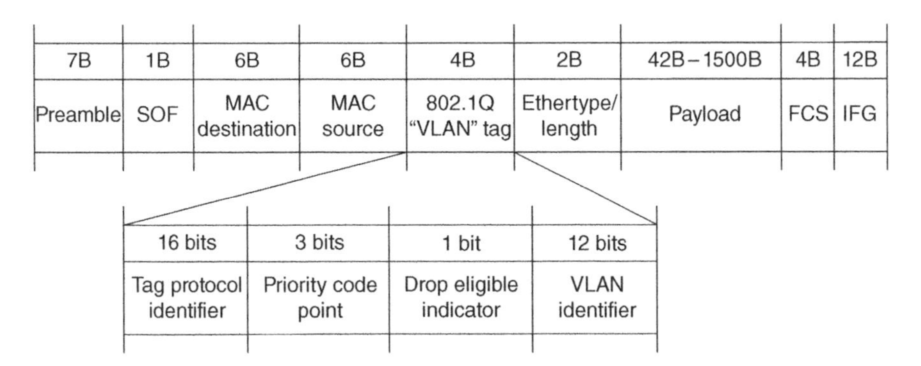

Figure 8.9 Ethernet frame format.

most promising candidate for such a network-level virtualization because of two main developments. First, automotive Ethernet physical layers are available, for example, Broad-R-Reach®, IEEE 100BASE-T1, and IEEE 1000BASE-T1. Second, the IEEE 802.1 audio/video task group has extended relevant Ethernet standards toward its use for real-time applications. Furthermore, the IEEE 802.1 Time-Sensitive Networking (TSN) task group continues standardization of robustness and fault-tolerant functionality in Ethernet-based networks as well as improved real-time communication. Deterministic Ethernet [8], TTTech's automotive Ethernet solution, implements IEEE AVB standards and IEEE TSN pre-standards as well as the SAE AS6802 standard, which provides fault-tolerant clock synchronization for safety-relevant applications. We continue with the discussion of some key features of Ethernet, IEEE AVB, IEEE TSN, and SAE AS6802 next.

Ethernet is standardized by the IEEE 802 standardization body. In particular the 802.3 working group defines the Ethernet media access control (MAC) layer and physical (PHY) layer, while the 802.1 working group defines Ethernet switch behavior (note: while switch is the well-known *terminus technicus* in industry, the standards use the term "bridge").

Ethernet messages are also called frames, and the format of an Ethernet frame is depicted in Figure 8.9. Traditionally, the IEEE enhanced the real-time properties of messages by using a priority field in the VLAN tag. Based on this priority, a switch can decide which one of the many messages to be transmitted next in case there are several messages waiting for forwarding. However, many use cases have been found in which the simple priority mechanism is insufficient. Thus, the IEEE 802.1 audio/video bridging (AVB) task group has developed real-time extensions to Ethernet that have been implemented in core IEEE standards in 2011 and include techniques such as bandwidth reservation, credit-based shaping, and a synchronization protocol, 802.1AS.

Although, AVB has enhanced Ethernet toward real-time capabilities (the goal is 2 ms latency over a network path of seven hops), the industrial and automotive industries have been pushing for better real-time performance as well as fault-tolerant features. Thus, the AVB task group has been renamed to the TSN task group and continues the standardization process. At the time of this writing, TSN is working on eight standardization projects (i.e., either defining a new standard document or modifying an existing standard). These projects address various areas, for example, time-triggered communication, improved synchronization, redundant communication, message preemption, and traffic policing and filtering. The first TSN standards are currently under finalization, most notably a basic form of time-triggered communication, called time-aware shaping.

A simple network and the operation IEEE 802.1Qbv time-aware shaper are depicted in Figure 8.10. Here, ECUs A and B send messages synchronized according a configured communication schedule (A1-2, B1-2), while ECUs C and D send messages unsynchronized (C1-3, D1-3). As depicted, switch 1 integrates the synchronized messages with the unsynchronized ones by selectively activating and deactivating outgoing queues of the switch: at those points in time when scheduled messages are arriving, the switch activates queue Q1 and deactivates queue Q2. If no synchronized messages are scheduled, then the switch activates queue Q2 and, thus, forwards the unsynchronized messages. Consequently, synchronized messages are transmitted through the network with minimal contention (possibly none) while unsynchronized messages may significantly queue up in the network.

The AVB and TSN shaping can be used to fine-tune the level of acceptable interference of messages belonging to different applications on each other. In case a medium level of interference is acceptable, then the messages can be sent as AVB traffic (i.e., being shaped by the credit-based shaper). On the other hand, if no interference (or only minimal interference) is acceptable, then the message transmissions should be scheduled. From a network virtualization point of view, the level of interference equals the quality level of the virtualization being in place—the less interference, the less the operational distinction of an application between running on a physically separated network from running on a virtualized network. Thus, synchronized communication (in the extreme, time-triggered communication, in which message transmissions are scheduled at each ECU and at each switch) allows a maximum of network virtualization quality.

The broader adoption of the IEEE Ethernet standards, with its layer 2 security features, resource prioritization and separation (e.g., VLANs), brings about important improvements in terms of security, with respect to the current automotive protocols, such as CAN Bus (e.g., no source address specified in the messages).

8.5.3 Emerging Use Case of Deterministic Fog Nodes in Automotive Applications—Vehicle-Wide Virtualization

Figure 8.11 depicts an example of VWV. In this simple example, we depict eight ECUs connected to each other by means of an automotive Ethernet network. In particular, ECUs 1–4 connect to a front switch, while ECUs 5–8 connect to a rear switch. The front and the back switches are connected to each other by means of an Ethernet link and together the switches and the link act as a backbone network that is shared for communication by all ECUs. The figure depicts ECU 1 and the backbone link

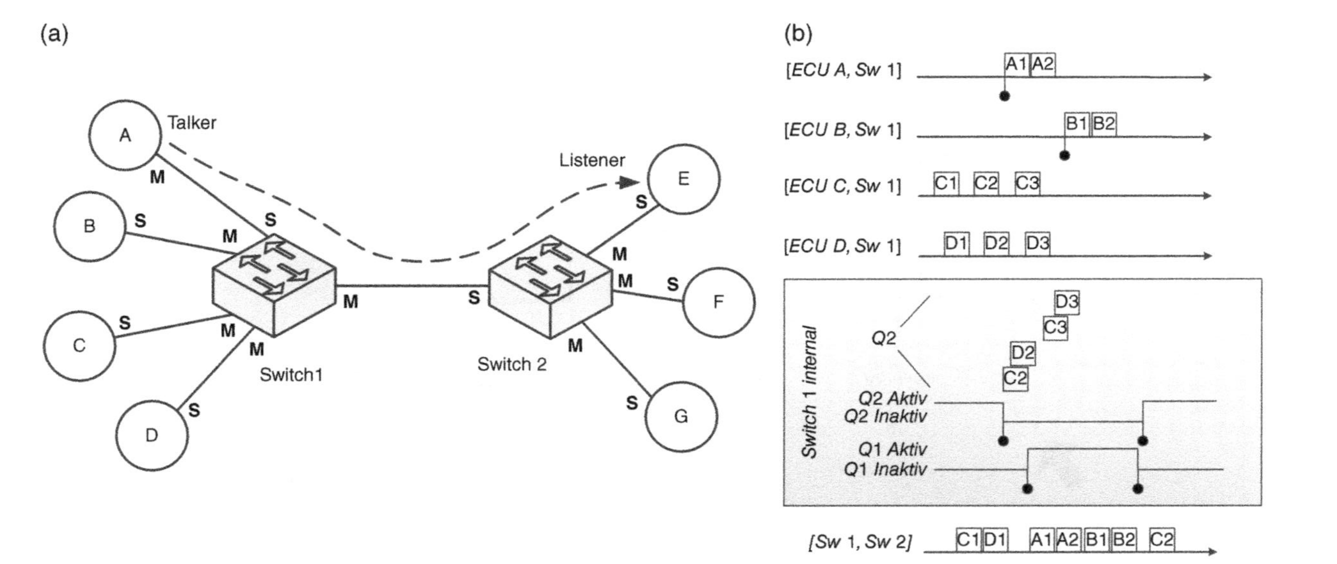

Figure 8.10 (a) Example of network and (b) communication scenario of the IEEE 802.1Qbv "time-aware shaper."

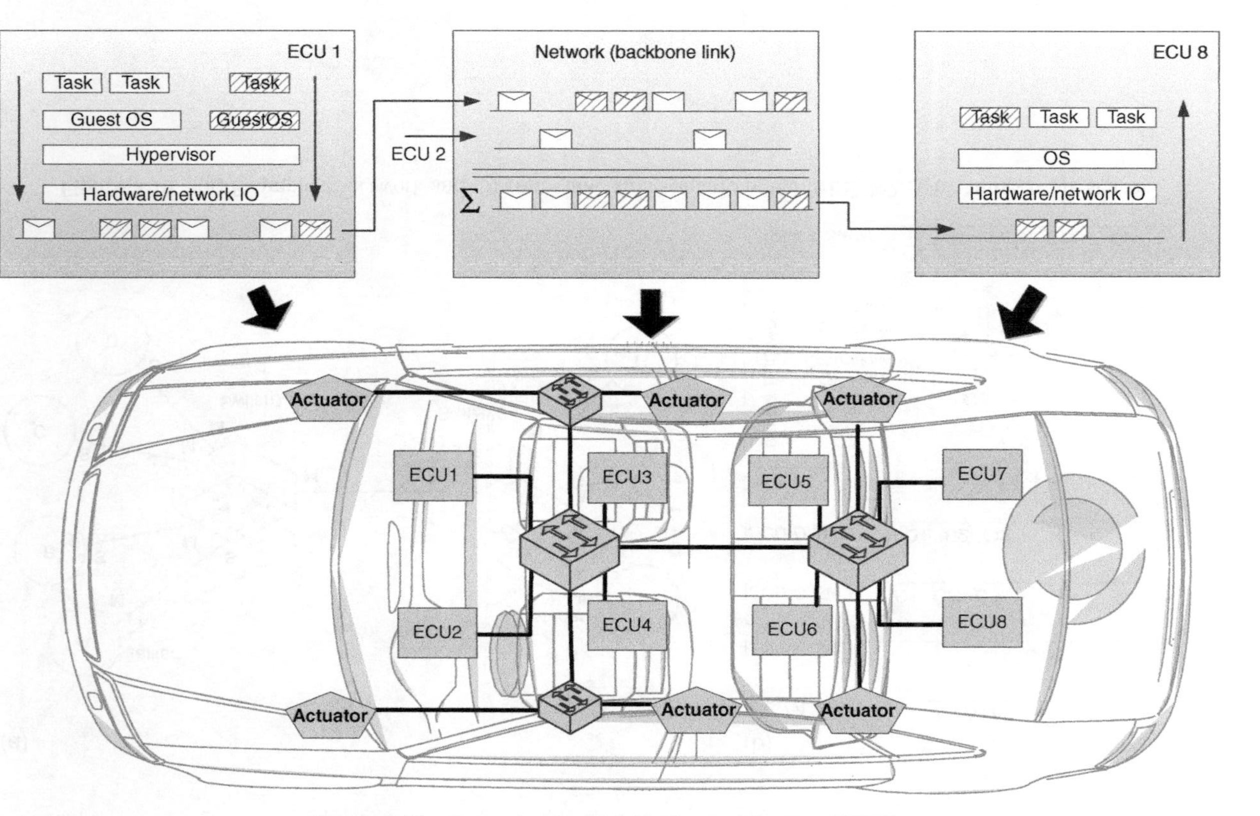

Figure 8.11 Example of vehicle-wide virtualization (VWV).

in more detail as well as the non-virtualized ECU 8. ECU 1 implements ECU virtualization in which a hypervisor allows the implementation of several (two in this example) guest OS. ECU 1 could actually be an automotive fog node. Applications run on top to the OS and consist of a number of tasks. As ECU 1 is connected to the shared Ethernet network, in particular to the front switch, the hypervisor needs to serialize the input from and output to the network. The network is then in charge to coordinate and serialize the messages from different ECUs on the Ethernet backbone link. In this example the network integrates the messages from ECU 1 with messages from ECU 2. Figure 8.11 furthermore depicts a simple ECU, ECU 8, as a consumer of the data produced by ECU 1 and transported by the network backbone link. We emphasize that VWV does not demand to construct all ECUs as virtualized devices. The concept rather suggests virtualization of ECUs as needed but allows the seamless interconnection also with non-virtualized ECUs.

This simple example illustrates the various scheduling decisions that can (but do not need to) be aligned within the distributed computer system. These scheduling decisions are (i) application-specific scheduler, (ii) task scheduler in the operating system, (iii) the hypervisor scheduler that controls the access of the virtual cores to shared hardware (e.g., to the network interface), and (iv) the network scheduler. It is easy to see that the better the individual scheduling routines are harmonized to each other, the more efficient the overall system may operate.

8.6 CONCLUSION

In this chapter, we discussed the challenges facing the current electronic automotive design and concluded that fog computing, in its most complete interpretation, provides an ideal infrastructure architecture responding to the fundamental challenges facing automotive electronic design. While this architecture makes software application design much more efficient and agile, the next challenge is exactly there.

More time-sensitive, secure, and functionally critical software will be needed to support future autonomous, electrically powered cars, based on a software engineering which needs to make important progress and is a topic for future discussion.

REFERENCES

1. F. Bonomi, R. Milito, J. Zhu, and S. Addepalli, "Fog computing and its role in the internet of things," in Proceedings of the First Edition of the MCC Workshop on Mobile Cloud Computing, MCC '12 (New York, NY, USA), pp. 13–16, ACM, 2012.
2. AUTOSAR, AUTOSAR Webpage, http://www.autosar.org/ (Accessed January 29, 2016; online).
3. Wired.com, http://www.wired.com/2015/07/hackers-remotely-kill-jeep-highway/ (Accessed January 29, 2016; online).
4. T. Zhang and L. Del Grossi, "Vehicle Safety Communications: Protocols, Security, and Privacy", John Wiley & Sons, Inc., Hoboken, NJ, October 2012.

5. P.C. Evans and M. Annunziata, “Industrial Internet: Pushing the Boundary of Minds and Machines,” 2012, http://www.ge.com/docs/chapters/Industrial_Internet.pdf (Accessed September 10, 2016).
6. H. Kopetz and G. Bauer, “The time-triggered architecture,” Proceedings of the IEEE, vol. 91, no. 1 (2003): 112–126.
7. G. Heiser, “The role of virtualization in embedded systems,” in Proceedings of the First Workshop on Isolation and Integration in Embedded Systems, EuroSys 2008 Affiliated Workshop, Glasgow, Scotland, UK, pp. 11–16, ACM, April 1, 2008. http://ess.cs.uni-dortmund.de/workshops/iies/2008/ (Accessed October 27, 2016).
8. TTTech, https://www.tttech.com/technologies/deterministic-ethernet/ (Accessed January 29, 2016; online).

9 Geographic Addressing for Field Networks

ROBERT J. HALL
AT&T Labs Research, Bedminster, NJ, USA

9.1 INTRODUCTION

Scenario. A team of firefighters is fighting a wildfire in a large wilderness area. They track each other on maps displayed on smartphone class devices to make sure they don't leave gaps in their lines through which the fire could move to get behind them. They monitor each other's biosensors in case someone is overtaken by smoke or heat or otherwise gets into trouble. An overflying drone helps by sending real-time images of the fire down to them. They collaborate by exchanging messages and map annotations with each other in real time to share intelligence about terrain, the fire, weather, or other exigent conditions. Oh, by the way, cell coverage is spotty at best and nonexistent in many parts of the wilderness.

Scenario. A group of young people is playing a large-scale game of *iTron* [1] in an area near their local high school. This is a new type of game involving both real-world athletic activity and virtual-world elements that add fun and imaginative content. Their person-worn devices (smartphones) maintain the state of the game, display a map with locations of both human and virtual elements, record scoring, and determine the outcomes of interactions. The cell coverage in this area is notoriously time dependent; rush hour on the nearby freeway can strongly impact or negate network availability and responsiveness unpredictably. There is no infrastructure Wi-Fi coverage.

9.1.1 Field Networking

Both of these scenarios, and many more, require *field networking*: data communications networks that enable devices in large-scale physical environments to communicate in order to support real-world tasks. In addition to emergency response and

Fog for 5G and IoT, First Edition. Edited by Mung Chiang, Bharath Balasubramanian, and Flavio Bonomi.

games, other applications of field networking ("field applications") include geosensing and data collection, process control, drone airspace awareness and control, military operations and force protection, military training, and connected and autonomous vehicles.

Field networking scenarios almost invariably involve fog networking subscenarios as well. The firefighters mentioned earlier need maps that come from servers in the cloud and may upload logged sensor data to the cloud to support after action reviews; the game players need maps as well, but they also store game results and records, such as high scores or league results which are computed in the field, in an Internet-accessible cloud service; geosensing applications, while often doing some local processing close to where the data collection occurs, still need to report the results up to cloud services for further processing and delivery to customers. Thus, field networking will necessarily need to support and interoperate with fog style networking applications. Also, since fog networking applications fundamentally involve some communications in the field, they need field networking support as well.

Field networks can be constructed using various wireless systems, including Wi-Fi (802.11), ad hoc Wi-Fi, and cellular data (e.g., 4G/LTE). These can be supplemented by wired networks, such as the Internet, when this is accessible. However, coverage by *infrastructure networks*, that is, those implemented using installed towers, relays, base stations, or access points, is not always available in the field. The military, for example, often operates in remote areas far from cell coverage or Wi-Fi hots pots. First responders often find themselves in areas where either there is no coverage at all or else the infrastructure network is not functioning either due to damage from the emergency situation or else simply from being overloaded with people calling and messaging the affected area. Thus, field networking must not be totally defeated by the (temporary or permanent) unavailability of infrastructure networks; rather, it should be able to continue operating among devices present and operating in the field, even if such operation is limited in some ways.

9.1.2 Challenges of Field Networking

Field networking brings many challenges that are not well met by traditional networking approaches based on Internet Protocol (IP) routing over infrastructure-based installations.

- *Device Mobility.* Field networking fundamentally involves movement of devices, whether such devices are smartphones or embedded devices. Even in nonmoving applications, such as sensor networks, the nodes simulate movement through temporal topology changes in the sense that from one moment to another, some devices may be sleeping to save power, and hence unavailable to participate in networking, or even failed due to running out of power. Movement (and other temporal topology changes) causes neighbor relationships ("links") among devices to change, and so viable routes through the network can change from moment to moment. The rate of change of topology is affected by factors including speed of movement, radio range of devices, and terrain complexity. Traditional IP networking relies on determining its routing tables by sending topology packets between devices and then caching the discovered information; each time links change, this information is degraded and will lead to lost packets and, worse, useless protocol retries. Thus, field networking is not well served by routing tables and routing based on cached next-hop information ("proactive" routing protocols). By contrast, the GA protocol discussed in Section 9.3 does not rely on topology information and discovers routes on the fly.

- *Spatial Density.* In some field applications, it is common to find large numbers of wireless devices near each other. This means that their packets must take turns using the common frequency. In such cases, it is easy for inefficient protocols to use up this common resource, leading to delays and failures. For example, consider distributing a video feed to all the devices in an area. If this must be done by sending a copy of the feed individually to each device, it doesn't take many devices to exhaust the available resource. A more scalable method takes advantage of the fact that one-to-many broadcast packets, as is possible in 802.11 Wi-Fi, naturally go to all in an area, so it is possible to send the feed as a sequence of broadcasts, without having to send a copy of each packet to each device. This symbiosis of broadcast and geographic addressing (GA) can be exploited (see Section 9.3) to gain algorithmic improvements [1,2].
- *Gaps in Coverage.* When operating in field conditions, it is common for devices to move out of range of cell towers. This severely limits the usefulness of applications implemented in a way that requires full-time access to a server in the cloud, because they stop working entirely when out of cell coverage. Instead, field applications should be able to continue operating, perhaps with reduced capability, even when out of range of infrastructure. For example, the firefighters should be able to continue tracking each other in the field even as they move into canyons or other areas not covered by cell data service. During those times, of course, they will not have access to cloud-based services (e.g., a map server or contact with central command) or even to tracking people who are far away from them, but they can still track and message each other locally, as long as the network can support peer-to-peer operations. The scalable geographic addressing framework (SGAF) multi-tier architecture [3] discussed in Section 9.5 integrates peer-to-peer wireless tiers with cloud-enabled long-range tiers in a way that (i) the two complement each other and provided redundancy when both are present and (ii) the system can seamlessly continue operating even if one of the tiers becomes unavailable.

This chapter overviews an approach to field networking, the SGAF, that meets these challenges better than traditional networking approaches. It is based on the observation that *GA is both a natural way to think about the communications underlying many field applications as well as an addressing paradigm that admits efficient and robust implementations under the stringent conditions experienced in the field.* The SGAF is demonstrated in a prototype known as the *AT&T Labs Geocast*

System (ALGS) [3]. ALGS supports prototype smartphone applications, described in Section 9.7, in the areas of athletic style games and field situational awareness and messaging, similar to the scenarios described mentioned earlier.

9.2 GEOGRAPHIC ADDRESSING

GA refers to communication protocols that allow a sender to specify the intended recipients of a message by where they are located in physical space. This is to be contrasted with traditional addressing schemes, such as IP addressing—where device addresses are specified by an integer that locates them within the Internet's hierarchical system of subnetworks—or even phone number addressing, which in olden times did actually indicate something about the location of the addressee but nowadays means little because most phones are mobile. Note that GA does not refer to a particular device or set of devices that is fixed for all time; typically, the sending device does not know in advance which, if any, devices lie in the area. The address refers exactly to the set of devices that are in the area at the time the message is transferred, a set whose extension changes rapidly with time. In the ALGS prototype, geographic addresses are specified as circles on the surface of the Earth, with a pair of integers specifying latitude and longitude in microdegrees and an integer representing the radius of the circle in meters. We will refer to the geographic address of a message as its *geocast region*.

One reason that GA is well suited to field applications is that their stereotypical communication patterns are often geographic. For example, a field situational awareness application allows devices to track the locations and sensor data of others in an area where the team is working. I will refer to this as *awareness messaging*. This naturally requires messages (e.g., queries and responses) flowing from one device to all others in the area. If the team is being monitored from a distant command center, once again messages flow between command and the area of operation. In the game applications, in addition to awareness messaging whereby the devices keep each other updated on location, score, health status, etc., there can be virtual weapon interactions, whereby a "shooter" can fire a simulated round by sending a message to all devices in the area that could be affected by the simulated round. A simulated sensing drone that virtually flies overhead to help search for opponents can send query messages to devices within its simulated area of sensing. Finally, in geosensing and data collection, a collection device (e.g., mounted in a drone overhead) can query the sensors that are currently awake within a particular field of interest to retrieve their stored data.

To handle extended transactions, involving more than one packet exchanged between nodes as they move around, a number of approaches are available. If GA is implemented over a normal IP network, then once awareness messaging discovers the identity of a desired transaction partner, a standard TCP/IP connection can be made. This will persist as long as IP connectivity persists. However, in networks either where the IP layer is not robust or even not present at all, a transaction can be retained by a technique known as *reverse path forwarding:* the initial GA packet discovers a route consisting of the sequence of relay nodes it passes through, and then replies and replies to replies can be source routed to follow the discovered path (or its reverse). In a highly dynamic network, it is necessary to periodically send a new GA message to rediscover the latest route, which may have changed.

In all the previously mentioned messaging patterns, a system based on traditional addressing schemes would first have to query some "geo-server" that was keeping track of all device locations and then formulate individual messages to send to each within the queried region. Moreover, as devices moved around, they must send updates to the geo-server with their new locations. This would be functionally equivalent to GA but forcing each application to maintain the service, at significant overhead, by itself. In a setting where many devices run many field applications (visualize a busy city at lunch hour), doing this over and over at the application level could easily swamp the available wireless network resources with overhead messages.

In some situations, there are even deeper problems. For example, when field applications must operate without coverage of infrastructure communications networks, such as cell data networks, it is not obvious where an application could locate its geo-server that would enable devices to continue functioning if some of them enter gaps in coverage. Instead, by building in the GA primitive into the core of the network, subsets of devices connected together by peer-to-peer networking can continue to function even when not in coverage.

Of course, there may still be applications, even wireless applications, that are better suited to traditional IP networking, such as watching a video from the Internet while in a local restaurant. GA is a tool well suited to many field applications, but it is not claimed to be the best tool for *all* wireless applications.

Outline. The rest of this chapter will overview an architecture, SGAF, for implementing GA within a large-scale heterogenous network environment. It will explain some of the scalability and other advantages, as well as some of the applications possible using GA. SGAF is composed of multiple network *tiers* that work together to implement end-to-end GA; each tier is characterized by how it implements GA. So the rest of the chapter is structured as follows. Section 9.3 describes scalable ad hoc geocast protocol (SAGP), a protocol for GA in peer-to-peer wireless networks. Section 9.4 describes a georouting approach to GA appropriate for tiers using wide area networking like the Internet. Section 9.5 describes how these are put together using bridging rules into a multi-tier large-scale GA service. Two final sections describe the ALGS prototype and some representative implemented and trialed applications, respectively.

Note that the full details of many of these building blocks are available in supporting papers, as cited; the purpose of this chapter is merely an overview.

9.3 SAGP: WIRELESS GA IN THE FIELD

This section gives a brief overview of the SAGP protocol; the reader is referred to Ref. [4] as the primary source for more details.

To meet the challenges of geographic density and operation in the absence of infrastructure networks, our first network tier is implemented over ad hoc Wi-Fi (802.11 in IBSS mode [5]). This is a peer-to-peer wireless network, sometimes known as a mobile ad hoc network *(MANET)* [6], where the individual end devices both source and maintain the network synchronization without the central coordination of a base station. Once attached to a given ad hoc SSID, a device will either maintain its synchronization when hearing other devices beaconing the same SSID or, when not hearing anyone else beacon it for a given interval, will transmit its own beacon to maintain the network (in case other devices are around). In this way, devices can communicate directly with each other even when there is no nearby base station controlling things. This mode is part of the 802.11 standards, and most commercial chip sets and devices are capable of participating, even though these days many smartphone operating system providers choose to disallow the creation or user selection of ad hoc SSIDs.

A key motivation for using 802.11 as basis for GA is that it allows devices to transmit broadcasts that can be heard and processed by any devices within range. This means that a single transmission can move information from a sender to everyone in the area covered by the radio range of the device. By *relaying* packets, that is, rebroadcasting received packets, in this way a GA packet can be routed to all recipients in the geocast region. I term this relaying of GA packets by repeated broadcasts a *geographic broadcast* or *geocast*. There are other ways to implement GA packet transmission that do not use this technique, so while all geocast protocols are GA implementations, not all GA implementations are geocast protocols.

9.3.1 SAGP Processing

The *SAGP* [4] is a GA protocol that can be implemented within any network that supports one-to-many broadcasts, such as ad hoc 802.11. It operates as follows. The originator formats a geocast packet that includes the geocast region in the form of three integers: latitude and longitude, in microdegrees, of center of circle, and radius in meters. The originator broadcasts the packet. This broadcast is heard by all devices within radio range.

For intuition, a smartphone held about four feet above ground seems to have about a 100–150 m range to another such smartphone in a clear flat terrain. This is strongly affected by terrain complexity, the model of phone, what is holding the phone, and how the phone is oriented.

Whenever an SAGP device receives a broadcast geocast packet, it carries out several operations concurrently:

- It determines whether it is within the packet's geocast region. If so, it delivers the packet to any higher layer components or applications that are waiting for geocast packets.
- Whether or not it is within the geocast region, it enqueues the packet for possible retransmission.

- For each copy of each geocast packet received, it records statistics about the geocast. The geocast—as opposed to the copies of the geocast—is identified by the originator device ID and its origination time recorded in the packet header. All copies of a geocast will have these two field values. The statistics include the nearest location of any transmitter of the geocast, the minimum distance of a transmitter of the geocast from the *center of geocast region (CGR)*, and the total number of transmissions (copies) of the geocast heard.
- It processes the head of the transmission queue. In particular, it dequeues the head packet and decides, based on the *geocast retransmission heuristics* described in the following text, whether to retransmit (i.e., rebroadcast) a new copy of the geocast.

9.3.2 SAGP Retransmission Heuristics

The heuristic decision is best thought of as a logical predicate P structured as

$$P \equiv F \wedge (M \vee T \vee \mathrm{CD})$$

where

- F is true iff the device is located within the *forwarding zone* defined by the packet's originator's location and geocast region. The forwarding zone is a configurable parameter of SAGP, but in the ALGS, the default for the SAGP tier is to use the union of a circle of radius one radio range surrounding the originator and a circle surrounding the CGR of radius one radio range larger than the radius of the geocast region.
- M is true iff the number of copies of the geocast heard by the device is less than the configurable parameter m. In the ALGS, it is typical to use $m = 2$ or $m = 3$ or 4 in complex terrain. The intuition is to ensure a minimum redundancy by transmitting at least m copies in any region containing a device.
- T is true iff the minimum distance of the transmitters of all copies heard so far has been at least t meters distant. The intuition here is that if the device is sufficiently far from all others it has heard, then there is a good chance someone even farther may benefit from hearing it transmit.
- CD is true iff the device is closer to the CGR than the transmitter of any previous copy. The intuition here is hill climbing: a device will decide to transmit if it appears to move closer to the CGR than other transmissions.

Figure 9.1 shows schematically the layout of sender, geocast region, and forwarding zone. The forwarding zone is the union of the two-dashed circles in the ALGS short-range tier.

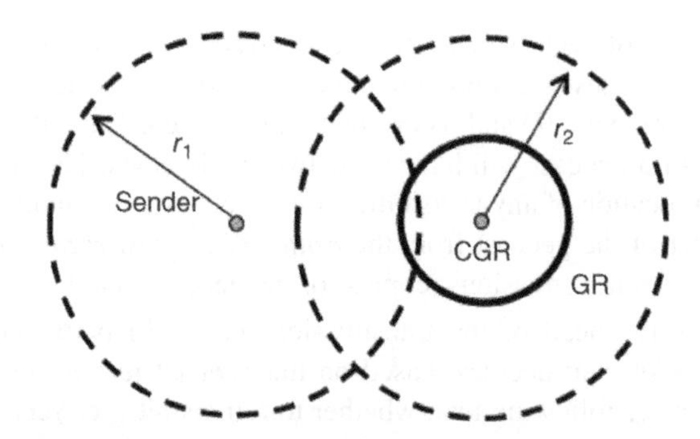

Figure 9.1 Sender, geocast region, and forwarding zone.

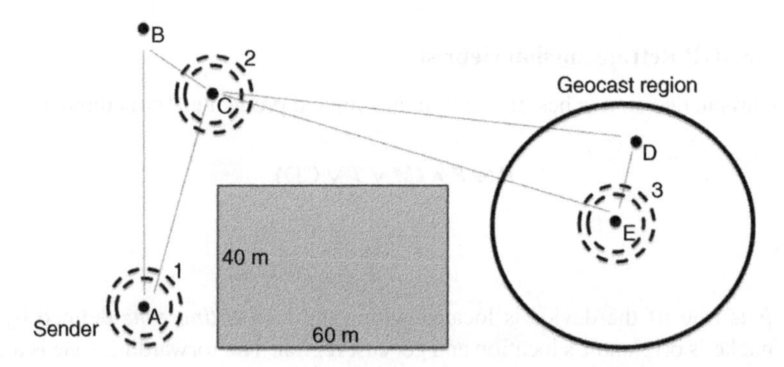

Figure 9.2 Example of a geocast propagated via SAGP.

9.3.3 Example of SAGP Packet Propagation

To illustrate how SAGP transfers a geocast packet, consider the diagram in Figure 9.2. Devices A, …, E are operating the SAGP protocol. A, the originator, transmits first (transmission 1). This is heard by B and C. C's M heuristic is satisfied, so it retransmits a copy of the geocast. B hears this, and since it is no closer to CGR than C, it has heard $m = 2$ copies, and since it is within the T heuristic's threshold distance ($t = 40m$) from C, it suppresses retransmission. Both D and E receive transmission 2 from C, which is their first copy, since the radio obstacle blocked transmission 1 from them. E, being closer to CGR than C was, retransmits based on the CD heuristic. (In this case, both T and M heuristics are also true.) D suppresses retransmission, because none of its heuristics are satisfied. Note that whether B or C decides to retransmit first is nondeterministic and based on low level factors. In fact, the ALGS implementation of SAGP uses a pseudorandom backoff delay, determined in part by node ID number, that helps avoid collisions between retransmissions.

Note that this scheme is significantly more efficient (measured in terms of fewer copies transmitted) than simple flooding. In simple flooding, every device who heard it would have retransmitted, leading to a total of five instead of the three resulting

Figure 9.3 Pictorial proof idea showing why SAGP only uses $O(\lg n)$ transmissions per geocast in dense scenarios. The originator is somewhere to the left of the diagram; the intuition is that, on average, successive transmissions will occur at devices approximately half way to the CGR.

from SAGP. In general, the difference between simple flooding and SAGP can be large, scaling as $n/\lg n$ with number n of participating devices.

For example, a critical class of situation is the case where all n devices are within radio range of each other without obstructions. In that case, each packet would be transmitted n times by simple flooding. SAGP, by contrast, would on the average only transmit $O(\lg n)$ copies of each. The proof is illustrated pictorially in Figure 9.3. Essentially, due to the CD heuristic, on the average, each successive transmission will take place approximately halfway closer to the CGR than all previous. In this dense scenario class, the T and M heuristics only increase the transmissions by constant amounts.

9.3.4 Followcast: Efficient SAGP Streaming

As has been observed by others, fully reactive routing, where each packet discovers its own routing path as in SAGP, can be inefficient in scenarios where neighbor relationships remain stable for times long compared with the inter-packet intervals.

A common case of sending many packets in a row to the same geocast region arises in streaming applications. In the firefighter example in Section 9.1, the drone could be streaming video down to team members; in that case many packets in succession would need to follow essentially the same route, since they would be sent much more quickly than humans walking will change neighbor relationships.

The key insight that allows optimizing this case is that we really only need to rediscover a geocast path on a timescale related to the changing device relationships, not as fast as packets are sent. The *followcast* extension to SAGP [2] has means for the network devices to remember paths taken by recent geocast and record whether they were *useful* relays. A relay was useful iff it was the transmission first received by some other device. When a followcast packet is sent following an earlier full geocast packet (which packet a followcast is following is indicated in the header), a device retransmits it if and only if its earlier transmission is recorded as useful.

Figure 9.4 illustrates how this works. The original geocast is originated by A, and its transmissions are shown as numbered dashed circles. The range of each transmission is indicated by the "RR" lines; the radius of the dashed circles is *not* the radio range here. The gray rectangle is an obstacle that blocks signals. Following the sequence, the transmissions indicated by arrows are the only ones that were useful: transmissions 3 and 5 were heard only by devices that had already heard at least one

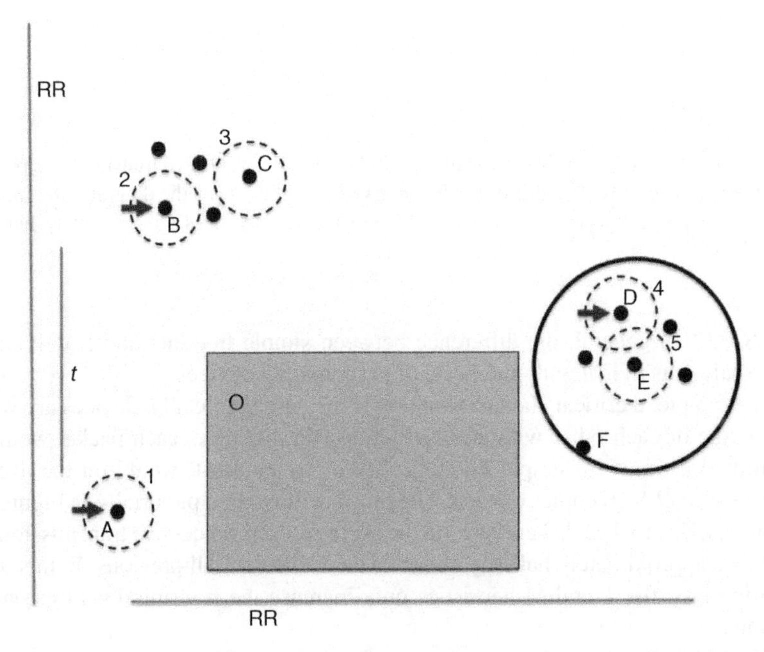

Figure 9.4 Example SAGP geocast propagation, with arrows showing the useful relays.

of the useful set $\{1,2,4\}$. The useful set is recorded in the geocast packet records at each device. They are retained as long as the packet records are retained, which in the ALGS implementation is 10 seconds. So followcast streaming sends a geocast once every *refresh interval*, which is chosen to be a bit less than 10 seconds, which is the lifetime of packet records. It then followcasts packets in between such geocasts. This typically results in large savings [2].

Note that followcast can be used both within streaming applications, where packet loss is tolerable within bounds, such as transmitting video or audio, or it can be used in reliable transfer, as long as retries (similar to those of TCP) are built into the higher level protocol that uses followcast as a stream primitive. Hall [2] describes these two types of application.

9.3.5 Meeting the Challenges

It is worth observing how SAGP meets the challenges of field networking discussed in Section 9.1.2.

Each geocast discovers (via the heuristics) a fresh path from sender to geocast region. Since this is done at the time the packet is sent, it is not relying on cached, possibly stale, routing tables. This minimizes packet losses due to mobility, since if a path exists at the time, SAGP will likely discover it, even if the discovered path has only very recently come into being. A second point relating to mobility is that traditional routing strategies rely on *routers*, which are distinguished nodes of the network that make decisions on transferring packets. SAGP can continue to operate even under arbitrary network partitioning resulting from mobility changes. If a path exists within the partition containing the source and destination devices, the SAGP can likely find it; however, if a device is partitioned from the routers in a traditional system, it cannot participate. Thus, SAGP meets the *device mobility* challenge better than traditional proactive routing strategies.

As discussed earlier, in a highly spatially dense scenario, SAGP typically uses about $O(\lg n)$ transmissions per geocast. This is fundamentally lower than the $\Omega(n)$ transmissions used either by simple flooding or by a traditional routed unicast-based approach. Even a traditional multicast group that sends packets over a spanning tree will use the same $\Omega(n)$ transmissions. So SAGP by exploiting one-to-many broadcasts can significantly reduce spectrum use in *spatially dense scenarios*.

Finally, since SAGP uses peer-to-peer ad hoc networking, it can operate (or continue to operate) when some or all of the devices move into *coverage gaps* of infrastructure networks, whether these be cellular data or Wi-Fi base stations. As long as the devices are within range and a chain of relays exist between sender and geocast region, SAGP continues to operate. In this sense, SAGP provides a way to extend the reach of wireless networking beyond the "edges" of infrastructure cloud networks.

9.4 GEOROUTING: EXTENDING GA TO THE CLOUD

Wireless ad hoc networks cannot support all requirements of field networking. In particular, remote servers accessible via the Internet may provide necessary services to users in the field. A good example is a map server holding a database of maps or satellite images that can be used as backgrounds to display tracking information or annotations on devices in the field. A long-range network connection can also allow remote users to communicate, monitor, or otherwise interact with team members operating in an area.

A second critical need is for covering long-range gaps between field users. For this purpose, wide area wireless networks such as cellular data (e.g., 4G/LTE) can reach much farther and can bridge gaps between devices too far for connectivity through a wireless ad hoc tier.

Of course, if devices do not have connections to long-range networks, then they only have access to the ad hoc tier. However, smartphone class devices typically have two radios, one for Wi-Fi that can be used for the ad hoc tier and one for a cellular data connection that can be used to connect to the Internet.[1] The long-range extension *(LRE)* tier of the SGAF architecture can provide GA that moves packets through the long-range tier seamlessly. It is "seamless" in the sense that devices still just address packets using a circle (center and radius) and the GA subsystem transfers

[1] Actually, most smartphones have a *third* communication medium as well: Bluetooth. It is typically used for short-range (a few meters) communications. There are cases where it may be useful to extend GA to this ultra-local tier, and the SGAF architecture can support this.

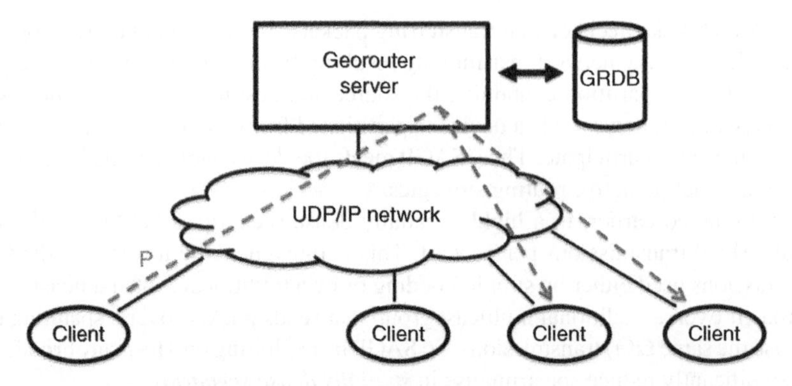

Figure 9.5 Packet transmission in the georouter tier.

the packet through whichever tiers are accessible as guided by transmission heuristics and bridging rules.

In SGAF and the ALGS, the LRE tier is implemented as a *georouter* tier. A georouter is a server in the network that is in contact with devices in the field and is capable of inferring which ones are in or near the geocast region of a packet and of transmitting a copy to some or all of them.

Figure 9.5 illustrates the process of transferring a GA packet using a georouter. The packet P is sent first via UDP/IP to the georouter server. The server access its *georouting database (GRDB)* that lets it map from a geocast region to a set of IP address and port pairs representing devices known to be in or near (how near is a configurable parameter) the geocast region.

The most obvious method would simply forward to *all* devices in or near the geocast region; however, in dense scenarios this could be inefficiently redundant. In SGAF, we allow different policies to be plugged in, on a per device basis if desired, depending on the capabilities of the devices and their characteristic speeds of movement. By limiting the number of devices each packet is sent to, we can let the scalability of the ad hoc tier transfer it from the few representatives chosen to all within the geocast region. In ALGS, when it is known that all devices have ad hoc tier capability, the georouter by default chooses four devices among all candidates to which to send copies of the packet; bridging rules then retransmit the packet across the ad hoc tier. See next section for more discussion of bridging.

Hall *et al.* [3] and Hall and Auzins [7] discuss the concept of multi-tiered geocast and the LRE, including georouting and bridging, in detail. It should be noted that the general idea of georouting is an old concept; however, the SGAF and ALGS add some new techniques for scalability and bridging with other tiers.

9.5 SGAF: A MULTI-TIERED ARCHITECTURE FOR LARGE-SCALE GA

The SGAF is an architecture framework that provides general techniques for combining different GA tiers to achieve an overall GA service that can cross multiple

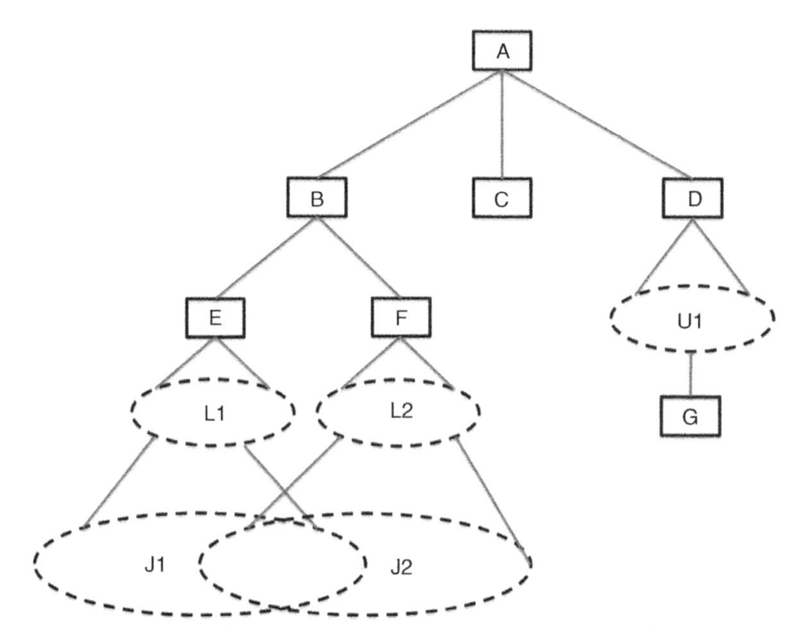

Figure 9.6 A notional example of a large-scale GA system built by bridging together many individual GA tiers. Rectangles represent georouter tiers, while dashed ovals represent geocast tiers.

heterogeneous networks. A *tier* is a set of devices that use a common communications medium to implement a GA service. A *geocast tier* is a tier that uses a geocast protocol (i.e., one based on wireless one-to-many broadcasts), such as SAGP, to implement GA. A *georouter tier* is a tier that does *not* use a geocast protocol; instead, it typically uses one or more georouters to implement GA.

Figure 9.6 shows a notional example of a large-scale GA system that is within the power of SGAF to express. It is composed hierarchically from multiple tiers. The rectangles represent georouter tiers, while the dashed ovals represent geocast tiers. Whenever a tree node contains multiple connections, it is connected by one or more bridge devices to the connected tier. As shown in the figure, there is no restriction on the types of connected tiers. A short-range geocast tier could be bridged to a longer-range geocast tier operating over longer-range radios, for example. Similarly, georouter tiers can be federated together so that different local georouter tiers could manage particular areas and a global georouter could connect them together.

9.5.1 Bridging Between Tiers

The SGAF is a parameterized framework that allows combining any number of tiers, of diverse types, together to implement a single overall GA service. Two tiers are connected by *bridging*, which occurs at *bridge devices*. A bridge device is one having multiple network interfaces, one to each tier in which it participates. The bridging behavior implemented at each bridging device is define by SGAF's *bridging rules*, which can be set differently for each tier interface type or even on a per bridge device level, if desired. Formally, bridging is governed by bridge functions:

$$BridgeFn_D : (GAPkt\ P, TierID\ T, Location\ L) \rightarrow Bool$$

where D is a bridge device and D transmits a received packet P on the tier T iff $BridgeFn_D$ returns true for P, T, at the D's current location L.

The simplest bridging function is identically true; that is, whenever a packet is received from one of its tiers, it always transmits it on its other tiers. Since a device will not reprocess a packet it has already processed, this cannot lead to infinite routing loops. However, it may be inefficient.

A different bridging rule, implemented in ALGS, acts as follows. Any field device having a long-range tier interface (e.g., LTE radio) will send the packet on the long-range tier iff the packet has not previously been sent on the long-range tier. The geocast header has a field recording this information that is updated each time the packet is retransmitted.

Figure 9.7 illustrates a multi-tier GA packet propagation. Device A transmits first, and then bridging device B transmits on both short-range and long-range tiers. The long-range tier transmission is sent to the georouter, which forwards it to bridging device C that happens to be located near the geocast region. C then transmits on the short-range tier, and D also retransmits, thereby enabling all the devices in the geocast region to receive it.

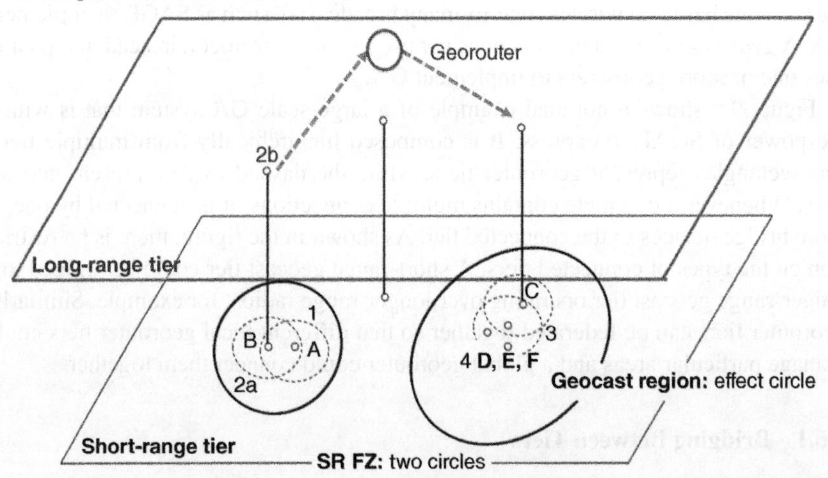

Figure 9.7 An example propagation across multiple tiers using bridging. Starting at A, the packet first traverses the geocast tier around A, then up through the georouter tier, and then back "down" into the georouter tier near the geocast region. Finally, it traverses that tier to reach the GR.

9.5.2 Hybrid Security Architecture

Network security is, of course, a major concern in large-scale networks. SGAF accommodates different approaches to security for each tier, because not all security techniques work in all tiers. For example, security in the long-range tier can be constructed using, for example, virtual private network (VPN) connections between devices and georouter servers. However, this style will not work in geocast tiers, due to the fact that packet propagation is fundamentally one to many. The one-to-one security relationships underlying VPNs scale quadratically in this context; that is, there would have to be a tunnel for every pair of devices. Not only is this too large a number to be manageable in large scenarios, the fact that devices move into and out of connectivity with each other rapidly means these tunnels would be difficult to maintain.

One way to secure a smaller-scale geocast tier is to use shared session keys managed centrally by a security administrator. That is, all devices in the group (e.g., responder team or game players) are issued the same session key, distributed from the administrator using individual key exchange keys. In this way, all geocast packets can be encrypted in the session key and read by all authorized devices for the purposes of secure and authentic packet transfer, while no outsiders can interfere or eavesdrop.

SGAF assumes that each bridging device is authorized for each tier in which it participates. Packets coming in securely through one tier interface are then resecured prior to bridging onto other tiers. This allows global-scale packet transfer across multiple heterogeneous security domains while still allowing localization of security administration per tier, effectively forming a web of trust for the purposes of GA packet transfer.

9.6 THE AT&T LABS GEOCAST SYSTEM

The ALGS is shown schematically in Figure 9.8. The name is a slight misnomer (in that it is not entirely a geocast tier) that has persisted for historical reasons. ALGS is actually a two-tier instance of the SGAF, with one ad hoc Wi-Fi (short-range geocast) tier and one LRE (georouter) tier. Some devices are Wi-Fi only and so connect only to the short-range tier. Examples of these include Wi-Fi-only sensor boards or tablets not having cellular service. Other devices are only long-range capable; for example, a cloud-resident information service that is accessible via GA only has an interface to the Internet. Still others, like smartphones, have both types of interfaces and take part in both tiers. Note that the cellular data system (GSM or 4G/LTE) is used to connect devices to the Internet as a key step in implementing the LRE.

Cloud-based services reside in the cloud on servers. These servers are given virtual–physical locations and report in to the georouting system exactly as if they were really located there. Thus, the programming abstraction is for GA-based applications to access cloud services by sending GA packets addressed to the area where the servers virtually reside. By convention, this location is in the Central Kalahari Game Reserve in Botswana. For example, the ALGS contains a map server

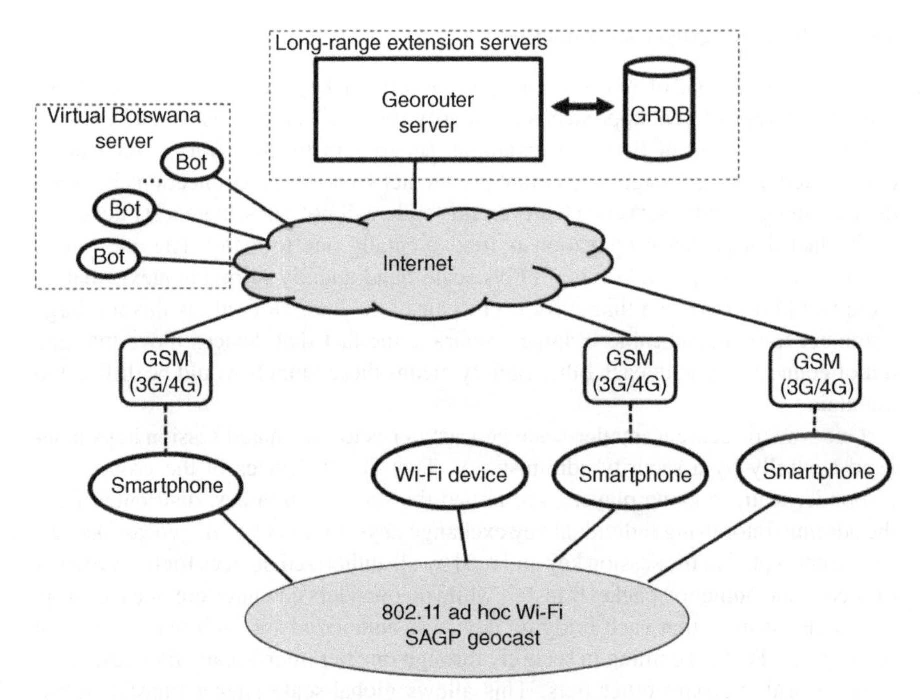

Figure 9.8 Schematic diagram of the AT&T Labs Geocast System.

resident in Botswana, and it is accessed via geotext messages, by which a device can request a new map by name. A requested map is transferred to the device using *geocast file transfer (GFT)* [2, 8], a reliable file transfer protocol built using GA and followcast as a primitive. In fact, the map can be transferred efficiently to *all* devices in an area in one GFT session, thereby avoiding the inefficiency of redundant individual file transfers. Other cloud-resident services include a game records server supporting the geocast games (see next section) and several bots, simulated entities used in gaming, demonstrations, and training.

The ALGS has been deployed and used for testing, trials, demonstrations, and other purposes for over 4 years. It supports both research and evaluations of new GA concepts and applications, as well as the applications described in the next section.

9.7 TWO GA APPLICATIONS

This section describes two field applications that are implemented and have been running in test/trial form for a few years now.

9.7.1 PSCommander

PSCommander is a smartphone application designed to support field operations teams. It provides a real-time map display showing positions, position histories,

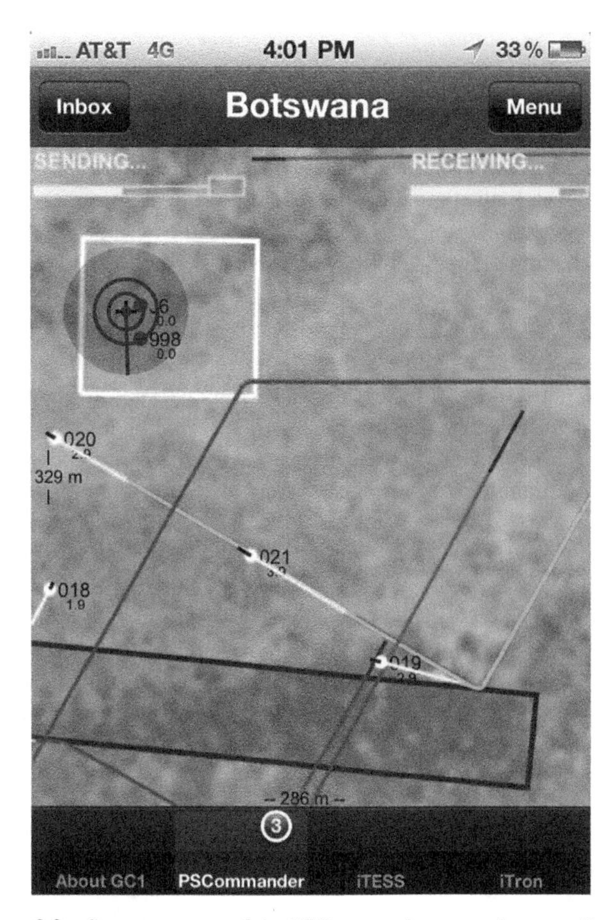

Figure 9.9 Screen capture of the PSCommander smartphone application.

and other data relating to other devices running PSCommander. An example screen display is shown in Figure 9.9.

PSCommander provides the following capabilities:

- *Live Location Tracking.* The user specifies one or more maps or satellite images of areas of interest. Figure 9.9 shows a sat image of part of Botswana. The system uses the GA-based *field common operating picture (FCOP)* algorithm [1] (see in the following text) to obtain location, location history ("tails"), and other data (in the figure: velocity information) from all devices in the areas covered by the maps. It draws the information using a color coding to indicate recency: green is the most recent, yellow is next most recent, etc. This is a critical aspect, because in field applications devices can drop in and out of contact, so in a safety support function, one needs to know how recent and reliable a report is in order to take actions.

- *Shared Map Annotations ("Zones").* Any user can create a rectangular zone on a map and attach a criticality level (indicated by color as green/yellow/red) and a text message. A different instance of the FCOP algorithm then propagates these zone definitions to all other devices in the area. When devices move into a zone, they get a displayed message and possibly an audible alarm, depending on criticality level. In the figure, the red rectangle is an alarm-level zone annotation with a message indicating operators should avoid entering. The green rectangle is an informational zone showing where virtual resources like the map server are located.
- *Geographic Message and File Transfer.* A geotext message is a text message addressed to a geocast region, and PSCommander supports these. However, it is useful also to be able to transfer larger files reliably, such as images. A user can, for example, use the smartphone's camera to take a picture, and then the GFT protocol [8] transfers it reliably to all operators in the designated geocast region. In the figure, light blue progress bars show incoming and outgoing transfers.

PSCommander has been trialed several times, and many scaling and other studies have been carried out, including a test involving attaching an SAGP-enabled smartphone to the side of radio-controlled airplane to measure relaying range and demonstrate feasibility of greatly improving connectivity among ground devices.

FCOP algorithm. The field monitoring problem is illustrated in Figure 9.10a. A device m wishes to request and then receive information from each of n devices

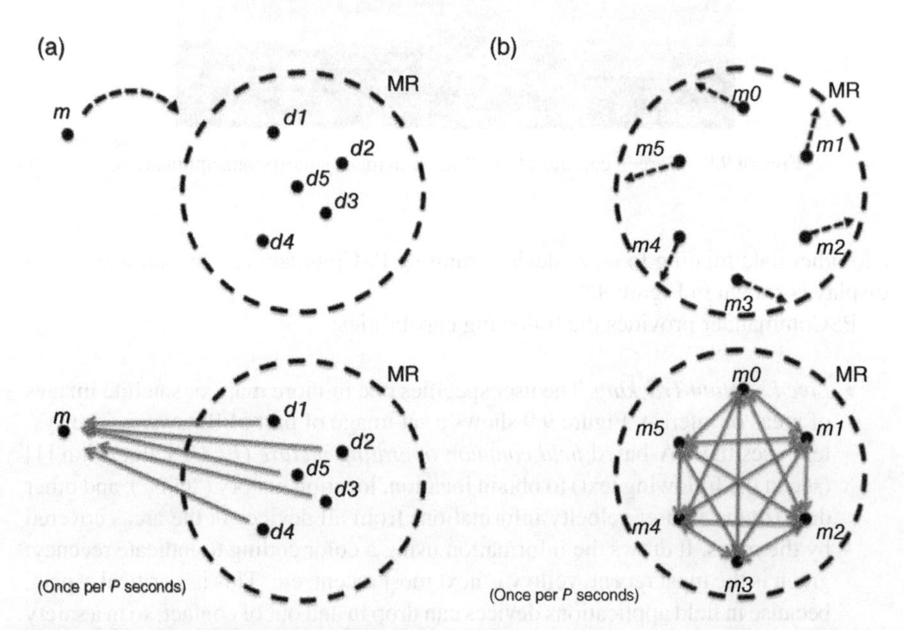

Figure 9.10 The FCOP problem illustrated. (a) The general monitoring problem and (b) the common operating picture special case.

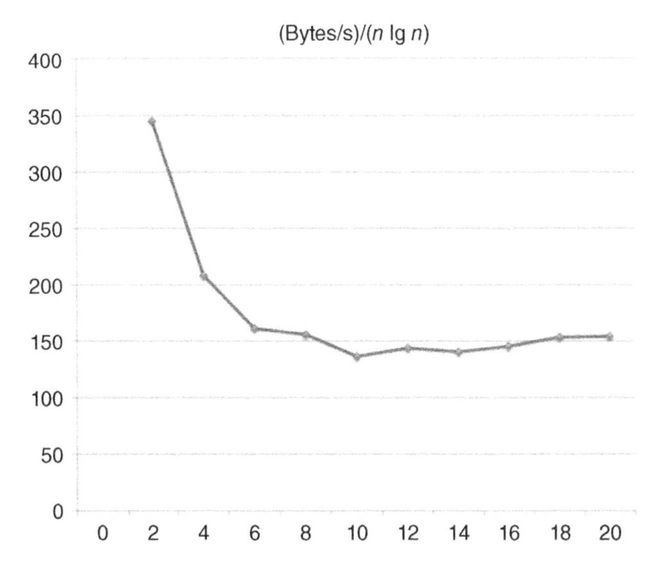

Figure 9.11 Graph of bytes per second used by the FCOP algorithm in a dense scenario, divided by $n \lg n$ versus number of devices n.

in a monitored region (MR). In the FCOP problem, Figure 9.10b, *every* device in the MR wishes to monitor MR. This results in $n(n-1)$ logical information flows, as shown in the lower diagram, because each device must report its information to each other. It only takes a few devices in the MR to swamp the available bandwidth.

However, using SAGP, we can do better. Using geocast queries and responses allows us in dense scenarios to use only $\lg n$ transmissions for each query and each response. The one-to-many property means that the response to a device's query can be received by other devices nearby. Thus, if a device has recently replied to a query from a device that is within the geocast region of a more recent query, it need not send another copy of the response, assuming that all devices record all responses they receive. The *FCOP algorithm* [1] is a distributed algorithm that sends two messages (one small query and one larger response) per device per interval of time, each using $O(\lg n)$ transmissions, so in total it gets its job done using only $O(n \lg n)$ transmissions, a significant savings for larger n over traditional approaches based on either simple flooding or unicast IP, which require $\Omega(n^2)$.

Figure 9.11 graphs FCOP bytes per second transmitted divided by $n \lg n$ versus number of devices. These measurements were taken of iPhones running PSCommander, with all n devices placed within radio range of each other in a clean 802.11 environment. The graph converges around 154, so the implemented FCOP algorithm transmits approximately $154n \lg n$ bytes per second. (The deviation for low values of n is due to lower order terms in SAGP's performance being significant for small n.)

The FCOP algorithm can be customized in various ways. In particular, the queries and responses can carry different information loads.

9.7.2 Geocast Games

Geocast games is another class of field application, whose members are games combining strenuous athletic activity with interesting virtual elements that augment the reality of the game. In the games implemented so far, the virtual elements and effects are displayed (visually or audially) using smartphones, which are carried or worn by the players. Figure 9.12 shows a simple game in progress, showing both the real-world view and the corresponding virtual view.

The prototype system offers three different geocast games:

- *The iTron Family.* In iTron, an arbitrary number of players move around in physical space, and where they move they leave a trail consisting of a virtual *wall.* This is similar to the classic *snake* video game but with real-world participants. The object is to be the last player who has not crossed any walls, either one's own, those of other players, or the edges of the bounding "arena." Figure 9.13 shows the final state of an iTron championship game I ran during a P.E. class at a local high school as the culmination of a multi-week teaching unit on iTron. The players used terrain features strategically and were forced to run at times to gain territory advantages. The networking was able to continue working even though this area has poor cellular network coverage, because the devices transferred messages over the peer-to-peer network.

  There are many variants of this game, including those with non-player virtual elements like pits and swamps, and noncompetitive artistic team-oriented variants. See Ref. [9] for a full description. Games can take place in arbitrarily interesting terrain, with game communications implemented using the GA system. The pattern is a variant of the FCOP algorithm but with scoring data in addition to location and trail information included in the messages.
- *iTESS.* The basis of iTESS [10] is hide and seek. However, each player is trying to locate all other players and tag them with virtual weapon fire. Each player's

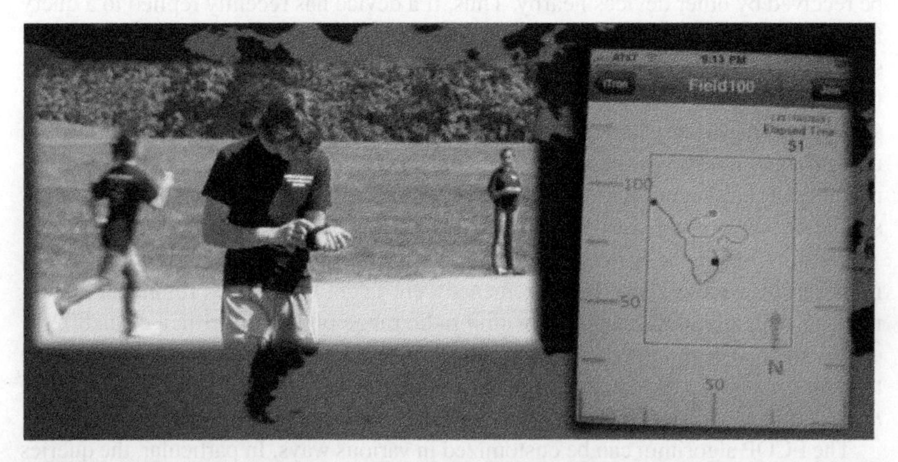

Figure 9.12 Typical iTron game, showing both real-world and virtual-world views.

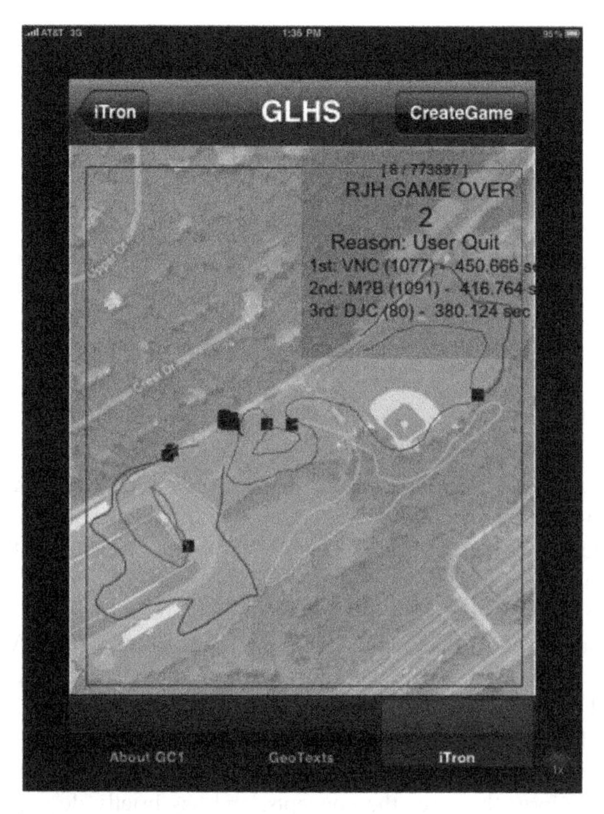

Figure 9.13 Final screenshot of the championship iTron match played at the culmination of a multi-week iTron teaching unit in a NJ high school Physical Education class. It was played at a larger scale, and players exploited different terrain types.

smartphone displays a map of the play area(s). Rather than simply displaying all positions of all players, however, the player commands a virtual "drone" that has a limited viewing area, so the player must direct it to search systematically. Of course, the player can also seek by looking as well, as the on-screen location matches the real-world location. In addition to this drone, players command an artillery weapon to drop on a circular target on the map; however, players are warned when they are within the target area of an incoming shell to give them time to react and run out of the way. There is also a point-and-shoot weapon [11] with similar properties.

iTESS's messaging patterns are for the most part direct geographically addressed messages to simulate the various elements. The drone simulation sends a GA message to the circle it can see; the weapon sends a GA message to the immediate vicinity of where it is targeted. Only game awareness messages (scoring, game state, and scoreboard information) are sent via an FCOP variant.

iTESS supports multiple play areas concurrently, allowing a player to switch views. This results in monitoring all the areas concurrently for game purposes. iTESS can be played over long distance, with games having been played between coasts of the United States, as long as all play areas have players with long-range capable devices within coverage of the cell network.

There are other geocast game families as well, such as *Butterflies* [10] and *Human Pong*. So far, after trials by hundreds of users of virtually all ages, there seems to be great interest in the potential of outdoor movement-style games that incorporate virtual elements. The current work has barely scratched the surface of this field networking application domain. GA has proven a valuable and scalable basis for game communications in outdoor venues.

9.8 CONCLUSIONS

Field applications have networking needs that are not well suited to traditional, centralized one-to-one style data communications. On the other hand, GA admits efficient and scalable implementations in wireless networks and meets the challenges brought by mobility, density, and operation out of coverage of infrastructure. GA is arguably a more natural building block for the communications patterns that arise in many field applications than in IP addressing. For example, the GA-based FCOP algorithm efficiently implements the awareness communication pattern that solves the FCOP and (more generally) field monitoring problems. Many field applications are built entirely or at least in critical part on an awareness component.

This chapter has introduced GA for field applications, described an implemented scalable GA platform that uses the concepts, and has briefly described some representative field applications that can use it to good advantage. While promising, the present work has only begun to achieve the great potential for benefits to society that can come from high quality field applications built on GA. I intend that follow-on work will continue to improve both the supporting technologies, like new high capacity wireless communications technologies supporting GA, and novel and exciting field applications that can help save lives and improve natural resource use, as well as to carry forward the vision of healthy and active outdoor field activities fundamentally augmented by engaging and useful elements from the virtual world.

REFERENCES

1. Hall, R.J., "A geocast based algorithm for a field common operating picture," in *Proceedings of 2012 IEEE Military Communications Conference* IEEE, 2012.
2. Hall, R.J., "An efficient protocol for geographically addressed streaming," in *Proceedings of 2014 IEEE Military Communications Conference* IEEE, 2014.
3. Hall, R.J., Auzins, J., Chapin, J., and Fell, B., "Scaling up a geographic addressing framework," in *Proceedings of 2013 IEEE Military Communications Conference* IEEE, 2013.
4. Hall, R.J., "An improved geocast for mobile ad hoc networks," *IEEE Transactions on Mobile Computing*, 10(2):254–266, 2011.

5. Gast, M., *802.11 Wireless Networks: The Definitive Guide*, O'Reilly, Sebastopol, 2002.
6. Murthy, C.S.R., and Manoj, B.S., *Ad hoc Wireless Networks: Architectures and Protocols*, Prentice Hall Upper Saddle River, 2004.
7. Hall, R.J., and Auzins, J., "A tiered geocast protocol for long range mobile ad hoc networking," in *Proceedings of 2006 IEEE Military Communications Conference* IEEE, 2006.
8. Hall, R.J., "A Geocast File Transfer Protocol," in *Proceedings of 2013 IEEE Military Communications Conference* IEEE, 2013.
9. Hall, R.J., "The iTron family of geocast games," *IEEE Transactions on Consumer Electronics*, 58(2):171–177, 2012.
10. Hall, R.J., "Software engineering challenges of multi-player outdoor smartphone games," Chapter 8 in *Computer Games and Software Engineering*, Cooper, K.M.L., and Scacchi, W., eds., CRC Press, Boca Raton, 2015.
11. Hall, R.J., "A point-and-shoot weapon design for outdoor multiplayer smartphone games," in *Proceedings of 2011 International Conference on the Foundations of Digital Games*. ACM, 2011.

10 Distributed Online Learning and Stream Processing for a Smarter Planet

DEEPAK S. TURAGA[1] and MIHAELA VAN DER SCHAAR[2]

[1] *IBM T. J. Watson Research Center, Yorktown, New York, NY, USA*
[2] *Electrical Engineering Department, University of California at Los Angeles, Los Angeles, CA, USA*

10.1 INTRODUCTION: SMARTER PLANET

With the world becoming ever more instrumented and connected, we are at the cusp of realizing a *Smarter Planet* [1], where insights drawn from data sources are used to adapt our environment and how we interact with it and with each other. This will enable a range of new services that make it easier for us to work, travel, consume, collaborate, communicate, play, entertain, and even be provided with care.

Consider the pervasiveness of the mobile phone. It is rapidly emerging as the primary digital device of our times—with over 6 (out of the 7) billion people in the world having access to a mobile phone [2]. We are witnessing the rapid emergence of services that use these phones (especially smartphones) as sensors of the environment and interfaces to people. For instance, it is now common with several map services (e.g., Google Maps) to be provided a live view of the traffic across a road network. This aggregate view is computed by processing and analyzing in real time the spatiotemporal properties of data collected from several millions of mobile phones. Applications such as Waze include adding crowd-sourced information to such data, where individual people use mobile phones to report traffic congestion, accidents, etc., and these are then transmitted to other users to inform and potentially alter their travel. While several of these applications are focused on aggregate information processing and dissemination, it is natural to expect more personalized applications, including *personal trip advisors*, that can provide dynamic routing as well as potentially combine multimodal transport options (e.g., car, train, walk, bus).

Fog for 5G and IoT, First Edition. Edited by Mung Chiang, Bharath Balasubramanian, and Flavio Bonomi.

Cities, which are responsible for providing several transportation related services, can use information from mobile phones, augmented with their own road sensors (loop sensors, cameras, etc.) and transport sensors (GPS on buses, trains, etc.), to optimize their transport grid in real time, provide emergency services (e.g., evacuations and dynamic closures) and real-time toll, modify public transport (e.g., allow for dynamic connections between bus/train routes based on current demand), and even control their traffic light systems. This ecosystem, including individual consumers and city infrastructure, is shown in Figure 10.1.

These types of applications have several unique characteristics driven by the distributed sources (often at large scale, i.e., with several thousands of sensors) of streaming data with limited communication and compute capabilities and the need for real-time low-latency analysis. This requires the computation to be distributed end to end, all the way from the data sources through the cloud—requiring the fog computing paradigm. Additionally, the streaming nature of the data necessitates the distributed computation to include streaming and online ways of preprocessing, cleaning, analyzing, and mining of the data, continuous adaptation to the time-varying properties of the data, and dynamic availability of resources. There need to be several advances in sensing and communication technology coupled with development of new analytic algorithms and platforms for these individual-centric and city-wide applications to become real and deliver value.[1] In this chapter, we introduce the emerging paradigm of *stream processing and analysis*, including novel platforms and algorithms, that support the requirements of these kinds of applications. We introduce distributed SPSs and propose a novel distributed online learning framework that can be deployed on such systems to provide a solution to an illustrative Smarter Planet problem. We believe that the recent arrival of new *freely available* systems for distributed stream processing such as InfoSphere Streams [3], Storm [4], and Spark [5] enables several new directions for advancing the state of the art in large-scale, real-time analysis applications and provide the academic and industrial research community the tools to devise end-to-end solutions to these types of problems and overcome issues with proprietary or piecemeal solutions.

This chapter is organized as follows. We start by defining a specific real-world transportation inspired problem that requires large-scale online learning, in Section 10.2. We then formalize the characteristics of such problems and their associated challenges in Section 10.3. We discuss distributed systems in Section 10.4 and how the emergence of SPSs allows us to build and deploy appropriate solutions to these problems. Following this, in Section 10.5, we propose a new framework for distributed, online ensemble learning that can naturally be deployed on a SPS to realize such applications, and we describe how to apply such a framework to a collision detection application. We conclude with a discussion on the several directions for future research enabled by this combination, in Section 10.6.

[1] While our description has been focused on transportation applications, there are several applications of these types in different domains ranging from healthcare, financial services, physical and cybersecurity, telecommunications, energy and utility, and environmental monitoring.

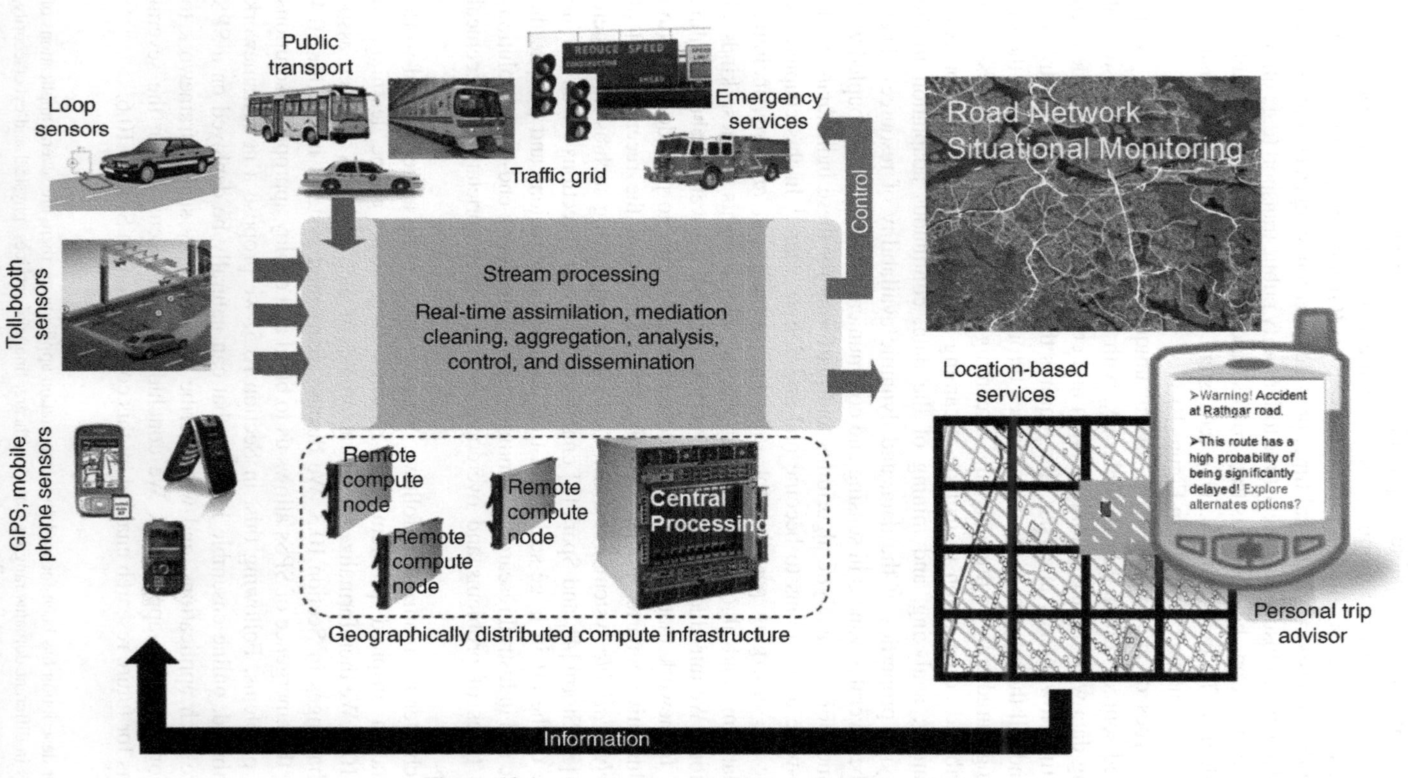

Figure 10.1 Smarter transportation: individuals and city.

10.2 ILLUSTRATIVE PROBLEM: TRANSPORTATION

In this section we define a concrete illustrative problem related to our transportation application domain that requires a joint algorithm–system design for online learning and adaptation. Consider a scenario where a city wants to modify its digital signage based on real-time predictions (e.g., 10 minutes in advance) of congestion in a particular zone. A visual depiction of this example is included in Figure 10.2, where the white spot—at the intersection of major road links—is the point of interest for congestion prediction.

Data for this real-time prediction can be gathered from many different types of sensors. In this example we consider cell phone location information from different points of interest and local weather information. This data is naturally distributed and may not be available at one central location, due to either geographical diversity or different cell phone providers owning different subsets of the data (i.e., their customers). In this simple example we consider geographic distribution of the data. The congestion prediction problem then requires deploying multiple distributed predictors that collect data from local regions and generate *local predictions* that are then merged to generate a more reliable *final prediction*. The local prediction can be about network conditions within a specific local region—for instance congestion within a particular part of the road network. The final prediction on the other hand is computed by combining these local predictions into a composite prediction about the state of the entire network.

An example with two distributed predictor applications—each depicted as a flow graph—is shown in Figure 10.2. The two different flowgraphs in this example look at different subsets of the data and implement appropriate operations for preprocessing (cleaning, denoising, merging, alignment, spatiotemporal processing, etc.), followed by operations for learning and adaptation to compute the local prediction. These are shown as the subgraphs labeled Pre-proc and Learner, respectively. In the most general case, a collection of different models (e.g., neural networks, decision trees, etc.), trained on appropriate training data, can be used by each predictor application.

The learners receive delayed feedback about their prediction correctness after a certain amount of time (e.g., if the prediction is for 10 minutes in advance, the label is

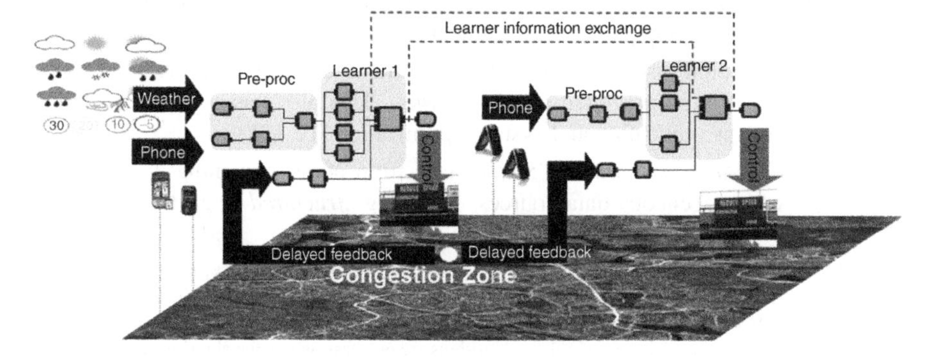

Figure 10.2 Distributed learning needed for real-time signage update.

available after 10 minutes) and can use it to modify their individual models and local aggregation. Additionally, these learners also need to exchange information about their predictions across distributed locations so that they can get a more global view of the state of the network and can improve their predictions. Prediction exchange between the learners is shown on the figure using dashed lines. Finally, the predictions from the learners can be used to update digital signage in real time and potentially alert or divert traffic as necessary.

We formalize the characteristics of this online data stream processing application (SPA) in the next section and discuss how developing it requires the design of online, distributed ensemble learning frameworks, while deploying it requires being able to instantiate these frameworks on a distributed system. In the following two sections, we show how we can leverage characteristics of modern SPSs in order to build and deploy general learning frameworks that solve distributed learning problems of this type. Our intent is to showcase how this enables a whole new way of thinking about such problems and opens up several avenues for future research.

10.3 STREAM PROCESSING CHARACTERISTICS

There is a unique combination of multiple features that distinguishes SPAs from traditional data analysis paradigms, which are often batch and offline. These features can be summarized as follows:

Streaming and In-Motion Analysis (SIMA). SPAs need to process streaming data on the fly, as it continues to flow, in order to support real-time, low-latency analysis and to match the computation to the naturally streaming properties of the data. This limits the amount of prior data that can be accessed and necessitates one-pass, online[2] algorithms [6–8]. Several streaming algorithms are described in [9, 10].

Distributed Data and Analysis (DDA). SPAs analyze data streams that are often *distributed*, and their large rates make it impossible to adopt centralized solutions. Hence, the applications themselves need to be distributed.

High Performance and Scalable Analysis (HPSA). SPAs require high-*throughput*, low-*latency*, and dynamic *scalability*. This means that SPAs should be structured to exploit distributed computation infrastructures and different forms of parallelism (e.g., pipelined data and tasks). This also means that they often require joint application and system optimization [11, 12].

Multimodal Analysis (MMA). SPAs need to process streaming information across heterogeneous data sources, including *structured* (e.g., transactions),

[2]Online algorithms are incremental in nature—processing data items as they arrive, without waiting for all the data to be available. In some cases a window of data can be processed together where the algorithm can keep a collection of items in memory and can make multiple passes over these items while they are in memory for analysis. A special class of online algorithms are one pass, where the algorithms make use of each data item exactly once.

unstructured (e.g., audio, video, text, image), and *semi-structured* data. In our transportation example this includes sensor readings, user-contributed text and images, traffic cameras, etc.

Loss-Tolerant Analysis (LTA). SPAs need to analyze lossy data with different noise levels, statistical and temporal properties, mismatched sampling rates, etc., and hence they often need appropriate processing to transform, clean, filter, and convert data and results. This also implies the need to match data rates, handle lossy data, synchronize across different data streams, and handle various protocols [7]. SPAs need to account for these issues and provide graceful degradation of results to loss in the data.

Adaptive and Time-Varying Analysis (ATA). SPAs are often long running and need to adapt over time to changes in the data and problem characteristics. Hence, SPAs need to support dynamic reconfiguration based on feedback, current context, and results of the analysis [6–8].

Systems and algorithms for SPAs need to provide capabilities that address these features and their combinations effectively.

10.4 DISTRIBUTED STREAM PROCESSING SYSTEMS

The signal processing and research community has so far focused on the theoretical and algorithmic issues for the design of SPAs but has had limited success in taking such applications into real-world deployments. This is primarily due to the multiple practical considerations involved in building an end-to-end deployment and the lack of a comprehensive system and tools that provide them the requisite support. In this section, we summarize efforts at building systems to support SPAs and their shortcomings and then describe how current SPSs can help realize such deployments.

10.4.1 State of the Art

Several systems, combining principles from data management and distributed processing, have been developed over time to support different subsets of requirements that are now central to SPAs. These systems include traditional databases and data warehouses, parallel processing frameworks, active databases, continuous query systems, publish–subscribe (pub-sub) systems, complex event processing (CEP) systems, and more recently the Map-Reduce frameworks. A quick summary of the capabilities of these systems with respect to the streaming application characteristics defined in Section 10.3 is included in Table 10.1.

More details on the individual systems and their examples can be obtained from Ref. [9]. However, as is clear, none of these systems were truly designed to handle all requirements of SPAs. Even the recent MapReduce paradigm does not support SIMA and ATA—critical to the needs of these types of applications. As a consequence, there is an urgent need to develop more sophisticated SPSs.

TABLE 10.1 Data Management Systems and Their Support for SPA Requirements

System	SIMA	DDA	HPSA	MMA	LTA	ATA
Databases (DB2, Oracle, MySQL)	No	Yes	Partly	No	Yes	No
Parallel processing (PVM, MPI, OpenMP)	No	Yes	Yes	Yes	Yes	No
Active databases (Ode, HiPac, Samos)	Partly	Partly	No	No	Yes	Yes
Continuous query systems (NiagaraCQ, OpenCQ)	Partly	Partly	No	No	Yes	Yes
Pub–sub systems (Gryphon, Siena, Padres)	Yes	Yes	No	No	Yes	Partly
CEP systems (WBE, Tibco BE, Oracle CEP)	Yes	Partly	Partly	No	Yes	Partly
MapReduce (Hadoop)	No	Yes	Yes	Yes	Yes	No

SIMA, Streaming and In-Motion Analysis; DDA, Distributed Data and Analysis; HPSA, High-Performance and Scalable Analysis; MMA, Multimodal Analysis; LTA, Loss-Tolerant Analysis; and ATA, Adaptive and Time-Varying Analysis.

10.4.2 Stream Processing Systems

While SPSs were developed by incorporating ideas from these preceding technologies, they required several advancements to the state of the art in algorithmic, analytic, and systems concepts. These advances include sophisticated and extensible programming models, allowing continuous, incremental, and adaptive algorithms and distributed, fault-tolerant, and enterprise-ready infrastructures or runtimes. These systems are designed to allow end-to-end distribution of real-time analysis, as needed in a fog computing world. Examples of early SPSs include TelegraphCQ, STREAM, Aurora–Borealis, Gigascope, and Streambase [9]. Currently available and widely used SPSs include IBM InfoSphere Streams [3] (streams) and open-source Storm [4] and Spark [5] platforms. These platforms are freely available for experimentation and usage in commercial, academic, and research settings. These systems have been extensively deployed in multiple domains for telecommunication call detail record analysis, patient monitoring in ICUs, social media monitoring for real-time sentiment extraction, monitoring of large manufacturing systems (e.g., semiconductor, oil, and gas) for process control, financial services for online trading and fraud detection, environmental and natural systems monitoring, etc. Descriptions of some of these real-world applications can be found in Ref. [9]. These systems have also been shown to scale rates of millions of data items per second, deployed across tens of thousands of processors in a distributed cluster, provide latencies of microseconds or lower, and connect with millions of distributed sensors.

While we omit a detailed description of stream processing platforms, we illustrate some of their core capabilities and constructs to support the needs of SPAs by outlining an implementation of the transportation application described in Section 10.2 in a stream programming language (SPL). In Figure 10.2, the application is shown as a flowgraph, which captures logical flow of data from one processing stage to another. Representing an application as a flowgraph allows for modular design and construction of the implementation, and as we discuss later, it allows SPSs to optimize the deployment of the application onto distributed computational infrastructures.

Each node on the processing flowgraphs in Figure 10.2 that consumes and/or produces a stream of data is labeled an *operator*. Individual streams carry data items or tuples that can contain structured numeric values, unstructured text, semi-structured content such as XML, and binary blobs. In Table 10.2 we present an outline implementation in SPL [3]. Note that this flowgraph actually implements the distributed learning framework that will be discussed in more detail in Section 10.5.

In this code, logical composition is indicated by an operator instance `stream <type> S3 = MyOp(S1;S2)`, where operator `MyOp` consumes streams `S1` and `S2` to produce stream `S3`, where type represents the type of tuples on the stream. Note that these streams may be produced and consumed on different computational resources—but that is transparent to the application developer. Systems like Streams also include multiple operators/tools that are required to build such an application. Examples include operators for:

- Data Sources and Connectors, for example, `FileSource`, `TCPSource`, `ODBCSource`
- Relational Processing, for example, `Join`, `Aggregate`, `Functor`, `Sort`
- Time Series Analysis, for example, `Resample`, `Normalize`, `FFT`, `ARIMA`, `GMM`
- Custom Extensions, for example, user-created operators in C++/Java or wrapping for MATLAB, Python, and R code

These constructs allow for the implementation of distributed and ensemble learning techniques, such as those introduced in the learning framework, within specialized operators. We discuss this more in Section 10.5. Stream processing platforms also include special tools for geo-spatial processing (e.g., for distance computation, map matching, speed and direction estimation, bounding box calculations) and standard mathematical processing. A more exhaustive list is available from Ref. [3].

Finally, programming constructs like the `Export` operator allows applications (in our case the congestion prediction application) to publish their output stream such that other applications, for example, other learners, can dynamically connect to receive these results. This construct allows for dynamic connections to be established and torn down as other learners are instantiated or choose to communicate. This is an important requirement for the learning framework in Section 10.5.

These constructs allow application developers to focus on the core algorithm design and logical flowgraph construction, while the system provides support for communication of tuples across operators, conversion of the flowgraph into a set of processes or processing elements (PEs), distribution and placement of these PEs across a distributed computation infrastructure, and finally necessary optimizations

TABLE 10.2 Example of Streams Programming to Realize Congestion Prediction Application Flowgraph

```
composite Learner1 {
graph
 stream <TPhone> PhoneStream = TCPSource()
    {param role: server; port: 12345u;}
 stream <TWeather> WeatherStream = InetSource()
    {param URIList: ["http://noaa.org/xx"];}
 stream <TPhone> CleanPhoneStream = Custom(PhoneStream){...}
 stream <TWeather> CleanWeatherStream =
     Custom(WeatherStream){...}
 stream <TFeature> FeatureStream = Join(CleanWeatherStream;
     CleanPhoneStream){...}
 stream <TPrediction> P1 = MySVM(FeatureStream){...}
 stream <TPrediction> P2 = MyNN(FeatureStream){...}
 stream <TPrediction> Learner1Pred = MyAggregation(P1;
    P2;Feedback;Learner2Pred){...}
 () as Sink2 = Export(Learner1Pred)
    {param properties: {name = "Learner1"}}
 stream <TPrediction> Learner2Pred = Import()
    {param properties: {name = "Learner2"}}
 stream <TFeedback> Feedback = Import()
    {param properties: {name = "Feedback"}}
}
```

for scaling and adapting the deployed applications. We present an illustration of the process used by such systems to convert a logical flowgraph to a physical deployment in Figure 10.3.

Among the biggest strengths of using a stream processing platform is the natural scalability and efficiency provided by these systems and their support for extensibility in terms of optimization techniques for *topology construction*, *operator fusion*, *operator placement*, and *adaptation* [9]. These systems allow users to formulate and provide additional algorithms to optimize their applications based on the application, data, and resource-specific requirements. This opens up several new research problems related to the joint optimization of algorithms and systems. For instance, consider a simple example application with two operators. Using the Streams composition language, these operators—shown as black and white boxes in Figure 10.4—can be arranged into a parallel (task parallel) or a serial (pipelined parallel) topology to perform the same task,[3] as shown in Figure 10.4. This topology can also be distributed across computational resources (shown as different sized CPUs in Figure 10.4). The right choice of topology and placement depends on the

[3]Note that with a stream processing platform, other forms of parallelism such as data parallel processing and hybrid forms of parallelism with arbitrary combinations of data-, task-, and pipelined–parallel processing are possible.

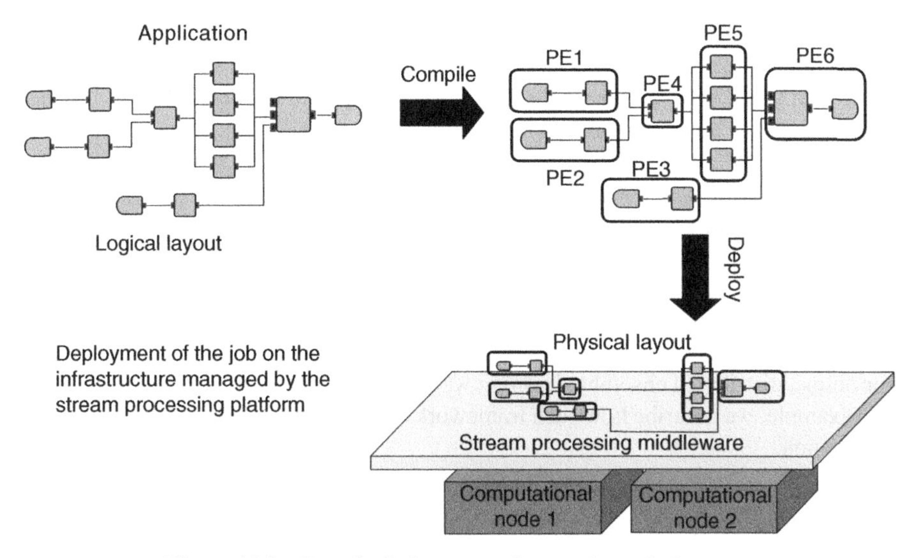

Figure 10.3 From logical operator flowgraphs to deployment.

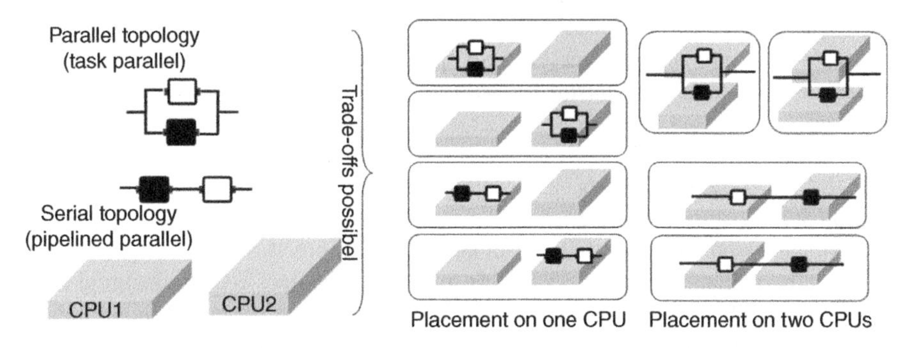

Figure 10.4 Possible trade-offs with parallelism and placement.

resource constraints and data characteristics and needs to be dynamically adapted over time. This requires solving a joint resource optimization problem whose solution can be realized using the controls provided by SPSs. In practice, there are several such novel optimization problems that can be formulated and solved—in this joint application–system research space. For instance, Refs. [11] and [13] discuss topology construction and optimization for non trivial compositions of operators for multi-class classification. Additionally, these systems provide support for design of novel meta-learning and planning-based approaches to dynamically construct, optimize, and compose topologies of operators on these systems [14].

This combination of systems and algorithms enables several other open research problems in this space of joint application–system research, especially in an online, distributed, large-scale setting. In the next section we propose a solution to build a large-scale online distributed ensemble learning framework that leverages the capabilities provided by SPSs (and is implemented by the code in Table 10.2) to provide solutions to the illustrative problem defined in Section 10.2.

10.5 DISTRIBUTED ONLINE LEARNING FRAMEWORKS

We now formalize the problem described in Section 10.2 and propose a novel distributed learning framework to solve it. We first review the state of the art in such research and illustrate its shortcomings. Then we describe a systematic framework for online, distributed ensemble learning well suited for SPAs. Finally, as an illustrative example, we describe how such framework can be applied to a collision detection application.

10.5.1 State of the Art

As mentioned in Section 10.3, it is important for stream processing algorithms to be online, one pass, adaptive, and distributed to operate effectively under budget constraints and to support combinations (or ensembles) of multiple techniques. Recently, there has been research that uses the aforementioned techniques for analysis, and we include a summary of some of these approaches next. We partition our review into *Ensemble methods*, *Diffusion adaptation*, and finally frameworks for distributed learning.

10.5.1.1 Ensemble Methods *Ensemble* techniques [15] build and combine a collection of base algorithms (e.g., classifiers) into a joint unique algorithm (classifier). Traditional ensemble schemes for data analysis are focused on analyzing stored or completely available datasets; examples of these techniques include bagging [16] and boosting [17]. In the past decade much work has been done to develop online versions of such ensemble techniques. An online version of AdaBoost is described in Refs. [18], and similar proposals are made in Refs. [19] and [20]. Minku and Xin [21] propose a scheme based on two online ensembles, one used for system predictions, and the other one used to learn the new concept after a drift is detected. Weighted majority [22] is an ensemble technique that maintains a collection of given learners, predicts using a weighted majority rule, and decreases in a multiplicative manner the weights of the learners in the pool that disagree with the label whenever the ensemble makes a mistakes. In Ref. [23] the weights of the learners that agree with the label when the ensemble makes a mistakes are increased, and the weights of the learners that disagree with the label are decreased also when the ensemble predicts correctly. To prevent the weights of the learners that performed poorly in the past from becoming too small with respect to the other learners, Herbster and Warmuth [24] propose a modified version of weighted majority adding a phase, after the multiplicative weight update, in which each learner shares a portion of its weight with the other learners.

While many of the ensemble learning techniques have been developed assuming no a priori knowledge about the statistical properties of the data—as is required in most of the SPAs—these techniques are often designed for a centralized scenario. In fact, the base classifiers in these approaches are not distributed entities; they all observe the same data streams, and the focus of ensemble construction is on the statistical advantages of learning with an ensemble, with little study of learning under communication constraints. It is possible to cast these techniques within the framework of distributed learning, but as is they would suffer from many drawbacks. For example, Refs. [18–21] would require an entity that collects and stores all the data recently observed by the learners and that tells the learners how to adapt their local classifiers, which are clearly impractical in SPAs that need to process real-time streams characterized by high data rates.

10.5.1.2 Diffusion Adaptation Methods *Diffusion adaptation* literature [25–32] consists of learning agents that are linked together through a network topology in a distributed setting. The agents must estimate some parameters based on their local observations and on the continuous sharing and diffusion of information across the network, and there is a focus on learning in distributed environments under communication constraints. In fact, Ref. [32] shows that a classification problem can be cast within the diffusion adaptation framework. However, there are some major constraints that are posed on the learners. First, in Refs. [25–32] all the learners are required to estimate the same set of parameters (i.e., they pursue a common goal) and combine their local estimates to converge toward a unique and optimal solution. This is a strong assumption for SPAs, as the learners might have different objectives and may use different information depending on what they observe and on their spatiotemporal position in the network. Hence, the optimal aggregation function may need to be specific to each learner.

10.5.1.3 Frameworks for Distributed Learning There has been a large amount of recent work on building frameworks for distributed online learning with dynamic data streams, limited communication, delayed labels and feedback, and self-interested and cooperative learners [7, 33–35]. We discuss this briefly next.

To mine the correlated, high-dimensional, and dynamic data instances captured by one or multiple heterogeneous data sources, extract actionable intelligence from these instances, and make decisions in real time as discussed previously, a few important questions need to be answered: which processing/prediction/decision rule should a local learner (LL) select? How should the LLs adapt and learn their rules to maximize their performance? How should the processing/predictions/decisions of the LLs be combined/fused by a meta-learner to maximize the overall performance? Most literature treats the LLs as black box algorithms and proposes various fusion algorithms for the ensemble learner with the goal of issuing predictions that are at least as good as the best LL in terms of prediction accuracy, and the performance bounds proved for the ensemble in these works depend on the performance of the LLs. In Ref. [34] the authors go one step further and study the joint design of learning algorithms for both the LLs and the ensemble. They present a novel systematic learning method (Hedge Bandits), which continuously learns and adapts the parameters of both the LLs and the ensemble, after each data instance, and provide both long-run (asymptotic) and short-run (rate of learning) performance guarantees. Hedge Bandits consists of a novel contextual bandit algorithm for the LLs and Hedge algorithm for the ensemble and is able to exploit the adversarial regret guarantees of Hedge and the data-dependent regret guarantees of the contextual bandit algorithm to derive a data-dependent regret bound for the ensemble.

In Ref. [7], the ensemble learning consists of multiple-distributed LLs, which analyze different streams of data correlated to a common classification event, and local predictions are collected and combined using a weighted majority rule. A novel online ensemble learning algorithm is then proposed to update the aggregation rule in order to adapt to the underlying data dynamics. This overcomes several limitations of prior work by allowing for (i) *different* correlated data streams with statistical dependency among the label and the observation being different across learners, (ii) data being processed incrementally, once on arrival leading to improved scalability, (iii) support for different types of local classifiers including support vector machine, decision tree, neural networks, offline/online classifiers, etc., and (iv) asynchronous delays between the label arrival across the different learners. A modified version of this framework was applied to the problem of collision detection by networked sensors similarly to the one that we discussed on Section 10.2 of this chapter. For details, please refer to Ref. [36].

A more general framework, where the rule for making decisions and predictions is general and depends on the costs and accuracy (specialization) of the autonomous learners, was proposed in Ref. [33]. This cooperative online learning scheme considers (i) whether the learners can improve their detection accuracy by exchanging and aggregating information, (ii) whether the learners improve the timeliness of their detections by forming clusters, that is, by collecting information only from surrounding learners, and (iii) whether, given a specific trade-off between detection accuracy and detection delay, it is desirable to aggregate a large amount of information or it is better to focus on the most recent and relevant information.

In Ref. [37], these techniques are considered in a setting with a number of speed sensors that are spatially distributed along a street and can communicate via an exogenously determined network, and the problem of detecting in real-time collisions that occur within a certain distance from each sensor is studied.

In Ref. [35], a novel framework for decentralized, online learning by many self-interested learners is considered. In this framework, learners are modeled as cooperative contextual bandits, and each learner seeks to maximize the expected reward from its arrivals, which involves trading off the reward received from its own actions, the information learned from its own actions, the reward received from the actions requested of others, and the cost paid for these actions—taking into account what it has learned about the value of assistance from each other learner. A distributed online learning algorithm is provided, and analytic bounds to compare the efficiency of these algorithms with the complete knowledge (oracle) benchmark (in which the expected reward of every action in every context is known by every learner) are established: regret—the loss incurred by the algorithm—is sublinear in time. These methods have been adapted in Ref. [38] to provide expertise discovery in medical environments. Here, an expert selection system is developed that learns online who is the best expert to treat a patient having a specific condition or characteristic.

In Section 10.5.2 we describe one such framework for online, distributed ensemble learning that addresses some of the challenges discussed in Section 10.3 and is well suited for the transportation problem described in Section 10.2. The presented methodology does not require a priori knowledge of the statistical properties of the data. This means that it can be applied both when a priori information is available and when a priori information is not available. However, if the statistical properties of the data are available beforehand, it may be convenient to apply schemes that are specifically designed to take into account the known statistical properties of the data. Moreover, the presented methodology does not require any specific assumption on the form of the loss or objective function. This means that any loss or objective function can be adopted. However, notice that the final performance of scheme depends on the selected function. For illustrative purposes, in Section 10.5.3 we consider a specific loss function, and we derive an adaptive algorithm based on this loss function, and in Section 10.5.4 we describe how the proposed framework can be adopted for a collision detection application.

10.5.2 Systematic Framework for Online Distributed Ensemble Learning

We now proceed to formalize the problem of large-scale distributed learning from heterogeneous and dynamic data streams using the problem defined in Section 10.2. Formally, we consider a set $\mathcal{K} = \{1, \ldots, K\}$ of *learners* that are geographically distributed and connected via *links* among pairs of learners. We say that there is a *link* (i, j) between learners i and j if they can communicate directly with each other. In the case of our congestion application, each learner observes part of the transportation network by consuming geographically local readings from sensors and phones within a region and is linked to other learner streams via interfaces like the export–import interface described in Section 10.4.

Each learner is an ensemble of *local classifiers* that observes a specific set of data sources and relies on them to make local classifications, that is, partition data items into multiple classes of interest.[4] In our application scenario, this maps to a binary classification task—predicting presence of congestion at a certain location within a certain time window. Each local classifier may be an arbitrary function (e.g., implemented using well-known techniques such as neural networks, decision trees, etc.) that performs classification for the classes of interest. In Table 10.2 we show an implementation of this in an SPL with two local classifiers, `MySVM` and `MyNN` operators, and an aggregate `MyAggregation` operator. In order to simplify the

[4]We present the ensemble learning in a classification setting, but the discussion is also applicable in a regression setting.

discussion, we assume that each learner exploits a single local classifier, and we focus on binary classification problems, but it is possible to generalize the approach to the multi-classifier and multi-class cases. Each learner is also characterized by a local *aggregation rule*, which is adaptive.

Raw data items in our application can include sensor readings from the transportation network and user phones, as well as information about the weather. These data items are cleaned, preprocessed, and merged, and features are extracted from them (e.g., see Table 10.2), which are then sent to the geographically appropriate learner. We assume a synchronous processing model with discrete time slots. At each time slot, each learner observes a feature vector. It first exploits the local classifier to make a local prediction for that slot, and then it sends its local prediction to other learners in its neighborhood and receives local predictions from the other learners, before it finally exploits the aggregation rule to combine its local predictions and the predictions from its neighbors into a final classification.

Consider a discrete time model in which time is divided into *slots*, but an extension to a continuous time model is possible. At the beginning of the n-th time slot, K multidimensional *instances* $\mathbf{x}_i^n \in \mathcal{X}_i$, $i = 1 \ldots K$, and K *labels* $y_i^n \in \{-1, +1\}$, $i = 1 \ldots K$, are drawn from an underlying and unknown joint probability distribution. Each learner i observes the instance $\mathbf{x}_i^n$, and its task is to predict the label y_i^n (see Figure 10.6). We shall assume that a label y_j^n is correlated with all the instances $\mathbf{x}_i^n$, $i = 1 \ldots K$. In this way, learner i's observation can contain information about the label that learner j has to predict. We remark that this correlation is not known beforehand.

Each learner i is equipped with a *local classifier* $f_i^n : \mathcal{X}_i \to \{-1, +1\}$ that generates the *local prediction* $s_i^n \triangleq f_i^n\left(\mathbf{x}_i^n\right)$ based on the observed instance $\mathbf{x}_i^n$ at time slot n. Our framework can accommodate both static pre-trained classifiers and adaptive classifiers that learn online the parameters and configurations to adopt [39]. However, the focus of this section will not be on classifier design, for which many solutions already exist (e.g., support vector machines, decision trees, neural networks, etc.); instead, we will focus on how the learners exchange and learn how to aggregate the local predictions generated by the classifiers.

We allow the distributed learners to exchange and aggregate their local predictions through multihop communications; however, within one time slot a learner can send only a single transmission to each of its neighbors. We denote by $\overline{s}_{ij}^n$ learner j's local prediction possessed by learner i before the aggregation at time instant n. The information is disseminated in the network as follows. First, each learner i observes $\mathbf{x}_i^n$ and updates $\overline{s}_{ii}^n = s_i^n = f_i^n(\mathbf{x}_i^n)$. Next, learner i transmits to each neighbor j the local prediction s_i^n and the local predictions $\overline{s}_{ik}^{n-1}$, for each learner $k \neq i$ such that the link (i,j) belongs to the shortest path between k and j. Hence, if transmissions are always correctly received, we have $\overline{s}_{ii}^n = s_i^n$ and $\overline{s}_{ij}^n = s_j^{n-d_{ij}+1}$, $i \neq j$, where d_{ij} is the distance in number of hops between i and j. For instance, Figure 10.5 represents the flow of information toward learner 1 for a binary tree network assuming that transmissions are always correctly received. More generally, if transmissions can be affected by communication errors, we have $\overline{s}_{ij}^n = s_j^m$, for some $m \leq n - d_{ij} + 1$.

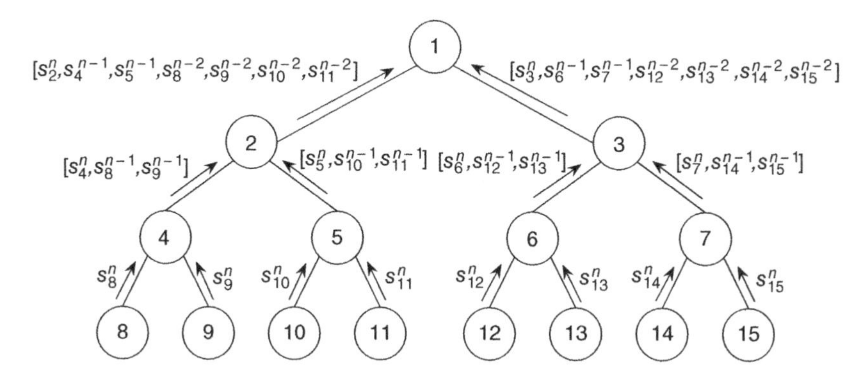

Figure 10.5 Flow of information toward learner 1 at time slot n for a binary tree network.

Each learner i employs a *weighted majority aggregation* rule to fuse the data it possesses and generates a final prediction $\hat{y}_i^n$ as follows:

$$\hat{y}_i^n \triangleq \text{sgn}\left(\sum_{j\in\mathcal{K}} w_{ij}^n \bar{s}_{ij}^n\right) \triangleq \begin{cases} +1, & \text{if argument of sgn is} \geq 0 \\ -1, & \text{otherwise} \end{cases} \tag{10.1}$$

where sgn(·) is the sign function. In the example earlier, we use a threshold of 0 on the value of the argument to separate the two classes. In practice this threshold can be arbitrary; however this does not affect the discussion next, as the threshold corresponds to a constant shift in the argument.

In the earlier construction, learner i first aggregates all possessed predictions $\{\bar{s}_{ij}^n\}$ using the weights $\{w_{ij}^n\}$ and then uses the sign of the fused information to output its final classification, $\hat{y}_i^n$. While weighted majority aggregation rules have been considered before in the ensemble learning literature [17–20], there is an important distinction in Equation (10.1) that is particularly relevant to the online distributed stream mining context: since we are limiting the learners to exchange information only via links, learners receive information from other learners with delay (i.e., in general $\bar{s}_{ij}^n \neq s_j^n$), as a consequence different learners have different information to exploit (i.e., in general $\bar{s}_{ij}^n \neq \bar{s}_{kj}^n$).

Each learner i maintains a total of K weights and K local predictions, which we collect into vectors:

$$\mathbf{w}_i^n \triangleq \left(w_{i1}^n, \ldots, w_{iK}^n\right); \quad \mathbf{s}_i^n \triangleq \left(\bar{s}_{i1}^n, \ldots, \bar{s}_{iK}^n\right). \tag{10.2}$$

Given the weight vector $\mathbf{w}_i^n$, the decision rule (10.1) allows for a geometric interpretation: the homogeneous hyperplane in $\mathfrak{R}^K$ that is orthogonal to $\mathbf{w}_i^n$ separates the positive prediction (i.e., $\hat{y}_i^n = +1$) from the negative predictions (i.e., $\hat{y}_i^n = -1$).

We consider an *online learning setting* in which the true label y_i^n is eventually observed by learner i. Learner i can then compare both $\hat{y}_i^n$ and y_i^n and use this information to update the weights it assigns to the other learners. Indeed, since we do not

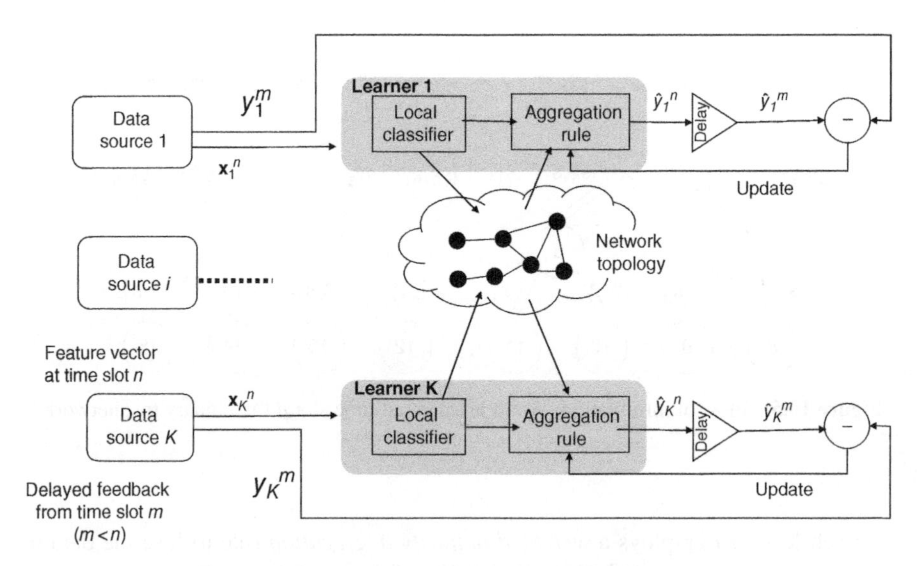

Figure 10.6 System model described in Section 10.5.2.

assume any a priori information about the statistical properties of the processes that generate the data observed by the various learners, we can only exploit the available observations and the history of past data to guide the design of the adaptation process over the network.

In summary, the sequence of events that takes place in a generic time slot n, represented in Figure 10.6, involves five phases:

1. *Observation.* Each learner i observes an instance $\mathbf{x}_i^n$ at time n.
2. *Information Dissemination.* Learners send the local predictions they possess to their neighbors.
3. *Final Prediction.* Each learner i computes and outputs its final prediction $\hat{y}_i^n$.
4. *Feedback.* Learners can observe the true label y_i^m that refers to a time slot $m \leq n$.
5. *Adaptation.* If y_i^m is observed, learner i updates its aggregation vector from $\mathbf{w}_i^n$ to $\mathbf{w}_i^{n+1}$.

In the context of the discussed framework, it is fundamental to develop strategies for adapting the aggregation weights $\{w_{ji}^n\}$ over time, in response to how well the learners perform. A possible approach is discussed next.

10.5.3 Online Learning of the Aggregation Weights

A possible approach to update the aggregation weights is to associate with each learner i an instantaneous loss function $\ell_i^n(\mathbf{w}_i)$ and minimize, with respect to the weight vector $\mathbf{w}_i$, the cumulative loss given all observations up to time slot n. In the following we consider this approach, adopting an instantaneous *hinge loss function* [40]:

For each time instant n, we consider the one-shot *loss function*

$$\ell_i^n(\mathbf{w}_i) \triangleq \begin{cases} -\alpha^{\text{MD}}\mathbf{w}_i \cdot \mathbf{s}_i^n & \text{if } \hat{y}_i^n = -1 \text{ and } y_i^n = 1 \\ \alpha^{\text{FA}}\mathbf{w}_i \cdot \mathbf{s}_i^n & \text{if } \hat{y}_i^n = 1 \text{ and } y_i^n = -1 \\ 0 & \text{if } \hat{y}_i^n = y_i^n \end{cases} \tag{10.3}$$

where the parameters $\alpha^{\text{MD}} > 0$ and $\alpha^{\text{FA}} > 0$ are the *mis-detection* and *false alarm* unit costs and $\mathbf{w}_i \cdot \mathbf{s}_i^n \triangleq \sum_{j\in\mathcal{K}} w_{ij}\overline{s}_{ij}^n$ denotes the scalar product between $\mathbf{w}_i$ and $\mathbf{s}_i^n$. The hinge loss function is equal to 0 if the weight vector $\mathbf{w}_i$ allows to predict correctly the label y_i^n; otherwise the value of the loss function is proportional to the distance of $\mathbf{s}_i^n$ from the separating hyperplane defined by $\mathbf{w}_i$, multiplied by α^{MD} if the prediction is -1 but the label is 1, we refer to this type of error as a *mis-detection*, or multiplied by α^{FA} if the prediction is 1 but the label is -1, we refer to this type of error as a *false alarm*.

The hinge loss function gives higher importance to errors that are more difficult to correct with the current weight vector. A related albeit different approach is adopted in AdaBoost [17], in which the importance of the errors increases exponentially in the distance of the local prediction vector from the separating hyperplane. Here, however, the formulation is more general and allows for the diffusion of information across neighborhoods simultaneously, as opposed to assuming each learner has access to information from across the entire set of learners in the network.

We can then formulate a global objective for the distributed stream mining problem as that of determining the optimal weights by minimizing the cumulative loss given all observations up to time slot n:

$$\left\{\mathbf{w}_i^{n+1}\right\}_{i=1}^K = \underset{\{\mathbf{w}_i\}_{i=1}^K}{\text{argmin}} \sum_{i=1}^K \sum_{m=1}^n \overline{\ell}_i^m \tag{10.4}$$

where $\overline{\ell}_i^m \triangleq \ell_i^m$ if y_i^m has been observed by time instant n, otherwise $\overline{\ell}_i^m \triangleq 0$.

To solve (10.4) learner i must store all previous labels and all previous local predictions of all the learners in the system, which is impractical in SPAs, where the volume of the incoming data is high and the number of learners is large. Hence, we adopt the stochastic gradient descent algorithm to incrementally approach the solution of (10.4) using only the most recently observed label. If label y_i^m is observed at the end of time instant n, we obtain the following update rule for $\mathbf{w}_i^n$:

$$\mathbf{w}_i^{n+1} = \begin{cases} \mathbf{w}_i^n + \alpha^{\text{MD}}\mathbf{s}_i^n & \text{if } \tilde{y}_i^m = -1 \text{ and } y_i^m = 1 \\ \mathbf{w}_i^n - \alpha^{\text{FA}}\mathbf{s}_i^n & \text{if } \tilde{y}_i^m = 1 \text{ and } y_i^m = -1 \\ \mathbf{w}_i^n & \text{if } \tilde{y}_i^n = y_i^n \end{cases} \tag{10.5}$$

where $\tilde{y}_i^m$ is the prediction that learner i would have made at time instant m with the current weight vector $\mathbf{w}_i^n$. This construction allows a meaningful interpretation.

It shows that learner i should maintain its level of confidence in its data when its decision agrees with the observed label. If disagreement occurs, then learner i needs to assess which local predictions lead to the misclassification: the weight w_{ij}^{n} that learner i adopts to scale the local predictions it receives from learner j is increased (by either α^{MD} or α^{FA} units, depending on the type of error) if the local prediction sent by j agreed with the label, otherwise w_{ij}^{n} is decreased.

[?] and [8] derive worst-case upper bounds for the misclassification probability of a learner adopting the update rule 10.5. Such bounds are expressed in terms of the misclassification probabilities of two benchmarks: (i) the misclassification probability of the best local classifiers and (ii) the misclassification probability of the best linear aggregator. We remark that the best local classifiers and the best linear aggregator are not known and cannot be computed beforehand; in fact, this would require to know in advance the realization of the process that generates the instances and the labels.

The optimization problem (10.4) can also be solved within the diffusion adaptation framework, as proposed in Ref. [32]. In this framework the learners combine information with their neighbors, for example, in the combine-then-adapt (CTA) diffusion scheme, they first combine their weights and then adopt the stochastic gradient descent [32]. Figure 10.7 illustrates the difference between our approach and the CTA scheme.

We remark that the framework described so far requires each learner to maintain a weight (i.e., an integer value) and a local prediction (i.e., a Boolean value) for each other learner in the network. This means that the memory and computational requirements scale linearly in the number of learners K. However, notice that the aggregation rule 10.1 and the update rule 10.5 only require basic operations such as add, multiply, and compare. Moreover, if the learners have a common goals, that is, they must predict a common class label y^{n}, it is possible to develop a scheme in which each learner keeps track only of its own local prediction and of the weight used to scale its local prediction and is responsible to update such a weight. In this scheme the learners exchange the weighted local predictions instead of the local predictions and the memory and computational requirements scale as a constant in the number of learners K. For additional details, we refer the reviewer to Ref. [8].

The framework discussed in this subsection naturally maps onto a deployment using an SPS. Each of the learners shown in Figure 10.6 maps onto the subgraph labeled learner in Figure 10.2, with the local classifiers mapping onto the shown parallel topology and the aggregation rule mapping to the fan-in on that subgraph. As mentioned earlier, the base classifiers may be implemented using the toolkits provided by systems like Streams that include wrappers for R and MATLAB. The feedback y_{i}^{n} corresponds to the delayed feedback in Figure 10.2, and the input feature vector $\mathbf{x}_{i}^{n}$ is computed by the Pre-proc part of the subgraph in Figure 10.2 and can include different types of spatiotemporal processing and feature extraction. Finally, the communication between the learners in Figure 10.6 is enabled by the learner information exchange connections in Figure 10.2. In summary, this online, distributed ensemble learning framework can naturally be implemented on a stream processing platform. This combination is very powerful, as it now allows the design, development, and deployment of such large-scale complex applications much more feasible, and it also

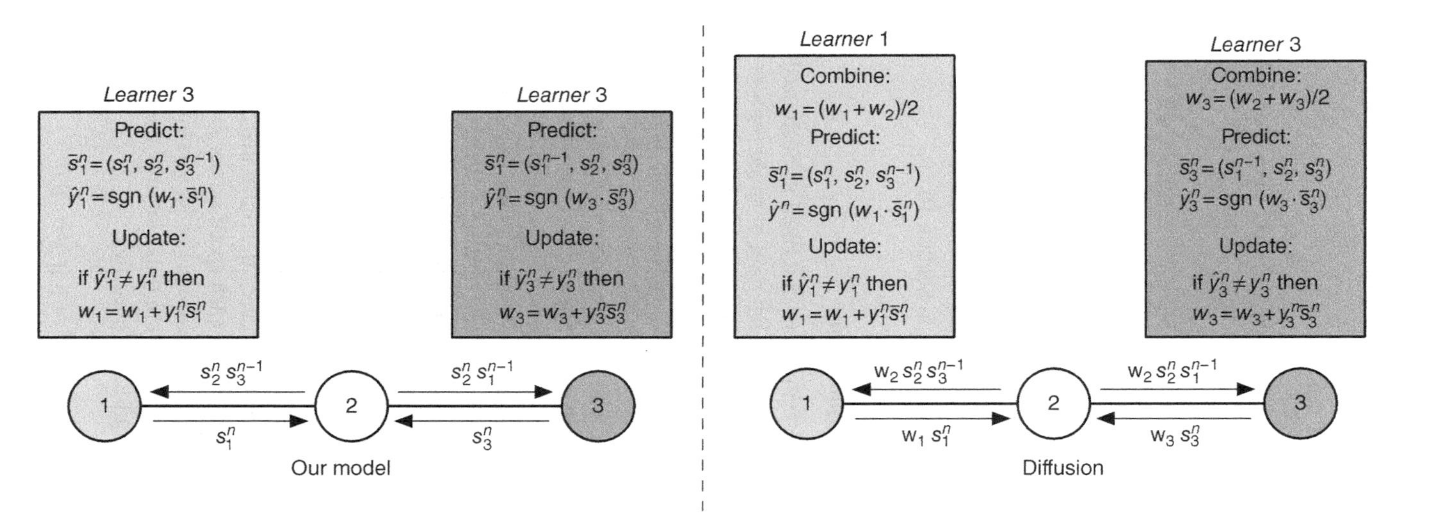

Figure 10.7 A comparison between the proposed algorithm and the combine-then-adapt (CTA) scheme in terms of information dissemination and weight update rule. Unlike diffusion, in the proposed approach the weight vectors do not need to be disseminated.

Algorithm 10.1

Initialization: $w_{ij}^n = 0,\ \forall i,j \in \mathcal{K}$
For each learner i and time instant n
  Observe $\mathbf{x}_i^n$ and update $\bar{s}_{ii}^n \leftarrow f_i^n(\mathbf{x}_i^n)$
  Send $\bar{s}_{ij}^n$ to all neighbors k where i is on path between j and k, $\forall j$
  Update $\bar{s}_{ij}^n \leftarrow \bar{s}_{kj}^n$ for each k and j such that $\bar{s}_{kj}^n$ is correctly received
  Predict $\hat{y}_i^n \leftarrow \mathrm{sgn}\left(\sum_{j\in\mathcal{K}} w_{ij}^n \bar{s}_{ij}^n\right)$
  For each time instant m such that y_i^m is observed
    If $\mathrm{sgn}\left(\sum_{j\in\mathcal{K}} w_{ij}^n \bar{s}_{ij}^m\right) \neq y_i^m$
      If $y_i^m = 1$ **do** $\mathbf{w}_i^n \leftarrow \mathbf{w}_i^n + \alpha^{MD}\mathbf{s}_i^n$
      Else do $\mathbf{w}_i^n \leftarrow \mathbf{w}_i^n - \alpha^{FA}\mathbf{s}_i^n$

enables a range of novel signal processing, optimization, and scheduling research. We discuss some of these open problems in Section 10.6.

10.5.4 Collision Detection Application

In this subsection we apply the framework described in Sections 10.5.2 and 10.5.3 to a collision detection application in which a set of speed sensors—which are spatially distributed along a street—must detect in real-time collisions that occur within a certain distance from them.

We consider a set $\mathcal{K} = \{1, \ldots, K\}$ of K speed sensors that are distributed along both travel directions of a street (see the left side of Figure 10.8). We focus on a generic sensor i that must detect the occurrence of *collision events* within z miles from its location along the corresponding travel direction, where z is a predetermined parameter. A collision event e_ℓ is characterized by an unknown *starting time* $t_{\ell,\text{start}}$, when the collision occurs, and an unknown ending time $t_{\ell,\text{end}}$, when the collision is cleared. The goal of sensor i is to detect the collision e_ℓ by the time $t_{\ell,\text{det}} = t_{\ell,\text{start}} + T_{\max}$, where $T_{\max}$ can be interpreted as the maximum time after the occurrence of the collision such that the information about the collision occurrence can be exploited to take better informed actions (e.g., the average time after which a collision is reported by other sources).

We divide the time into *slots* of length T. We write $y_i^n = +1$ if a collision occurs at or before time instant n and is not cleared by time instant n, whereas we write $y_i^n = -1$ to represent the absence of a collision. Figure 10.9 illustrates these notations.

At the beginning of the n-th time slot, each speed sensor j observes a speed value $x_j^n \in \Re$, which represents the average speed value of the cars that have passed through sensor j from the beginning of the $(n-1)$-th time slot until the beginning of the n-th time slot. We consider a threshold-based classifier:

$$s_j^n = f_j^n(x_j^n) \triangleq \begin{cases} -1 & \text{if } x_j^n > \beta_j v_j^n \\ +1 & \text{if } x_j^n \leq \beta_j v_j^n \end{cases} \tag{10.6}$$

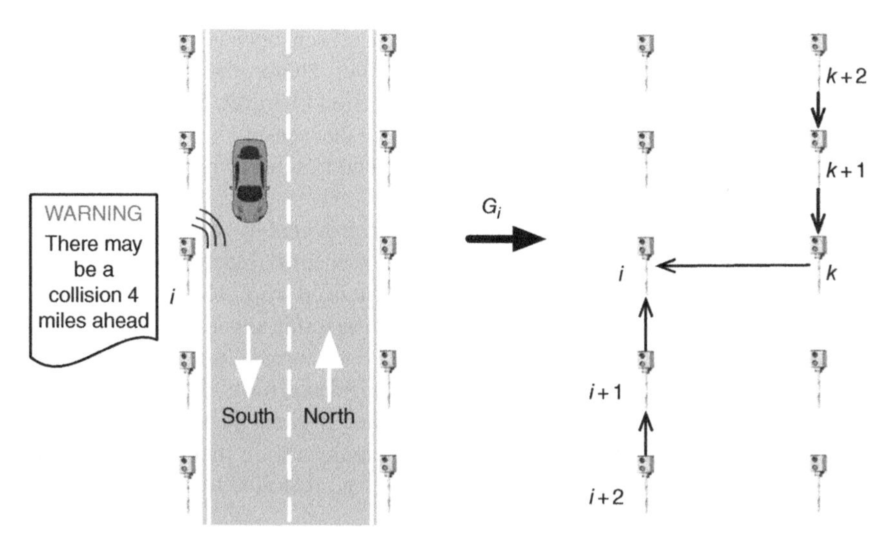

Figure 10.8 A generic speed sensor i must detect collisions in real time and inform the drivers (left). To achieve this goal sensor i receives the observations from the other sensors, and the flow of information is represented by a directed graph $\mathcal{G}_i$ (right).

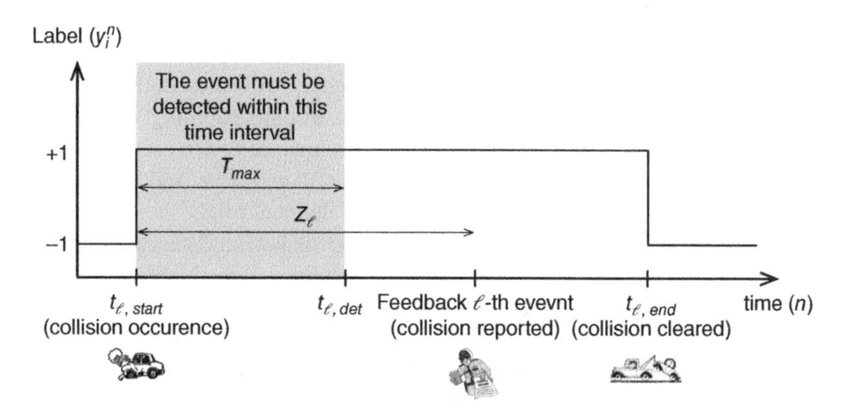

Figure 10.9 Illustration of the considered notations.

where v_j^n is the average speed observed by sensor j during that day of the week and time of the day, and $\beta_j \in [0, 1]$ is a threshold parameter.

If sensor j is close to sensor i, the speed value x_j^n and the local prediction s_j^n are correlated to the occurrence or absence of the collision events that sensor i must detect. For this reason, to detect collisions in an accurate and timely manner, the sensors must exchange their local predictions. Specifically, we denote the sensors such that sensor i precedes sensor $i + 1$ in the direction of travel. In order to detect whether a collision has occurred within z miles from its location, sensor i requires the observations of the subsequent sensors in the direction of travel (e.g., sensors $i+1$, $i+2$, etc.) up to the sensor that is far from sensor i more than z miles. Hence, the information flows in the opposite direction with respect to the direction of travel. Such a scenario is represented by the right side of Figure 10.8. Notice that sensor $i+1$ is responsible to collect the observation from sensor $i+2$ and to send to sensor i both its observation and the observation of $i+2$.

Figure 10.8 shows also the flow of information provided by one side of the street to the other side of the street (i.e., from sensor k to sensor i). Indeed, the fact that the observations on one side of the street are not influenced by a collision on the other side of the street can be extremely useful to assess the traffic situation and distinguish between collisions and other types of incidents. For example, the sudden decrease of the speed observed by some sensors in the considered travel direction may be a collision warning sign; however, if at the same time instants the speed observed by the sensors in the opposite travel direction decreases as well, then an incident that affect both travel directions may have occurred (e.g., it started to rain) instead of a collision.

We can formally define the flow of information represented by the right side of Figure 10.8 with a directed graph $\mathcal{G}_i \triangleq (\mathcal{K}_i, \mathcal{L}_i)$,[5] where $\mathcal{K}_i \subset \mathcal{K}$ is the subset of sensors that send their local predictions to sensor i (included i itself) and $\mathcal{L}_i \subset \mathcal{K}_i \times \mathcal{K}_i$ is the set of links among pairs of sensors.

Now both the local classifiers and the flow of information are defined, learner i can adopt the framework described in Sections 10.5.2 and 10.5.3 to detect the occurrence of collisions within z miles from its location. Specifically, learner i maintains in memory K_i weights (K_i is the cardinality of $\mathcal{K}_i$) that are collected in the weight vector $\mathbf{w}_i^n$; it predicts adopting 10.1 and it updates the weights adopting 10.5 whenever a feedback is received. The feedback about the occurrence of the collision event e_ℓ can be provided, for example, by a driver or by a police officer, and it is in general received with delay. In Figure 10.8 such a delay is denoted by Z_ℓ.

We have evaluated the proposed framework over real-word datasets. Specifically, we have exploited a dataset containing the speed readings of the loop sensors that are distributed along a 50 mile segment of the freeway 405 that passes through Los Angeles County and a collision dataset containing the reported collisions that occurred along the freeway 405 during the months of September, October, and November 2013. For a more detailed description of the datasets, we refer the reader to Ref. [41]. Our illustrative results show that the considered framework is able to detect more than half of the collisions occurring within a distance of 4 miles from a specific sensors while generating false alarms in the order of one false alarm every 100 predictions. The results show also that by setting the ratio $\alpha^{\text{MD}}/\alpha^{\text{FA}}$, mis-detections and false alarms can be traded off.

[5] We remark that we focus on a particular sensor i to keep the discussion and the notations simple. However, all the sensors may be required to detect collisions that are within z miles from their location. Hence, each sensor applies the same scheme we propose in this chapter for a generic sensor i. This means that there are many directed graphs (e.g., $\mathcal{G}_1$, $\mathcal{G}_2$, etc.) representing how information flows among different sensors.

10.6 WHAT LIES AHEAD

There are several open research problems at this application–algorithm systems interface—needed for fog computing—that are worth investigating. First, there is currently no principled approach to decompose an online distributed large-scale learning problem into a topology/flowgraph of streaming operators and functions. While standard engineering principles of modularity, reuse, atomicity, etc. apply, there is no formalism that supports such a decomposition.

Second, there are several optimization problems related to mapping a given processing topology onto physical processes that can be instantiated on a distributed computation platform. This requires a multi-objective optimization where communication costs need to be traded off with memory and computational costs while ensuring efficient utilization of resources. Also, given that resource requirements and data characteristics change over time, these optimization problems may need to be solved incrementally or periodically. The interaction between these optimizations and the core learning problem needs to be also formally investigated.

Third, there are several interesting topology configuration and adaptation problems that can be considered: learners can be dynamically switched on or off to reduce system resource usage or improve system performance; the topology through which they are connected can adapt to increase parallelism; the selectivity operating points of individual classifiers can be modified to reduce workloads on downstream operators; past data and observations can be dynamically dropped to free memory resources; etc. The impact of each individual adaptation and of the interaction among different levels of adaptation is unclear and needs to be investigated. Some examples of exploiting these trade-offs have been considered in Ref. [11], but this is a fertile space for future research.

Another important extension is the use of active learning approaches [42] to gather feedback in cases where it is sparse, hard, or costly to acquire.

Fourth, there is need to extend the meta-learning aggregation rule from a linear form to other forms (e.g., decision trees) to exploit the decision space more effectively. Additionally, meta-learners may themselves be hierarchically layered into multiple levels—with different implications for learning, computational complexity, and convergence.

Fifth, in the presence of multiple learners, potentially belonging to different entities, these ensemble approaches need to handle noncooperative and in some cases even malicious entities. In Ref. [43], a few steps have been taken in this direction. This work studies distributed online recommender systems, in which multiple learners, which are self-interested and represent different companies/products, are competing and cooperating to jointly recommend products to users based on their search query as well as their specific background including history of bought items, gender, and age.

Finally, while we have posed the problem of distributed learning in a supervised setting (i.e., the labels are eventually observed), there is also a need to build large-scale online algorithms for knowledge discovery in semi-supervised and unsupervised settings. Constructing online ensemble methods for clustering, outlier detection, and frequent pattern mining are interesting directions. A few steps in these directions have been taken in Refs. [44] and [45], where context-based unsupervised ensemble learning was proposed and clustering, respectively.

More discussion of such complex applications built on a stream processing platform, open research problems, and a more detailed literature survey may be obtained from Ref. [9]. Overall, we believe that the space of distributed, online, large-scale ensemble learning using stream processing middleware is an extremely fertile space for novel research and construction of real-world deployments that have the potential to accelerate our effective use of streaming Big Data to realize a Smarter Planet.

ACKNOWLEDGMENT

The authors would like to acknowledge the Air Force DDAS Program for support.

REFERENCES

1. "IBM Smarter Planet," http://www.ibm.com/smarterplanet (accessed September 24, 2016), retrieved October 2012.
2. "ITU Report: Measuring the Information Society," http://www.itu.int/dms_pub/itu-d/opb/ind/D-IND-ICTOI-2012-SUM-PDF-E.pdf (accessed September 24, 2016), 2012.
3. "IBM InfoSphere Streams," www.ibm.com/software/products/en/infosphere-streams/ (accessed September 24, 2016), retrieved March 2011.
4. "Storm Project," http://storm-project.net/ (accessed September 24, 2016), retrieved October 2012.
5. "Apache Spark," http://spark.apache.org/ (accessed September 24, 2016), retrieved September 2015.
6. Y. Zhang, D. Sow, D. S. Turaga, and M. van der Schaar, "A fast online learning algorithm for distributed mining of big data," in *The Big Data Analytics Workshop at SIGMETRICS*, Pittsburgh, PA, USA, June 2013.
7. L. Canzian, Y. Zhang, and M. van der Schaar, "Ensemble of distributed learners for online classification of dynamic data streams," *IEEE Transactions on Signal and Information Processing over Networks*, vol. 1, no. 3, pp. 180–194, 2015.
8. J. Xu, C. Tekin, and M. van der Schaar, "Distributed multi-agent online learning based on global feedback," *IEEE Transactions on Signal Processing*, vol. 63, no. 9, pp. 2225–2238, 2015.
9. H. Andrade, B. Gedik, and D. Turaga, *Fundamentals of Stream Processing: Application Design, Systems, and Analytics*. Cambridge University Press, Cambridge, 2014.
10. C. Aggarwal, Ed., *Data Streams: Models and Algorithms*. Kluwer Academic Publishers, Norwell, 2007.
11. R. Ducasse, D. S. Turaga, and M. van der Schaar, "Adaptive topologic optimization for large-scale stream mining," *IEEE Journal of Selected Topics in Signal Processing*, vol. 4, no. 3, pp. 620–636, 2010.

12. S. Ren and M. van der Schaar, "Efficient resource provisioning and rate selection for stream mining in a community cloud," *IEEE Transactions on Multimedia*, vol. 15, no. 4, pp. 723–734, 2013.
13. F. Fu, D. S. Turaga, O. Verscheure, M. van der Schaar, and L. Amini, "Configuring competing classifier chains in distributed stream mining systems," *IEEE Journal of Selected Areas in Communications*, vol. 1, no. 4, pp. 548–563, 2007.
14. A. Beygelzimer, A. Riabov, D. Sow, D. S. Turaga, and O. Udrea, "Big data exploration via automated orchestration of analytic workflows," in *USENIX International Conference on Automated Computing*, June 26–28, 2013, San Jose, CA, 2013.
15. Z. Haipeng, S. R. Kulkarni, and H. V. Poor, "Attribute-distributed learning: Models, limits, and algorithms," *IEEE Transactions on Signal Processing*, vol. 59, no. 1, pp. 386–398, 2011.
16. L. Breiman, "Bagging predictors," *Machine Learning*, vol. 24, no. 2, pp. 123–140, 1996.
17. Y. Freund and R. E. Schapire, "A decision-theoretic generalization of on-line learning and an application to boosting," *Journal of Computer and System Sciences*, vol. 55, no. 1, pp. 119–139, 1997.
18. W. Fan, S. J. Stolfo, and J. Zhang, "The application of AdaBoost for distributed, scalable and on-line learning," in *Proceedings of ACM SIGKDD*, San Diego, CA, USA, August 1999, pp. 362–366.
19. H. Wang, W. Fan, P. S. Yu, and J. Han, "Mining concept-drifting data streams using ensemble classifiers," in *Proceedings of ACM SIGKDD*, Washington DC, USA, August 2003, pp. 226–235.
20. M. M. Masud, J. Gao, L. Khan, J. Han, and B. Thuraisingham, "Integrating novel class detection with classification for concept-drifting data streams," in *Proceedings of ECML PKDD*, Bled, Slovenia, September 2009, pp. 79–94.
21. L. L. Minku and Y. Xin, "DDD: A new ensemble approach for dealing with concept drift," *IEEE Transactions on Knowledge and Data Engineering*, vol. 24, no. 4, pp. 619–633, 2012.
22. N. Littlestone and M. K. Warmuth, "The weighted majority algorithm," *Information and Computation*, vol. 108, no. 2, pp. 212–261, 1994.
23. A. Blum, "Empirical support for winnow and weighted-majority algorithms: Results on a calendar scheduling domain," *Machine Learning*, vol. 26, no. 1, pp. 5–23, 1997.
24. M. Herbster and M. K. Warmuth, "Tracking the best expert," *Machine Learning*, vol. 32, no. 2, pp. 151–178, 1998.
25. A. Ribeiro and G. B. Giannakis, "Bandwidth-constrained distributed estimation for wireless sensor networks-part I: Gaussian case," *IEEE Transactions on Signal Processing*, vol. 54, no. 3, pp. 1131–1143, 2006.
26. A. Ribeiro and G. B. Giannakis, "Bandwidth-constrained distributed estimation for wireless sensor networks-part II: Unknown probability density function," *IEEE Transactions on Signal Processing*, vol. 54, no. 7, pp. 2784–2796, 2006.
27. J.-J. Xiao, A. Ribeiro, Z.-Q. Luo, and G. B. Giannakis, "Distributed compression estimation using wireless sensor networks," *IEEE Signal Processing Magazine*, vol. 23, no. 4, pp. 27–41, 2006.
28. J. B. Predd, S. R. Kulkarni, and H. V. Poor, "Distributed learning for decentralized inference in wireless sensor networks," *IEEE Transactions on Signal Processing*, vol. 23, no. 4, pp. 56–69, 2006.

29. H. Zhang, J. Moura, and B. Krogh, "Dynamic field estimation using wireless sensor networks: Tradeoffs between estimation error and communication cost," *IEEE Transactions on Signal Processing*, vol. 57, no. 6, pp. 2383–2395, 2009.
30. S. Barbarossa and G. Scutari, "Decentralized maximum-likelihood estimation for sensor networks composed of nonlinearly coupled dynamical systems," *IEEE Transactions on Signal Processing*, vol. 55, no. 7, pp. 3456–3470, 2007.
31. A. Sayed, S.-Y. Tu, J. Chen, X. Zhao, and Z. Towfic, "Diffusion strategies for adaptation and learning over networks: An examination of distributed strategies and network behavior," *IEEE Signal Processing Magazine*, vol. 30, no. 3, pp. 155–171, 2013.
32. Z. J. Towfic, J. Chen, and A. H. Sayed, "On distributed online classification in the midst of concept drifts," *Neurocomputing*, vol. 112, pp. 139–152, 2013.
33. L. Canzian and M. van der Schaar, "Timely event detection by networked learners," *IEEE Transactions on Signal Processing*, vol. 63, no. 5, pp. 1282–1296, 2015.
34. C. Tekin and M. van der Schaar, "Active learning in context-driven stream mining with an application to image mining," *IEEE Transactions on Image Processing*, vol. 24, no. 11, pp. 3666–3679, 2015.
35. C. Tekin and M. van der Schaar, "Distributed online learning via cooperative contextual bandits," *IEEE Transactions on Signal Processing*, vol. 63, no. 14, pp. 3700–3714, 2015.
36. L. Canzian, U. Demiryurek, and M. van der Schaar, "Collision detection by networked sensors," *IEEE Transactions on Signal and Information Processing over Networks*, vol. 2, no. 1, pp. 1–15, 2016.
37. J. Xu, D. Deng, U. Demiryurek, C. Shahabi, and M. van der Schaar, "Mining the situation: Spatiotemporal traffic prediction with big data," *IEEE Journal on Selected Topics in Signal Processing*, vol. 9, no. 4, pp. 702–715, 2015.
38. C. Tekin, O. Atan, and M. van der Schaar, "Discover the expert: Context-adaptive expert selection for medical diagnosis," *IEEE Transactions on Emerging Topics in Computing*, vol. 3, no. 2, pp. 220–234, 2015.
39. D. Shutin, S. R. Kulkarni, and H. V. Poor, "Incremental reformulated automatic relevance determination," *IEEE Transactions on Signal Processing*, vol. 60, no. 9, pp. 4977–4981, 2012.
40. L. Rosasco, E. D. Vito, A. Caponnetto, M. Piana, and A. Verri, "Are loss functions all the same?" *Neural Computation*, vol. 16, no. 5, pp. 1063–1076, 2004.
41. B. Pan, U. Demiryurek, C. Gupta, and C. Shahabi, "Forecasting spatiotemporal impact of traffic incidents on road networks," in *IEEE ICDM*, Dallas, TX, USA, December 2013.
42. M.-F. Balcan, S. Hanneke, and J. W. Vaughan, "The true sample complexity of active learning," *Machine Learning*, vol. 80, no. 2–3, pp. 111–139, 2010.
43. C. Tekin, S. Zhang, and M. van der Schaar, "Distributed online learning in social recommender systems," *IEEE Journal of Selected Topics in Signal Processing*, vol. 8, no. 4, pp. 638–652, 2014.
44. E. Soltanmohammadi, M. Naraghi-Pour, and M. van der Schaar, "Context-based unsupervised data fusion for decision making," in *International Conference on Machine Learning (ICML)*, Lille, France, July 2015.
45. D. Katselis, C. Beck, and M. van der Schaar, "Ensemble online clustering through decentralized observations," in *CDC*, Los Angeles, CA, USA, December 2014.

11 Securing the Internet of Things: Need for a New Paradigm and Fog Computing

TAO ZHANG, YI ZHENG, RAYMOND ZHENG, and HELDER ANTUNES

Corporate Strategic Innovation Group, Cisco Systems, Inc., San Jose, CA, USA

11.1 INTRODUCTION

The emerging Internet of things (IoT) will interconnect a significantly larger number and broader range of things (devices) than today's Internet. These additional devices will range from simple sensors to wearable devices on humans and animals; to consumer goods such as clothes and parcels; to complex endpoints such as automobiles, trains, bicycles, drones, smart appliances, and commercial and consumer robots that will each contain multiple-networked subsystems; to sophisticated systems such as industrial control systems, connected transportation systems, smart buildings and cities, oil and gas systems, and smart energy grids (smart grids).

Industries and academia have been devoting tremendous efforts to building market consensus and developing enabling technologies and standards. The IEEE P2413 (Draft Standard for an Architectural Framework for the Internet of Things Working Group) is developing an IoT architectural framework [1]. The International Telecommunication Union (ITU) Study Group 20 is developing IoT standardization requirements focusing initially on smart city applications [2]. The Object Management Group (OMG) is developing standards for modeling and managing data and devices in the IoT [3]. The oneM2M consortium is defining standards for a common machine-to-machine (M2M) service layer to connect devices with M2M application servers, targeting business domains, such as connected transportation, healthcare, and utilities, and industrial automation [4]. The Industrial Internet Consortium (IIC) is working to accelerate IoT development and adoption in the industrial sectors to interconnect machines, business flows, intelligent analytics, and people

Fog for 5G and IoT, First Edition. Edited by Mung Chiang, Bharath Balasubramanian, and Flavio Bonomi.

at work [5]. The Open Interconnect Consortium (OIC) is defining connectivity and interoperability requirements for connecting billions of devices [6]. The OpenFog Consortium is developing an open fog computing architecture for distributing computing services and resources close to users and endpoints to meet growing demands for local computing in IoT [7].

The benefits of the IoT rely critically on its ability to provide adequate security. Securing the IoT brings a vast range of new challenges, which have attracted wide attention over the past few years. The National Institute of Standards and Technology (NIST) released a framework for improving critical infrastructure cybersecurity [8]. Major industries have been studying their specific IoT security issues, such as security challenges in connected transportation [9–11], industrial control systems [12], and smart grids [13]. Significant efforts have also been devoted to addressing IoT security issues that apply across different industries, such as how to secure billions of heterogeneous devices [14, 15], how IP-based security protocols may apply to resource-constrained devices [16], how to secure multimedia traffic in the IoT [17], how to secure cyber–physical systems that are becoming increasingly abundant in the IoT [18], and how cloud-based services can help address security issues in the IoT [10, 11, 19].

However, there has been far from adequate clarity on two fundamental questions: (i) what IoT security challenges are unique compared to protecting conventional enterprise networks and consumer electronics in today's Internet and why the existing security paradigm cannot adequately address these challenges, and (ii) what fundamental changes to the existing security paradigm will be needed to address these new challenges.

While the answers to these questions will differ from industry to industry, many fundamental challenges that will cause significant changes to the future security paradigm apply across multiple industries. For example, IoT devices will vary widely in their abilities to support security operations; many will have long life spans but highly constrained resources that are impractical to upgrade. Systems in many industry verticals will be distributed and have to operate in physically unprotected environments. In the growing number of cyber–physical systems that perform mission-critical tasks, such as industrial control systems, equipment uptime will take highest priority. Once installed, these devices are expected to operate nonstop in the field with minimal human intervention. These IoT characteristics make the conventional security technologies, which rely predominately on perimeter prevention to prevent security threats from penetrating the system, impractical and incomplete. Furthermore, conventional incident response mechanisms depend heavily on disruptive human interventions to remediate security compromises, which is often unacceptable in mission-critical cyber–physical systems.

Therefore, this chapter will first examine several categories of unique new IoT security challenges including those mentioned previously and discuss required fundamental changes to the existing security paradigm to address these new challenges. Our goal is not to enumerate all potential security challenges in IoT for every industry vertical but to focus on several major challenges that apply across multiple industry verticals and show how they will change the future security paradigm.

This chapter will then describe a new security paradigm to address these unique challenges. We will focus on the following key pillars for enabling this new security paradigm: (i) How to assess in trustworthy ways whether a very large number of devices are operating securely? (ii) How to secure a wide range of resource-constrained devices? (iii) How to dynamically adapt responses to security compromises based on the requirements of the system and the risk levels of the compromises?

Fog computing plays an essential role in this new security paradigm. Being physically and logically close to the endpoints, a fog system is best positioned to provide security services to the vast range of IoT devices, especially resource-constrained devices, to, for example, help enable scalable and trustworthy monitoring of a large number of devices and perform time-critical and resource-intensive security tasks on behalf of the endpoints.

11.2 NEW IOT SECURITY CHALLENGES THAT NECESSITATE FUNDAMENTAL CHANGES TO THE EXISTING SECURITY PARADIGM

Cyber security solutions for today's Internet, designed primarily for protecting enterprise networks, data centers, and consumer electronics, have focused on providing perimeter-based protections. Systems under protection are placed behind firewalls, which work with intrusion detection systems and intrusion prevention systems to prevent security threats from reaching the protected systems. Perimeter-based protection has been further extended onto individual hosts. Such host-based protection calls for sophisticated and typically resource-intensive security functions, such as threat detection, to be implemented on each individual host to prevent security threats from breaking into the host. More recently, cloud-based security services have been developed to off-load resource-intensive security protection capabilities from hosts onto resource-rich clouds [20]. Existing cloud-based security services continue to focus on providing perimeter-based protection. Examples of such services include redirecting e-mail and Web traffic to the clouds for threat detection and redirecting access control requests to the clouds for authentication and authorization processing. Cloud-based security services can introduce intolerable delays for many systems and applications and require impractically high long-haul communication bandwidth.

Should threats penetrate these protections; a system or host under protection will typically have limited and primitive capabilities to fight against the compromises. For example, when malware infects a device or system, the common practices have been for human operators to take the system offline, clean up or replace compromised files and devices, and then put the system back online. After an attack, forensic analysis, which often requires intensive human involvement, will be carried out to understand what has happened. Results from such analysis will be used to harden existing protection mechanisms and develop future remediation measures [21].

This existing security paradigm will no longer be adequate for protecting the diverse range of IoT devices, systems, and applications. To illustrate why, we will next discuss four categories of unique IoT security challenges that the existing security paradigm cannot adequately address without fundamental changes.

11.2.1 Many Things Will Have Long Life Spans but Constrained and Difficult-to-Upgrade Resources

Many things in the IoT will have highly constrained resources and will be incapable of supporting processing-intensive host-based threat protection mechanisms, such as malware signature scanning and complex intrusion detection and protection mechanisms. Even when the devices have sufficient resources, implementing sophisticated security capabilities on a large number of devices can be cost prohibitive. Keeping security installations on these devices up to date over time can become overly complex and difficult to manage, especially when the devices are distributed over large geographical areas.

Further complicating the matter is the fact that the hardware or software on many devices will be infeasible or impractical to upgrade, and yet, the devices must be secure over their very long life spans. For example, replacing any hardware on cars, which have already been sold to consumers, can create significant inconvenience to vehicle owners and result in heavy costs and reputation damages to carmakers. However, over a car's long life span that averages over 11 years [22], security threats will become significantly more advanced, and the amounts of resources required to combat the fast evolving threats will increase accordingly.

In many industrial control systems, hardware and software updates are also difficult because these systems must operate continuously for long periods of time. Taking a system offline for any reason can cause significant business loss and disruptions and therefore must be planned days, weeks, and often months in advance [23]. A nuclear reactor, for example, typically runs on 18-month cycles, and any downtime can cause tens of thousands of dollars [24]. Due to the high costs of industrial control and manufacturing systems, employing redundant systems to ensure production continuity can also be impractical in some cases. Therefore, unlike the routers, switches, laptop computers, and smartphones in today's Internet, the security hardware and software in an industrial manufacturing or control system often cannot be upgraded timely every time security vulnerabilities are discovered and patches are required.

Therefore, the existing security paradigm that relies heavily on each individual endpoint or network device to use its built-in security mechanisms to defend itself will be impractical in many IoT environments. Instead, many devices in the IoT will need external or off-board security services to help protect them, allowing the hardware and software on the devices to be simple and require no or rare updates.

11.2.2 Putting All IoT Devices Inside Firewalled Castles Will Become Infeasible or Impractical

While the existing security paradigm relies extensively on firewalled castles, many things in the IoT cannot be easily placed inside firewalled castles and yet have to operate in physically unprotected or highly vulnerable environments. Examples of such things include connected cars, drones, bicycles, sensors, wearable devices, smart appliances, and communication devices deployed along roadsides and at traffic intersections to support smart cities. These devices can often be accessed by anyone with relative ease physically or via wired or wireless local network connections.

For example, a modern car consists of tens of microcomputers or electronic control units (ECUs) interconnected by multiple types of in-vehicle networks. Anyone can physically attach tools to these onboard networks via many attach points to eavesdrop on in-vehicle communications and inject false data into the onboard networks. Furthermore, today's vehicles provide a standardized onboard diagnostics (OBD) interface, with standard communications protocol, to allow complete access into each vehicle's internal networks for vehicle diagnosis, repair, and ECU firmware and software update. The right-to-repair laws in the United States and similar laws in many other countries require that automakers provide the same information to all independent repair shops as they do to automaker-authorized car dealerships to create a fair vehicle repair marketplace. As a result, anyone can use low-cost and readily available tools to access a vehicle's internal networks through the OBD ports, which makes it essentially impossible to put all ECUs inside a firewalled castle. To make matters worse, OBD ports with built-in wireless access are becoming increasingly common for supporting a wide range of applications from over-the-air firmware updates to remote data collection and to telematics applications. Aftermarket wireless dongles that can be attached to the OBD ports to enable remote access are also widely available now. Placing sophisticated firewalls, intrusion detection, and prevention mechanisms on every individual ECU is not a plausible solution either because it can result in prohibitively high costs and unmanageable complexity for managing and updating these firewalls, besides the fact that many ECUs are highly resource-constrained and incapable of supporting sophisticated firewalls or advanced encryption technologies.

In a smart grid, power usage meters in residential homes, commercial buildings, and other power consumption sites can often be physically accessed with relative ease [25]. The many data collectors used to collect data from these smart meters are also deployed in highly vulnerable environments, typically close to the meters, and can be readily accessed by many people. It will be impractical to put all these devices behind firewalls. Firewalling each individual meter and data collector will not be more feasible either. The hardware or software of these meters and data collectors could be tampered with. They could also be replaced with fake devices. Adversaries could also introduce additional bogus meters or data collectors that pretend to be legitimate devices. Furthermore, adversaries could compromise the data transmitted over the air between the meters and the data collectors.

Therefore, today's perimeter-based security paradigm alone will no longer be adequate for securing the diverse range of IoT devices and systems and their operating environments.

11.2.3 Mission-Critical Systems Will Demand Minimal-Impact Incident Responses

Today's incident response solutions rely predominately on brute-force mechanisms such as shutting down a potentially compromised system, reinstalling and rebooting its software, or replacing its components and subsystems. Such maximal responses, which largely disregard how severe the compromises actually are, can cause intolerable disruptions to mission-critical systems. However, maintaining uninterrupted and safe operation, even when the system is compromised, is often a top priority for mission-critical systems such as industrial manufacturing and control systems, connected vehicles, drones, and smart grids. For example:

- An electric power generator may be infected by a malware that merely seeks to steal power for unauthorized use. Shutting down the power generator could cause severe disruptions to the smart grid and excessive power outages.
- Industrial control systems often have little tolerance for downtime. Manufacturing operations can also have critical safety implications. As a result, manufacturers usually value uninterrupted operation and safety over system integrity. This means that hardware and software updates can only be installed during a system's scheduled downtimes, which have to be short and far between, rather than every time any security compromise is detected.
- A connected car can be infected by malware that can become active while the car is in motion. While the malware can do a range of damages to the vehicle and can put the driver and passengers in harm's way, abruptly shutting down the engine each time any malware is detected could be an even quicker and surer way to cause deadly traffic accidents.
- If a drone flying midair is abruptly turned off just because a security compromise is detected, it can crash from the sky onto people, houses, and other properties to cause serious damages. Instead, safe landing or safe return-home mechanisms will be essential for responding to such security threats that can compromise a drone's flight.
- A server in a data center may be infected by a spyware that seeks to steal commercial secrets. While allowing such a compromised server to continue to operate could give the attacker access to some sensitive data, it may not directly impact the data center's mission-critical services. If we shut down the server, or halt the execution of the malware-infected files to wait for the malware to be removed, the system downtime could cause significantly more damage, including causing vast economic losses to the data center operator, business disruptions to those who count on the data centers to operate their businesses, and inconvenience to other users of the data center.

Therefore, today's highly disruptive incidence response paradigm will no longer be adequate for securing the many mission-critical systems in the emerging IoT.

11.2.4 The Need to Know the Security Status of a Vast Number of Devices

IoT will support a vast number of heterogeneous devices and distributed systems. A manufacturer, for example, may need to network many manufacturing plants. A smart city will consist of many smart buildings and devices deployed along the roads to control traffic signals and communicate with vehicles. An oil and gas company often has hundreds of remote sites (such as oil rigs, exploration sites, refineries, and pipelines) that need to be connected to its corporate networks. A smart grid will consist of networked subsystems for metering, data collection, data aggregation, energy distribution, and demand response in multiple geographical areas. A large carmaker will need to ensure the security of tens of millions of cars on the road in a large country such as the United States.

Therefore, a fundamental issue is how security administrators get to know, in a trustworthy way, whether the many devices and systems, many of which will be operating in widely distributed and highly vulnerable environments, are functioning securely.

Today's security health monitoring systems can collect historical, real-time, or event-triggered security status reports and log data from individual devices. However, many devices in the IoT, especially those operating in vulnerable environments, can be compromised and used to send false information [10, 11]. A wide range of solutions have been developed for filtering out abnormal data from multiple data sources. These solutions typically rely on the majority of the data sources to be honest, that is, uncompromised and not malfunctioning. In many IoT scenarios, however, adversaries can easily cause the compromised devices to form local majorities to fool the detection system. Consider, for example, vehicles using vehicle-to-vehicle communications to inform each other of their current locations to support collision avoidance applications. Vehicles can be compromised to send false information to other vehicles to cause traffic disasters. To detect such malicious vehicles, vehicles can be required to report suspicious messages, or suspected malicious vehicles, to a central security system that can analyze the information from multiple vehicles to detect malicious vehicles. A very small group of hackers can modify their own vehicles, then drive these compromised vehicles to easily form a local majority against an innocent vehicle on the road, and then report to the central security system that the innocent vehicle is sending malicious messages.

Consider smart grids as another example, where smart meters and data collectors are often deployed in highly vulnerable environments and can be physically accessed by hackers with relative ease [25]. Hackers can compromise these meters and data collectors, replace them with fake ones, add additional fake meters, or hack the communication links from the meters and data collectors to the rest of the system to send false energy usage data to the electricity distribution and demand response systems to disrupt energy distribution or cause excessive amount of power to be distributed to the wrong cites. These compromised or fake meters and data collectors can easily dominate an area so that the messages from them will appear to represent the "normal" power usage in the area, making the compromises difficult to detect using today's anomaly detection mechanisms.

Many IoT devices and systems, such as cars, manufacturing machines, and smart grids, will rely heavily on sensory input. Adversaries can also compromise sensory and other input data to a device to cause the device to send false information using its valid security credentials, making the false messages even more difficult to detect [9–11]. Furthermore, adversaries can implement advanced evasive technology to hide their malicious behaviors.

As a result, existing approaches for collecting security status reports or log data from remote devices or systems will become inadequate.

Remote attestation mechanisms have been developed to allow a device to prove its trustworthiness to a remote verifier [26,27]. A device makes a claim about certain properties of its hardware, software, or runtime environment to the verifier and uses its security credentials (e.g., a hardware-based root of trust and public key certificates) to vouch for these properties. The verifier then cryptographically verifies these claims.

However, existing remote attestation methods have focused on enabling an individual device to attest to its own trustworthiness. A wide range of resource-constrained devices in the IoT will not be able to support the often processing-intensive remote attestation algorithms and protocols. Furthermore, requesting a large number of devices to perform remote attestation can result in prohibitively high cost and management complexity.

Therefore, we need new ways to determine, in a trustworthy manner, whether a large number of distributed and diverse devices or systems are operating securely.

11.3 A NEW SECURITY PARADIGM FOR THE INTERNET OF THINGS

Addressing the unique challenges described in Section 11.2 will necessitate fundamental changes to the existing security paradigm. Such changes will include:

1. *Providing External Help to the Less Capable Devices.* Rather than implementing comprehensive security capabilities on every individual device, we need ways to reduce the security complexity and costs on individual devices and compensate the devices' constrained security capabilities with off-board security services. These external security services must be provided in highly scalable, timely, resource-efficient, and easy-to-manage manners.
2. *Trustworthy Ways to Monitor a Large Number of Diverse Devices and Systems.* We can no longer expect to know the security status of a large number of heterogeneous and distributed devices and systems by relying solely on retrieving security log data from every device or forcing all devices to perform remote attestation. Instead, we will need significantly more scalable and trustworthy ways to monitor the security status of large distributed systems, many of which have to operate in highly vulnerable environments.
3. *Dynamic Risk-and-Benefit Proportional Protection.* Instead of human-intensive and highly disruptive incident responses, we need to provide dynamic and adaptive protection that can respond to security compromises based on their risk levels to enable uninterrupted and safe operation of a system even when the system is compromised. Such adaptive protection represents a new thinking: rather than focusing primarily on preventing threats from entering a system, we will treat security compromises as a normal way of life and provide ways for a system to fight against the compromises automatically while achieving a proper balance between the benefits of continuous safe system operation and potential risks of the security compromises.

These capabilities will form the key pillars for the new security paradigm required to protect the emerging IoT. In the rest of this section, we will discuss each of these three pillars in greater detail.

11.3.1 Help the Less Capable with Fog Computing

In IoT, many applications will demand computing, data processing, and smart networking services to be moved from the clouds closer to the endpoints. Such local intelligence will be necessary to meet stringent latency requirements, reduce processing load and conserve battery power on the endpoints, overcome bandwidth and cost constraints for long-haul communications, and support local communication needs. Fog computing [28] can meet such demands by distributing computing, data processing, and advanced networking services close to or sometimes onto the endpoints and anywhere along the continuum from the clouds to the endpoints.

As illustrated in Figure 11.1, fog nodes—functional nodes used to provide fog services—can be deployed between endpoints and the cloud. One of the key tenets of fog nodes and fog systems is that they work together with the cloud to form an end-to-end system to serve the end users. Fog nodes and systems can also operate autonomously or collaborate with each other to ensure non-interrupted services to the end users even when they lose contact with the cloud.

With these properties, fog computing is in an ideal position to provide a wide range of new security services to help protect resource-constrained endpoints and network devices. The following are some examples of such fog-based security services:

Update of Security Credentials and Software for Endpoints. Fog systems can assist endpoints, and resource-constrained network edge devices, in acquiring and updating their security credentials (e.g., security keying materials and public key certificates), security software, and security configurations. Keeping the security

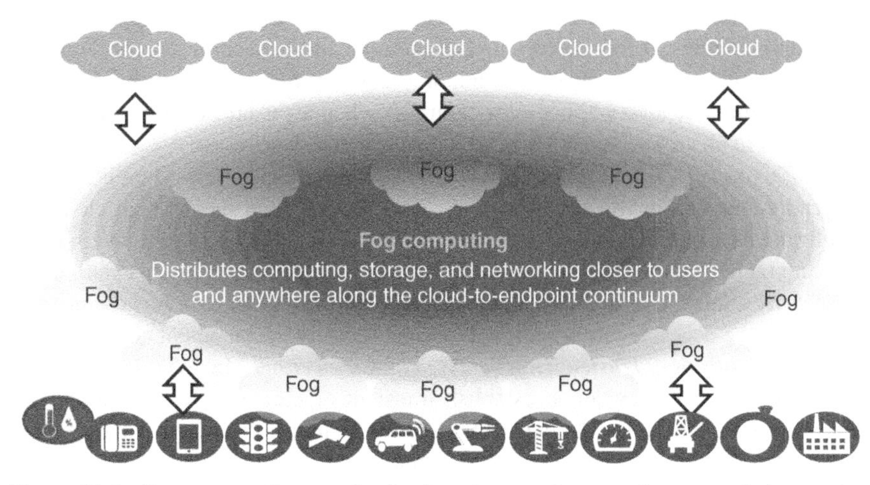

Figure 11.1 Fog computing and fog-based security services to help protect resource-constrained devices and systems.

credentials and software on a large number of devices up to date is a growing challenge. Today's approach of requiring every endpoint to update its security credentials and software from centralized remote servers often proves impractical. Consider a car as an example again. Many of the ECUs on a car are highly constrained in processing and networking capabilities. Forcing every ECU to directly communicate with an off-board authority for security credential and software updates will be excessively complex and costly and hence impractical.

A fog node or system can serve as a proxy for the endpoints to retrieve the security credentials and software updates from remote sources. It can then use lighter-weight protocols and procedures to distribute the updates to the endpoints locally. In the example of the car, the fog node can be a security gateway or software client running on the car or a device deployed along the roadside that communicates with the car. The fog node may, for example, use the Let's Encrypt framework [29] to automatically acquire public key certificates from remote certificate authorities on behalf of the endpoints and then distribute the certificates to the endpoints locally. As a result, the many resource-constrained endpoints will no longer need to run complex procedures and protocols to acquire security credentials and software updates directly from remote servers.

Authentication of Endpoints. Adequate authentication of endpoints and other network devices is an essential requirement to ensure system security and safe operations. Consider a car again; ECU authentication will be an important step to help prevent compromised, rogue, and other authorized ECUs from being installed onto vehicles. However, having every ECU on a car to directly authenticate with remote servers can be overly complex and impractical. Similarly, requiring every sensor in a smart grid or on each oil rig to directly authenticate with remote security systems will also be impractical.

A fog system can perform local authentication of the endpoints so that each individual endpoint no longer needs to authenticate directly with remote security servers. Where required, a fog node may maintain sufficient knowledge to locally authenticate endpoints without relying on help from remote authentication servers. This will enable local authentication when the endpoints and the fog systems lose connectivity to remote authentication servers—for example, when a car breaks down and has to be repaired in an area where there is no wireless network coverage.

Monitor and Report Security Status of Endpoints. Fog systems, being close to user endpoints, can prove to be more effective and efficient means than clouds in monitoring the security status of the endpoints. It is usually unnecessary, and in some cases impractical, to send all status data to remote servers. A fog system can monitor the security status of local devices and systems, perform local data processing, and report only the necessary information to remote security management centers.

Help Establishing Trusted Transient Local Connections. In IoT, there will be increasing needs for devices to establish transient or temporary connections with other nearby devices. For example, technicians on factory floors may need to connect to manufacturing robots to perform on-site repairs and maintenance. In an emergency response situation, medical personnel may need to connect to patients' imbedded pacemakers for timely diagnosis and treatment. In such transient situations, it is often impractical to pre-configure universally interoperable security credentials on all these devices. Fog systems, being close to the endpoints, can assist the endpoints in establishing trusted local transient connections. For the first example previously mentioned, a fog system can authenticate a technician's device that wishes to connect to a manufacturing robot and then provide it with the necessary temporary security keys for communicating with the robot. For the second example previously mentioned, a fog system deployed along roadside, on police vehicles, or on emergency response vehicles can help authenticate and authorize devices and establish the required temporary security keys needed to connect emergency responders' devices to a patient's pacemaker in a traffic accident scene.

Protection of Endpoints. Due to their close proximity to the endpoints, fog systems are often the most or the only logical choices to implement conventional perimeter-based protections, such as firewalls, to defend endpoints against threats that come from outside of the security perimeters. More importantly, however, many security functions that are traditionally implemented on endpoints in today's Internet can be moved to fog systems. For example, comprehensive malware detection and protection mechanisms can be moved out of the endpoints and onto fog systems. Endpoints only need to detect suspicious files, which may come from local or remote communication channels, and forward these files to a fog system for malware detection. Mechanisms for detecting whether a file is suspicious are usually significantly simpler and need to be updated much less frequently than the mechanisms for determining whether the file contains malware.

Often the fog system may have already seen some of the suspicious files. These could be the case, for example, when a file has been previously reported by some endpoints or has been indicated by a more powerful cloud-based security system to be compromised. This means that an endpoint may often just need to send a short representation, such as a hash value, of a suspicious file to the fog system.

A group of fog nodes can collectively carry out security defense functions for the endpoints in a distributed and collaborative manner. For example, all or a subset of the fog nodes can form a dynamic or pre-configured defense cluster to carry out malware defense on behalf of the endpoints. With the fog nodes' collective defense capabilities, each individual fog node within the cluster does not have to implement the entire spectrum of malware defense capabilities. Instead, the malware defense capabilities will be distributed across multiple fog nodes within the cluster where each fog node can have the same or complementary threat defense capabilities. Such a fog-based distributed defense system can not only make efficient use of shared resources but also improve overall security by making it more difficult for attackers to disrupt the security operations. Some sample ways of fog-based distributed malware examination include:

- Some fog nodes can support signature-based malware scanning, while other fog nodes support heuristic-based malware detection mechanisms.
- Some fog nodes can be responsible for detecting malware targeted to Windows operating systems, while other fog nodes can handle malware targeted to Linux operating systems.

- Some fog nodes can maintain more comprehensive malware signature databases because of their higher storage capacities and abilities to communicate with the centralized cloud services more frequently, while other fog nodes that have limited storage may have only subsets of the signature database pertaining to only the latest updates.
- Multiple fog nodes can offer the same protection capability with other fog nodes for reasons of load balancing and/or backup.

Using the aforementioned approach, each endpoint only needs to detect files that are suspicious, which can be significantly simpler than detecting whether a file contains malware. For example, when an endpoint starts with a known set of authorized files or a "golden image," any file that deviates from the authorized files will be considered suspicious. Examples of suspicious files include files that have changed in size and new files that have not been digitally signed by authorized parties. To detect such suspicious files, a thin software client can reside in a device with the specific purpose to detect new and changed files, which may enter the endpoint through any interface, local or remote, wired or wireless. Upon detecting suspicious files, the thin client will send either the metadata or copies of the files to the fog node cluster for further analysis and assessment for malware. The fog nods in the cluster will then work collaboratively and collectively to detect whether the file is infected by malware. A more advanced approach may also include cleaning up or discarding the infected files and sending clean files back to the endpoint.

11.3.2 Scale Security Monitoring to Large Number of Devices with Crowd Attestation

To address the need to verify the trustworthiness of large number of devices and systems, we propose a new approach that enables a system to attest to its trustworthiness without requiring every individual device to attest to its own trustworthiness. We refer to this approach as *crowd attestation*. With crowd attestation, rather than requiring every individual device to vouch for its own trustworthiness, a subset of the devices in a system will act as attesters. Each attester will not only attest to its own trustworthiness but will also monitor, evaluate, and vouch for the trustworthiness of selected other devices. The set of attesters collectively will *cover* all the devices in the system. In other words, every device, including every attester, will be monitored and attested to by at least one attester. Verifiers, which can be cloud-based or fog-based security management servers, will collect the attestation reports from the attesters and use these reports to evaluate the trustworthiness of the overall system.

Crowd attestation can be provided as a service by a fog system. That is, the attesters can be fog nodes, and the crowd attestation functions can be a subset of the fog services provided by these fog nodes.

The following key technologies will be essential for enabling crowd attestation:

- Methods for assessing the trustworthiness of other devices (monitored devices)
- Methods for ensuring trustworthy attestation
- Methods for scaling the solution to monitor very large and distributed systems

11.3.2.1 Assess the Trustworthiness of Monitored Devices A key to enabling crowd attestation is the technology for an attester to tell how trustworthy a monitored device is. An attester may use different metrics, in the cyber and the physical domains, to assess the trustworthiness of other devices. These metrics can range from externally observable behaviors (e.g., network traffic patterns, application layer behaviors, device locations and movements, devices' external temperatures) to intrinsic properties that require more intrusive methods to collect (e.g., the device's file system status, currently running processes, memory access patterns, and electrical current draws). Different attesters may attest to different properties of a device. The set of devices and their properties an attester should attest to can be determined based on security policies and the capabilities of the attesters. The set of attesters will collectively provide a holistic view of the entire system.

Table 11.1 shows three possible categories of approaches for assessing the trustworthiness of a monitored device.

The first category of approaches, which we refer to as "black-box" monitoring, analyzes the monitored device's external behaviors and characteristics in the cyber domain (e.g., patterns of network traffic to and from the device), in the physical domain (e.g., physical movement patterns of the device), or in both the cyber and the physical domains. The main advantage of black-box monitoring is that there is

TABLE 11.1 Ways to Determine the Trustworthiness of Another Device

Category of Monitoring Methods	Description
Black-box	The attester is assumed to have no knowledge about the internal characteristics of the monitored device and cannot rely on the monitored device to assist in the monitoring. The attester will rely on the monitored device's externally observable behaviors and characteristics to detect any anomaly and will need to understand what externally observable behaviors and parameters are normal
Clear-box	The attester is assumed to be able to obtain trustworthy measurements of some internal characteristics of the monitored device. Examples of such characteristics may include profiles of authorized program files, device temperature change patterns, electric current patterns, memory access and usage patterns, and radio-frequency signal patterns. The attester can use its knowledge on these characteristics to identify anomalies caused by compromises such as malware
Gray-box	Compromised devices could fake their externally observable behaviors and send false measurements of their internal characteristics to an attester. Gray-box monitoring is a new way of thinking about detecting compromises of a monitored device. It operates on the assumption that the attester can take actions to provoke or cause a monitored device to react in ways that will reveal potential compromises to the device, such as if a malware is running on it

no need to implement any special software or hardware on the monitored devices just for the purpose of monitoring. However, a compromised device can fake its external behaviors to evade detection while causing damage to the device. Such compromises are especially dangerous in the many cyber–physical systems, where computer technologies are used to control physical equipment and processes, such as vehicles, trains, road traffic control systems, and industrial control systems such as manufacturing systems, smart grids, and smart building control systems. In a cyber–physical system, a cyber-domain compromise can be used to damage the physical processes while keeping the system's cyber behavior appear to be normal. For example, the famous Stuxnet attack on the Iranian nuclear facility was masqueraded with normal status sent back to the system administers, while the attack was carried out to spin a nuclear reactor out of control [30]. As another example, an adversary could compromise a manufacturing robot and have it do random actions while maintaining a normal communication pattern with external entities so the network monitoring system cannot detect the compromise. Malware on a drone, which is flying out of its controller's line of sight, can cause the drone to veer off its planned path while telling its controller that it is still flying as planned.

The second category of approaches is "clear-box" monitoring. Clear-box monitoring goes beyond observing only the external behaviors to monitor selected internal parameters of a monitored device. Examples of such internal parameters may include profiles of authorized program files, device temperature change patterns, electric current patterns, memory access and usage patterns, and radio-frequency signal patterns. Clear-box monitoring will typically require the monitored device to report the values of the selected internal parameters to the attester. For example, an agent inside a monitored device can measure the internal parameters and report the results to the attester. Compared with black-box monitoring, clear-box monitoring makes it harder for adversaries to circumvent the monitoring. However, a compromised device could still falsify measurement reports, by falsifying the input to the agent or compromising the agent itself.

The third category of approaches, gray-box monitoring, assumes that the attester cannot fully trust the external behaviors of the monitored device nor the data reported by the monitored device. Instead, an attester will challenge a monitored device to force it to react in ways that will reveal potential compromises. Several mechanisms for enabling such gray-box monitoring have been developed [31]. However, to support the broad spectrum of IoT devices, significantly more scalable and effective gray-box attestation mechanisms, which require minimal help from the monitored devices, will be necessary.

11.3.2.2 Ensure Trustworthy Attestation The collection of the attesters for a system must be able to report the security status of the system in ways that cannot be substantially tampered by adversaries. To achieve this goal, the first step is to ensure that all or a critical mass of the attesters have not been tampered with. Therefore, all or a carefully selected subset of the attesters should attest to its own trustworthiness in different situations: when the attester powers up, when any new software is loaded up to run, when any change to the attester's own device is detected (e.g., a change to the file system), or when requested by a verifier. Attestation to other devices' trustworthiness can be performed periodically, upon requests from a verifier, and upon detecting any significant changes to the behaviors or properties of the other devices.

Even when an attester is not tampered with, an adversary could still cause it to send false attestation reports by compromising the input data that the attester relies on for generating its attestation reports. For instance, an adversary could falsify speed and location sensory inputs to the communication module on a car to cause it to send erroneous speed and location information to neighboring cars [9–11]. Such false information can disrupt vehicle collision avoidance applications, which rely on vehicle-to-vehicle communications, and cause fatal accidents. In many other industries such as smart grids, sensors use automatic power-off cycles to conserve battery power. During a sensor's powered-off cycle, an attacker could physically replace the sensor with a rogue one to send false information to an attester, which may not be able to detect the changes by only monitoring the communications with the sensor.

Therefore, a second step to ensure trustworthy attestation is that each attester should be covered by at least one other attester. When the majority of the attesters can be trusted, the verifier can use the attestation reports from all the attesters to detect false attestation claims using voting mechanisms. In scenarios where voting mechanisms are inapplicable or ineffective (e.g., when there is only a single attester), the system must provide additional means (e.g., hardware-based root of trust) to ensure the attesters' own trustworthiness.

To further increase the trustworthiness and accuracy of crowd attestation, multiple attesters can collaborate with each other to jointly assess the security status of a monitored device by correlating their observations on the same or different sets of parameters regarding the monitored device. For example, the external behaviors observed by some attesters can be correlated with the results of clear-box or gray-box monitoring by other attesters to piece together a more comprehensive picture of the security status of a monitored device or system.

Select attesters. The selection of attesters must meet several important requirements. First, there should be as few attesters as possible to reduce system complexity and costs. Second, the attesters collectively must cover all the devices, including all attesters, in the system. Each attester should be covered by at least one other attester because attesters can be compromised and used to send erroneous attestation reports.

This attester selection problem can be formulated as a connected vertex cover problem: finding a minimal subset A of nodes in a graph that represents the system to be monitored, so that (i) nodes in subset A form a connected graph and (ii) any node outside subset A is connected to at least one node inside A. Polynomial time algorithms are available for this problem [32].

Additional practical considerations should also be factored into attester selection. For example, it will be desirable to select the attesters in a way that ensures that even when a given number of the attesters are compromised, the entire set of attesters can still collectively produce trustworthy attestations to the security status of the system. Furthermore, attesters, which have higher levels of trustworthiness, better visibility into the behaviors of the overall system, and more interactions with other devices, can be preferred attesters. Such preferred attesters may include local gateways, routers, and servers. Methods to select attesters that can meet these practical requirements are yet to be developed.

Attestation graph and trustworthiness scores. A crowd attestation graph shows who attests to whose behaviors. Each node in the graph represents a device in the real-life system. Each node will have a native trustworthiness value. If a device is capable of performing remote attestation, its trustworthiness value will be based on its attestation to its own hardware, software, and runtime behavior. For a device that is incapable of performing remote attestation, its trustworthiness value will be determined based on the last known and trusted status about the device.

A trustworthiness score is then calculated by the verifiers for each node based on the node's native trustworthiness value and other attesters' attestation reports.

11.3.2.3 Scale to Support Very Large Systems with Hierarchical Crowd Attestation To scale crowd attestation up to support large and distributed systems, attesters can be arranged in a hierarchy as illustrated in Figure 11.2. Layer-1 attesters, those at the bottom layer of a crowd attestation hierarchy, can each attest to the security status of a subset of the devices in a system. Higher-layer attesters can then use the attestation reports from Layer-1 attesters to assess the security status of the overall system.

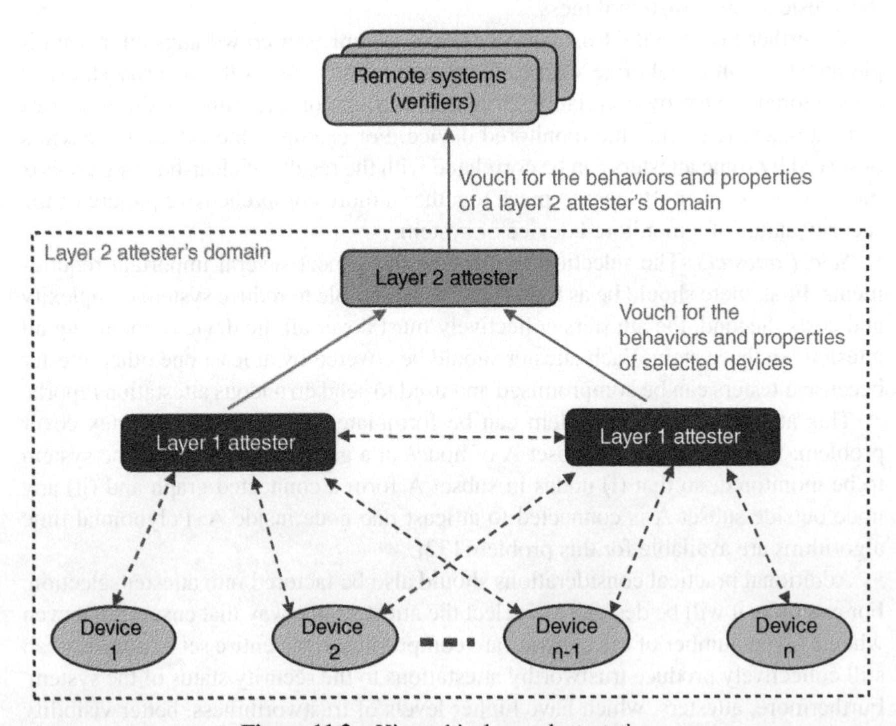

Figure 11.2 Hierarchical crowd attestation.

Layer-1 attesters may often need to be deployed close to the devices they monitor, for example, on the same local networks as the monitored devices, in order to collect the necessary information required to evaluate the monitored devices' security status. This means that a large number of Layer-1 attesters may still be needed for monitoring a large number of geographically distributed systems, even though crowd attestation already reduced the total number of devices required to perform attestation. In such scenarios, cost considerations may cause the Layer-1 attesters to be constrained in computing resources.

Therefore, resource-constraint Layer-1 attesters can just vouch for the raw data they have observed without judging whether the data represents malicious activities. Higher-layer attesters can then assess the security status of the system based on the raw data from the Layer-1 Attesters.

More generally, each attester at any layer of a crowd attestation hierarchy can attest to the raw observations it collects from the monitored other devices and the attestation reports it has received from lower-layer attesters, without having to judge whether the data reveals any compromises to the system. With this approach, a rich set of data can be made available to the verifiers to assess the security status of a system. However, reporting the raw observations of all devices in a large system can require the attesters to send an excessively large amount of data to the verifiers.

Therefore, all or selected attesters at each layer of the crowd attestation hierarchy may analyze its own observations and the information contained in the attestation reports it has received from lower-layer attesters to draw conclusions on the security status of the devices it covers and then attest to its conclusions. This can significantly reduce the amount of data that has to be sent to the verifiers. When a verifier needs further information, it can request the attesters to submit the detailed observations they have collected.

A properly designed hierarchical attestation approach can greatly increase the scalability of a crowd attestation implementation.

11.3.3 Dynamic Risk–Benefit-Proportional Protection with Adaptive Immune Security

Some existing computerized systems already implement rudimentary adaptive incident response mechanisms, which allow the systems to revert to a safe mode of operation to run only a subset of its functions when parts of the system malfunction for any reason. For example, when subsystems on some modern cars malfunction, the cars can automatically restrict their speeds while allowing the vehicles to continue to run to allow time for the drivers to bring the vehicles to a safe stop [33]. Such techniques, however, will not be sufficient to address the needs of the broad spectrum of mission-critical systems in the IoT.

Protecting computer systems against security threats is analogous to the ways our human bodies respond to diseases such as virus infections. The conventional security paradigm is similar to having only the innate human immune system that seeks to prevent viruses from entering human bodies. Conventional incident response approaches are analogous to having doctors apply maximal intrusive treatments every time a sign of illness is detected, regardless how serious the illness is. Specifically, upon detecting any suspected compromise of any severity, system administrators, as doctors for our computer systems, will put the patient (computer system) in general anesthesia (shutting the system down) and then surgically remove the suspicious body masses or parts (suspicious files or devices).

Humans have a much more elegant and effective way to defend against diseases. There is no need to involve doctors every time a person gets sick. Instead, the human body has an adaptive immune system that can self-defend against and self-heal from viruses that have entered the body without relying on external intervention. This adaptive immune system tracks the disease, learns its nature and behavior, and activates counter measures that are dynamically adjusted in proportion to the severity of the disease. In the process, the human body cultivates immunity to the disease and will be able to respond to and heal from similar attacks much faster in the future. External intervention is necessary only when the adaptive immune capabilities are overwhelmed by a disease.

Addressing the new security challenges in IoT will demand a new security paradigm similar to the human immune system. This new paradigm will add a crucial missing component—the adaptive immune security—to today's security paradigm. When a compromise is detected, the security system will, in real time, diagnose the threats, evaluate the risks, garner available resources at its disposal, and activate counter measures proportional to the risks to fight the threats. A compromised file (or device) may be allowed to continue to run if the risk–benefit analysis results indicate that deleting, quarantining, or denying the file's execution will cause significantly more harm. Meanwhile, the system's adaptive immune process will monitor the execution and movement of the compromised files, continue risk–benefit analysis based on newly available information, and adjust counter measures accordingly. The system may automatically reduce its functionalities and modify its behaviors in the cyber domain (e.g., limiting the network traffic destinations to contain the spread of the malware-infected files) and in the physical domain (e.g., constraining a machine's movement speeds). Once the threat is removed, the system will resume its normal operations. By monitoring in real time the firsthand behaviors of the running files, devices, and communications in a real environment (as compared with emulated environment) and making conscious decisions on how to respond to an attack, the system develops a more accurate visibility to how the attack unfolds. This knowledge can then be fed back into the system's self-diagnosing and self-healing functions to train and improve its immune capabilities, making the system more resilient after each compromise.

Applying human adaptive immune principles to computer system security has been studied before [34–36]. However, prior studies have focused on using artificial immune techniques to *detect* compromises [35], and there have not been sufficient efforts on applying adaptive immune mechanisms to *respond* to compromises.

Next, we discuss the main steps of an adaptive immune security system, using malware infection as an example of compromise. Here, we will focus on how to respond to a compromise rather than how to detect a compromise. The main steps for responding to a compromise include self-monitoring, self-diagnosing, self-defending, self-healing, and self-learning:

Self-monitoring. The devices in a system will collaborate to track, in real time, the activities of a compromise such as a malware. Examples of such compromise-related activities include:

- The behaviors of the malware-infected programs and devices (e.g., communication patterns, functions performed, and even physical behaviors such as device movement patterns)
- The movement of the malware from one file to another file
- The movement of each malware-infected file from one device to another or from one part of a device to another part of the same device
- The movement of data generated by malware-infected programs and devices (e.g., is the data sent to unauthorized locations?)

Self-diagnosing. As the compromises (e.g., malware and the malware-infected files and devices) are tracked, the compromised devices and the devices that interact with the compromised devices will conduct risk–benefit trade-off evaluations. These devices can perform the analysis locally and can also send selected compromise-related information to resource-rich fog systems or cloud-based servers that can carry out deeper and more comprehensive analysis on behalf of the devices.

The risk–benefit evaluation should take into consideration factors such as:

- The types and functions of the compromised devices. For example, shutting down a mission-critical function or device can result in higher risks and costs than shutting down a less mission-critical function or device.
- The nature of the compromise. Different types of compromises, such as different types of malware, can cause vastly different degrees of damage to a system.
- The behaviors of the compromised devices (or compromised applications on a device). Compromises to a device can often cause the device's behaviors in the cyber domain or in the physical domain to change. Examples of cyber-domain behaviors include the patterns of data to and from the device and memory usage patterns. Examples of physical-domain behaviors include a device's physical movements and the amount of electrical current a device draws.
- The paths along which the compromises have been progressing through the system. For example, if a malware can only spread inside a non-mission-critical portion of a system, then the risks of allowing the malware in the system will be lower than when the malware can spread to mission-critical portions of system.

Self-defending. When the benefits outweigh the risks, a compromised application or device can be allowed to operate. Whether to allow a compromised application or device to run also depends on how well the compromised application or device can be controlled. For example, we may only allow a compromised application to continue to run if we can shut it down when necessary.

When a compromised application or device is allowed to continue to operate, counter measures matching the levels of the risks should be activated on the compromised devices and on other devices that have high risks of being compromised by this known ongoing compromise. Examples of the counter measures include:

- Limit devices' functions (e.g., only perform a minimum set of mission-critical functions).
- Constrain devices' operating ranges (e.g., maximum engine rotations, temperature range, and the amount of electricity draws).
- Restrain or quarantine compromised devices' communications (e.g., capping the maximum amount traffic it can send and receive and limiting the set of devices it can communicate with).
- Identify vulnerable files and devices and verify their integrity. For example, a malware may likely spread among files of the same type. Therefore when a file is detected to be infected by malware on one device, files of the same type on other devices can be reexamined for malware.
- Challenge a compromised device to attest to the set of the programs running on the device.
- Track the movements of the compromises (e.g., malware, malware-infected files, and malware-infected devices) through the system and activate counter measures along the paths.
- Notify system operators so they can take proper actions.

Different devices can take different counter measures, depending on factors such as how mission-critical a device is, what security protections are required for a device, what capabilities a device has, and how vulnerable a device is.

At any time, if the risks of letting a compromised file or device to operate exceed the risks of preventing the file or device from running, the compromised file or device will not be allowed to execute. This can be done on an individual case-by-case basis: a compromised program can be removed from one device but still allowed to execute on a different device, depending on the specific benefits and risks of running the compromised file on each specific device.

The risk–benefit trade-off, the decision on whether to allow a compromised file or device to continue to operate, and the counter measures will be based on the system requirements and the system's operational environment and context. An adaptive immune security system seeks to detect the presence of compromises, automate the decision-making to select proper forms of response to a compromise, carry out the selected incident responses automatically, and bring human into the loop when necessary.

Self-healing and Self-learning. As a compromise runs through its course in a system, the self-healing procedure tracks what counter measures have been activated and can later stop the counter measures and restore the system back to trusted states when the compromises are removed.

As the attack's kill chain is revealed and the vulnerabilities of the system uncovered, such threat intelligence information can be fed back to the self-monitoring and self-diagnosing processes and converted into proper self-defense and remediation actions for the future. For example, counter measures that have been found effective can be distributed to other devices that may need similar defense capabilities (i.e., to vaccinate similar systems), and patches for newly discovered vulnerabilities can be developed and sent to the relevant devices.

It is important to point out that not all network nodes or endpoints need to implement all the adaptive immune security capabilities described previously. Such capabilities are typically required only on devices that must maintain continuous safe operation in the face of security compromises. These devices are effectively behaving as fog nodes, where the adaptive immune security functions are subsets of the fog services deployed on the nodes.

11.4 SUMMARY

Securing the emerging IoT imposes a range of unique challenges, which the existing cyber security paradigm cannot adequately address without fundamental changes. The large-scale, highly distributed, and adaptive real-time defense requirements for IoT systems demand a new security paradigm. This chapter discussed several categories of such new challenges and outlined a new security paradigm for addressing them. This new paradigm calls for off-board security services to help secure resource-constrained endpoints, scalable and trustworthy ways to assess whether a large number of distributed devices and systems are operating securely, and dynamic risk-and-benefit-proportional real-time responses to security compromises to minimize disruptions. This chapter further identified key technologies that need to be developed to enable this new paradigm. Fog computing, which distributes computing, storage, and smart networking closer to end users, is a key enabler for this new security paradigm.

ACKNOWLEDGMENT

The authors would like to thank Jack Cham and Anoop Nannra for the insightful discussions on topics covered in this chapter.

REFERENCES

1. https://standards.ieee.org/develop/project/2413.html (Accessed February 18, 2016).
2. http://www.itu.int/en/ITU-T/studygroups/2013-2016/20/Pages/default.aspx (Accessed February 18, 2016).
3. http://www.omg.org (Accessed February 18, 2016).
4. http://www.onem2m.org (Accessed February 18, 2016).

5. http://www.iiconsortium.org (Accessed February 18, 2016).
6. http://openinterconnect.org (Accessed February 18, 2016).
7. http://www.openfogconsortium.org (Accessed February 18, 2016).
8. National Institute of Standards and Technology (NIST) version 1.0, "Framework for Improving Critical Infrastructure Cybersecurity," February 12, 2014, https://www.cisecurity.org/images/frame.pdf (Accessed September 10, 2016).
9. Luca Delgrossi and Tao Zhang, "Vehicle Safety Communications: Protocols, Security, and Privacy," John Wiley & Sons, Inc., Hoboken, NJ, 2012.
10. Tao Zhang, Helder Antunes, and Siddhartha Aggarwal, "Defending Connected Vehicles against Malware: Challenges and a Solution Framework," IEEE Internet of Things Journal, Vol. 1, Issue 1, February 2014.
11. Tao Zhang, Helder Antunes, and Siddhartha Aggarwal, "Securing Connected Vehicles End to End," SAE 2014 World Congress and Exhibition, Detroit, MI, USA, April 8–10, 2014.
12. National Institute of Standards and Technology (NIST) Special Publication 800-82, Revision 2, "Guide to Industrial Control Systems (ICSs) Security," May 2015; http://csrc.nist.gov/publications/drafts/800-82r2/sp800_82_r2_second_draft.pdf (Accessed October 26, 2016).
13. Ye Yan, Yi Qian, Hamid Sharif, and David Tipper, "A Survey on Cyber Security for Smart Grid Communications," IEEE Communications Surveys & Tutorials, Vol. 14, Issue 4, January 30, 2012.
14. Gang Gan, Zeyong Lu, and Jun Jiang, "Internet of Things Security Analysis," 2011 International Conference on Internet Technology and Applications (iTAP), Wuhan, China, August 16–18, 2011.
15. Rodrigo Roman, Jianying Zhou, and Javier Lopez, "On the Features and Challenges of Security and Privacy in Distributed Internet of Things," Computer Networks, Vol. 57, Issue 10, pp. 2266–2279, July 5, 2013.
16. Tobias Heer, Oscar Garcia-Morchon, René Hummen, Sye Loong Keoh, Sandeep S. Kumar, and Klaus Wehrle, "Security Challenges in the IP-based Internet of Things," Wireless Personal Communications, Vol. 61, Issue 3, pp. 527–542, December 2011.
17. Liang Zhou and Han-Chieh Chao, "Multimedia Traffic Security Architecture for the Internet of Things," IEEE Networks, Vol. 25, Issue 3, May 2011.
18. Huansheng Ning and Hong Liu, "Cyber-Physical-Social Based Security Architecture for Future Internet of Things," Advances in Internet of Things, Vol. 2, pp. 1–7, 2012, http://dx.doi.org/10.4236/ait.2012.21001 (Published online January 2012; http://www.SciRP.org/journal/ait (Accessed September 10, 2016).
19. Ibbad Hafeez, Aaron Yi Ding, Lauri Suomalainen, Seppo Hätönen, Valtteri Niemi, and Sasu Tarkoma, "Demo: Cloud-based Security as a Service for Smart IoT Environments," 2015 Workshop on Wireless of the Students, by the Students, and for the Students, New York, NY, September 11, 2015.
20. Jayant Shukla, US Patent Application US 20100031361 A1, "Fixing Computer Files Infected by Virus and Other Malware," priority date: July 21, 2008.
21. https://support.symantec.com/en_US/article.TECH122466.html# (Accessed February 26, 2016).
22. Polk (Online), "Polk Finds Average Age of Light Vehicles Continues to Rise," August 2013, https://www.polk.com/company/news/polk_finds_average_age_of_light_vehicles_continues_to_rise (Accessed October 29, 2013).

23. National Institute of Standards and Technology (NIST), U.S. Department of Commerce, Special Publication 800-82, "Guide to Industrial Control Systems (ICS) Security," June 2011. http://webcache.googleusercontent.com/search?q=cache:x_Y5pLdaBEAJ:http://csrc.nist.gov/publications/nistpubs/800-82/SP800-82-final.pdf%2B Guide+to+Industrial+Control+Systems+(ICS)+Security&safe=strict&hl=en-IN&gbv=2 &ct=clnk (Accessed October 26, 2016).
24. Warwick Ashford, "Industrial Control Systems: What Are the Security Challenges?," October 15, 2014, http://www.computerweekly.com/news/2240232680/Industrial-control-systems-What-are-the-security-challenges (Accessed January 28, 2016).
25. Kris Ardis, "7 Serious Smart Meter Security Threats That Do Not Involve Hacking the Network," July 28, 2014, http://www.smartgridnews.com/story/7-serious-smart-meter-security-threats-do-not-involve-hacking-network/2014-07-28 (Accessed January 28, 2016).
26. Ruizhong Chen, Lihao Wei, Hong Zou, and Meijie Zhai, "A TCM-Based Remote Anonymous Attestation Protocol for Power Information System," The International Power, Electronics and Materials Engineering Conference 2015 (IPEMEC 2015), May 16–17, 2015, Dalian, China.
27. Aurelien Francillon, Quan Nguyen, Kasper B. Rasmussen, and Gene Tsudik, "A Minimalist Approach to Remote Attestation," Conference on Design, Automation & Test in Europe (DATE), Dresden, Germany, March 24–28, 2014.
28. Flavio Bonomi, Rodolfo Milito, Jiang Zhu, and Sateesh Addepalli, "Fog Computing and Its Role in the Internet of Things," The First Edition of the MCC Workshop on Mobile Cloud Computing, 2012, New York, NY.
29. https://github.com/letsencrypt/acme-spec (Accessed January 23, 2016).
30. Nicolas Falliere, Liam O. Murchu, and Eric Chien, "W32.Stuxnet Dossier," Symantec Security Response, Version 1.4, February 2011, https://www.symantec.com/content/en/us/enterprise/media/security_response/whitepapers/w32_stuxnet_dossier.pdf (Accessed September 10, 2016).
31. Raghunathan Srinivasan, Partha Dasgupta, and Tushar Gohad, "Software Based Remote Attestation for OS Kernel and User Applications," IEEE International Conference on Privacy, Security, Risk, and Trust, and IEEE International Conference on Social Computing, Boston, MA, October 2011.
32. Bruno Escoffier, Laurent Gourves, and Jerome Monnot, "Complexity and Approximation Results for the Connected Vertex Cover Problem in Graphs and Hypergraphs," LNCS (Lecture Notes in Computer Science), Vol. 4769, pp. 202–213, Springer, Heidelberg, 2007.
33. Clutch & Transmission Technicians, "Transmission 'Limp Mode' or 'Fail Safe' Mode," June 6, 2014, http://ctttransmissions.com/techtalk/transmission-limp-mode-or-fail-safe-mode/ (Accessed October 12, 2015).
34. Anil Somayaji, Steven Hofmeyr, and Stephanie Forrest, "Principles of a Computer Immune System," 1997 Workshop on New Security Paradigms (NSPW'97), pp. 75–82, ACM, New York, NY, USA, 1997.
35. P.K. Harmer, P.D. Williams, G.H. Gunsch, and G.B. Lamont, "An Artificial Immune System Architecture for Computer Security Applications," IEEE Transactions on Evolutionary Computation, Vol. 6, Issue 3, pp. 252–280, June 2002.
36. Jungwon Kim, Peter J. Bentley, Uwe Aickelin, Julie Greensmith, Gianni Tedesco, and Jamie Twycross, "Immune System Approaches to Intrusion Detection—A Review," Natural Computing, Vol. 6, Issue 4, pp. 413–466, December 2007 (First online January 12, 2007).

INDEX

Fog for 5G and IoT, First Edition. Edited by Mung Chiang, Bharath Balasubramanian, and Flavio Bonomi.

WILEY SERIES ON INFORMATION AND COMMUNICATION TECHNOLOGY

Series Editors: T. Russell Hsing, Vincent K. N. Lau, and Mung Chiang

The Information and Communication Technology (ICT) book series focuses on creating useful connections between advanced communication theories, practical designs, and end-user applications in various next generation networks and broadband access systems, including fiber, cable, satellite, and wireless. The ICT book series examines the difficulties of applying various advanced communication technologies to practical systems such as WiFi, WiMax, B3G, etc., and considers how technologies are designed in conjunction with standards, theories, and applications. The ICT book series also addresses application-oriented topics such as service management and creation and end-user devices, as well as the coupling between end devices and infrastructure.

T. Russell Hsing, PhD, is the Executive Director of Emerging Technologies and Services Research at Telcordia Technologies. He manages and leads the applied research and development of information and wireless sensor networking solutions for numerous applications and systems. Email: thsing@telcordia.com

Vincent K.N. Lau, PhD, is Associate Professor in the Department of Electrical Engineering at the Hong Kong University of Science and Technology. His current research interest is on delay-sensitive cross-layer optimization with imperfect system state information. Email: eeknlau@ee.ust.hk

Mung Chiang, PhD, is the Arthur LeGrand Doty Professor of Electrical Engineering at Princeton University, the Director of the Keller Center for Innovation in Engineering Education, and the Chair of Princeton Entrepreneurship Council, USA. Dr. Chiang founded the Princeton EDGE Lab in 2009 and a co-founder of OpenFog Consortium in 2015. He is the recipient of the 2013 Alan T. Waterman Award by US National Science Foundation.

- Smart Data Pricing
by Soumya Sen, Carlee Joe-Wong, Sangtae Ha, and Mung Chiang
- WiFi, WiMAX and LTE Multi-hop Mesh Networks: Basic Communication Protocols and Application Areas
by Hung-Yu Wei, Jarogniew Rykowski, and Sudhir Dixit
- Networks and Services: Carrier Ethernet, PBT, MPLS-TP, and VPLS
by Mehmet Toy
- Vehicle Safety Communications: Protocols, Security, and Privacy
by Tao Zhang and Luca Delgrossi
- RF Circuit Design, 2nd Edition
by Richard C. Li
- Equitable Resource Allocation: Models, Algorithms and Applications
by Hanan Luss
- Fundamentals of Wireless Communication Engineering Technologies
by K. Daniel Wong

• The Fabric of Mobile Services: Software Paradigms and Business Demands
by Shoshana Loeb, Benjamin Falchuk, and Thimios Panagos

• RF Circuit Design
by Richard C. Li

• Digital Signal Processing Techniques and Applications in Radar Image Processing
by Bu-Chin Wang

• Wireless Internet and Mobile Computing: Interoperability and Performance
by Yu-Kwong Ricky Kwok and Vincent K.N. Lau

• Digital Signal Processing for RFID
by Feng Zheng and Thomas Kaiser

Printed and bound by CPI Group (UK) Ltd, Croydon, CR0 4YY

16/04/2025

14658364-0003